T0092355

Springer Theses

Recognizing Outstanding Ph.D. Research

For further volumes:
http://www.springer.com/series/8790

Aims and Scope

The series "Springer Theses" brings together a selection of the very best Ph.D. theses from around the world and across the physical sciences. Nominated and endorsed by two recognized specialists, each published volume has been selected for its scientific excellence and the high impact of its contents for the pertinent field of research. For greater accessibility to non-specialists, the published versions include an extended introduction, as well as a foreword by the student's supervisor explaining the special relevance of the work for the field. As a whole, the series will provide a valuable resource both for newcomers to the research fields described, and for other scientists seeking detailed background information on special questions. Finally, it provides an accredited documentation of the valuable contributions made by today's younger generation of scientists.

Theses are accepted into the series by invited nomination only and must fulfill all of the following criteria

- They must be written in good English.
- The topic should fall within the confines of Chemistry, Physics, Earth Sciences, Engineering and related interdisciplinary fields such as Materials, Nanoscience, Chemical Engineering, Complex Systems and Biophysics.
- The work reported in the thesis must represent a significant scientific advance.
- If the thesis includes previously published material, permission to reproduce this must be gained from the respective copyright holder.
- They must have been examined and passed during the 12 months prior to nomination.
- Each thesis should include a foreword by the supervisor outlining the significance of its content.
- The theses should have a clearly defined structure including an introduction accessible to scientists not expert in that particular field.

Alireza Faed

An Intelligent Customer Complaint Management System with Application to the Transport and Logistics Industry

Doctoral Thesis accepted by
Curtin University, Australia

Author
Dr. Alireza Faed
School of Information Systems
Curtin University
Perth, WA
Australia

Supervisor
Prof. Dr. Elizabeth Chang
School of Information Systems
Curtin University
Perth, WA
Australia

ISSN 2190-5053 ISSN 2190-5061 (electronic)
ISBN 978-3-319-00323-8 ISBN 978-3-319-00324-5 (eBook)
DOI 10.1007/978-3-319-00324-5
Springer Cham Heidelberg New York Dordrecht London

Library of Congress Control Number: 2013937191

© Springer International Publishing Switzerland 2013
This work is subject to copyright. All rights are reserved by the Publisher, whether the whole or part of the material is concerned, specifically the rights of translation, reprinting, reuse of illustrations, recitation, broadcasting, reproduction on microfilms or in any other physical way, and transmission or information storage and retrieval, electronic adaptation, computer software, or by similar or dissimilar methodology now known or hereafter developed. Exempted from this legal reservation are brief excerpts in connection with reviews or scholarly analysis or material supplied specifically for the purpose of being entered and executed on a computer system, for exclusive use by the purchaser of the work. Duplication of this publication or parts thereof is permitted only under the provisions of the Copyright Law of the Publisher's location, in its current version, and permission for use must always be obtained from Springer. Permissions for use may be obtained through RightsLink at the Copyright Clearance Center. Violations are liable to prosecution under the respective Copyright Law.
The use of general descriptive names, registered names, trademarks, service marks, etc. in this publication does not imply, even in the absence of a specific statement, that such names are exempt from the relevant protective laws and regulations and therefore free for general use.
While the advice and information in this book are believed to be true and accurate at the date of publication, neither the authors nor the editors nor the publisher can accept any legal responsibility for any errors or omissions that may be made. The publisher makes no warranty, express or implied, with respect to the material contained herein.

Printed on acid-free paper

Springer is part of Springer Science+Business Media (www.springer.com)

Parts of this thesis have been published in the following articles:

1. Faed, A. R., Ashouri, A, Wu, C., "Maximizing Productivity Using CRM Within the Context of M-Commerce", International Journal of Information Processing Management, Human and Sciences Publication, **2010**. 2 (1), pp. 33–43.
2. Faed, A.R., Hussain, O.K., Chang, E.,"A Methodology to Map Customer Complaints, and Measure Customer Satisfaction and Loyalty" Service oriented and computing and application. Springer, **2012**, Under review.
3. Faed, A.R., Omar, K. Hussain, "Linear Modelling and Optimization to Generate Customer Satisfaction and Loyalty", The 9th IEEE International Conference on e-Business Engineering, **2012**.
4. Faed, A. R., "Adaptive Neuro-Fuzzy Inference System Modelling to Predict and Ascertain Customer Satisfaction" Submitted to Second World Conference on Soft Computing, 3-5 December, **2012**.
5. Faed, A.R., Ashouri, A, Wu, C., "The Impact of Trust and Interactivity on Intensifying Customer Loyalty for CRM", International Conference on Knowledge Management and Information Sharing (KMIS 2010), IC3k, **2010**. 113–120.
6. Faed, A. R., "A conceptual framework for E- Loyalty in Digital Business Environment", the 4th 2010 IEEE International Conference on Digital Ecosystems and Technologies (IEEE DEST 2010), IEEE, **2010**. 547–552.
7. Faed, A. R., Wu, C.,"The Relationship between Interactivity, Complaint and Expectation in the CRM Environment", International Conference on Computer Information Systems and Industrial Management Applications," CISIM 2010. IEEE.
8. Faed, A. R., Wu, C., Chang, E., "I-CRM on the cloud", the 13th International Conference on Network-Based Information System (NBIS 2010), IEEE, **2010**. 216–223.

Dedicated to our parents,
my wife and my son

Supervisor's Foreword

The work of Dr. Alireza Faed represents a first attempt to formally address the issue of customer complaints in the context of Customer Relationship Management (CRM). Though, most organizations currently have some customer complaints handling processes, a rigorous approach in CRM that allows them to address these complaints, find out their root causes and define their impact factors has been lacking so far. Also, complaints data is largely unstructured, and there has been no systematic approach which could help organizations with the management of such data. Plenty of statistical and data analytics methods have been available, yet there has been no systematic methodology or tool that could make it possible to use them to manage customer complaints. Further, with the existing processes it has not been possible to turn the higher level of uncertainty and imprecision of the complaints into a clear interpretation of root causes. Most importantly, there has been no systematic method so far to turn the complaints into useful information, in order to improve both internal and external relationships and to design future strategies for process improvement and business opportunities development.

To address the aforementioned issues, the author carried out a formal data collection process. Data were collected through extensive field studies and qualitative interviews with customers (operators, authorities, large and small business enterprises). Complaints data was then analysed through a new proposed framework of intelligent CRM (I-CRM). This new methodology, which is integrated with text-mining, categorisation, type mapping, analytical hierarchy process, structural equation modelling, linear statistical analysis and fuzzy approach, allowed to find root causes and underlying problems, carry out sensitivity analysis, prioritise significance of the issues, define impact factors, identify key customers in the context of business values and map the significance of the problems to the customers types. The methodology is unique in its kind and applicable to various industries regardless of the services they provide. It helps organizations to transform the customer complaints into organizational opportunities, hence leading to business improvement.

As the first work to formally examine these issues and propose solutions, the framework the thesis presents will offer a sound basis for future research in CRM, and more advanced results can hopefully be generated on the basis of and by taking inspiration from it. In addition to its novel approach, the thesis provides readers with a comprehensive survey of the current literature on CRM and of existing CRM approaches, together with a critical discussion of their pros and cons.

Australia, Perth, April 2013 Prof. Dr. Elizabeth Chang

Acknowledgments

I wish to acknowledge the contribution of everyone whose support has made possible the completion of this thesis.

First and foremost, I wish to express my sincere gratitude to Prof. Elizabeth Chang, my main supervisor. Her boundless support, enthusiasm and guidance kept me on track and gave me the encouragement to meet many challenges and overcome hardships. Without her, this work would never have been completed.

I take this opportunity to acknowledge and extend my thanks to Prof. Tharam S. Dillon and Dr. Omar Hussain. They have generously given of their time to provide insightful comments, support, understanding and precious knowledge.

It would not have been possible for me to undertake this doctoral thesis without the support of my family.

I owe my deepest gratitude to my parents Parisima Soltani and Hamid Faed for their unconditional and endless support, inspiration and love. They have contributed both spiritually and financially to ensure that I accomplish my goals. I am most fortunate to have them as my role models in life. They have made sacrifices for me without expectations and have enabled me to study with peace of mind. There are no words to express my appreciation and the extent of their help which I will always treasure. My sincere gratitude goes to my grandmother, Parvaneh, for her endless love and prayers. She is the living symbol of sacrifice, dignity and sincerity.

I owe loving thanks to my wife, Afsaneh for her love, prayers and immense support throughout the Ph.D. journey. I am so blessed to have her and my son, Sayman who is my pride and joy. Without Afsaneh's encouragement and understanding, it would have been impossible for me to finish this work. I would like to express my special heartfelt thanks to my brother, Mehdi, who supported me and my whole family during this challenging time.

I would like to extend my gratitude to all my friends and colleagues. Without their friendship, and support, this undertaking would have been much more difficult.

Contents

Chapter 1
Introduction

1.1 Overview of the Thesis

This thesis presents a new conceptual framework and practical solution for advanced customer relationship management (CRM), customer satisfaction and customer complaint management. Companies are competing with each other to not only continue to be viable, but to win customers and develop their relationships with them. CRM has emerged as a technology to help companies maximise their products, services and sales, thereby establishing good relationships with their customers. CRM is a set of disciplinary business strategies to create and sustain long-term, cost-effective customer relationships by satisfying the customers [1, 2]. Also, using CRM, companies can prioritise their long term survival strategies ahead of their profits. In order to better achieve this aim, this thesis introduces the notion of intelligent customer relationship management (I-CRM), and defines and develops an I-CRM framework. To completely understand CRM, we need to highlight its components and examine its impact on business activity.

In this chapter, we discuss the key concepts of customer relationship management (CRM), CRM Systems, following this with a detailed discussion of academic research on CRM, CRM adoption, CRM challenges, understanding CRM in general, and the benefits of CRM. We identify the research gap and the existing issues in CRM systems. This enables us to determine our research objectives and define the scope of the research. At the end of the chapter, we illustrate the structure of the thesis.

1.2 Why Customer Relationship Management?

To date, CRM has been acknowledged and utilised as a productive method for gathering, analysing, and transforming customer data into knowledge and managerial performances [2]. CRM systems and strategies are focused on professional interactions in order to generate value for customers, and develop beneficial

A. R. Faed, *An Intelligent Customer Complaint Management System with Application to the Transport and Logistics Industry*, Springer Theses, DOI: 10.1007/978-3-319-00324-5_1,
© Springer International Publishing Switzerland 2013

strategies for both company and customers. The majority of organizations that utilise CRM systems receive a massive amount of customer feedback which includes both compliments and complaints. If the feedback is addressed by customer relationship expertise and the customers' issues are satisfactorily resolved, the company will advance itself through continuing improvement. However, existing CRM technology does not provide an automated or semi-automated customer feedback analysis system. Companies consistently attempt to gather customer feedback and store this in order to address the complaints using CRM systems and strategies. For example, the Department of Transport in Western Australia has more than two hundred telephone operators and handles 5,000 to 25,000 calls a day, most of which are complaints. Due to the large volume of calls, the department has increased the number of operators in the call centre to respond to and deal with the calls. The CEO of this company states that they cannot keep employing new staff, but instead need to examine the nature of the complaints. What are the major issues among the customers? How many repeated calls does the department receive each day and are any of these complimentary? However, existing CRM systems and technologies have not provided computerised and automated analysis tools to help the organisation analyse the customer feedback data. Hence, the following are required:

(1) Automated data and text mining tool, coupled with semantics reasoning tools to help analyse customer feedback and address the issues in a timely fashion.
(2) Scientific analytic tools to determine customer satisfaction, loyalty, retention and feedbacks. These include statistical tools coupled with fuzzy analysis to provide business intelligence.
(3) Tools and approaches that can manage data types and temporary data sets, such as collecting customer complaints by various operators, since inadequate responses or failure to address such complaints could harm the future of the company.
(4) An integrated real-time system to categorise and address customer feedback both internally and externally using a complete mathematical approach.

While CRM vendors provide significant concepts and methods in the management of customers for business advantages, the lack of a technology-supported system as mentioned above has hindered the efficiency of CRM systems.

1.3 CRM Adoption in Industry

To date, companies that have implemented customer relationship management systems and strategies feel that they are not able to address customers' issues adequately, and doing so would help them to enhance the business and effectively satisfy the customers [3, 4]. It is always difficult to manage relationships, especially the company-customer relationship. It is also hard to plan or prepare a CRM given the varieties of customer types in the CRM processes. In addition, there is no evidence that existing state-of-the-art CRM systems in organizations have provided substantial assistance to customers or business and there is no evidence that such systems have led to profitability [5, 6]. In addition, companies which implement CRM indicate that the acquisition and retention of customers are their biggest concerns and that companies have difficulty building value for the company through CRM; moreover, CRM does not provide them with enough knowledge to recognize customer values and needs [7, 8]. Furthermore, CRM to many small-media enterprises is an ill-defined, misconceived and ambiguous subject; in other words, it is obscure [9, 10]. Most companies that use a CRM system believe that it does not provide them with significant information; CRM does not, for example, differentiate between customer behaviors or determine the level of customer satisfaction [11]. Failures occur even with powerful CRM implementations [12]. There are also cases of failure in customer data transmission and the acknowledgement that CRM has not brought in more business or improved sales [13, 14]. CRM is not always a cost-effective method for some companies [15]. As an example, although more than half of the hotels in London implement some aspect of e-CRM, they still resist the notion of CRM [16]. In some companies, the concepts of CRM have not attracted any attention and the companies do not wish to apply customer relationship models as a framework in their operations [17, 18]. The capturing of false customer information, the inability to integrate customer management processes into business, and the use of misleading metrics or inappropriate measurement approaches, are major issues [19]. The disparity between customer opinion and CRM strategies shows that CRM still faces some hurdles [20]. There are also various managerial problems in spite of substantial investments in CRM, with difficulties and blurring in management causing many discontented customers to receive less returns on investments [14, 21]. CRM does not lead to long-term relationships with the customers, acceptable customer loyalty and desirable customer acquisition due to some weaknesses in managerial methods [22]. Moreover, they do not have a robust methodology and means of analysis to deal with all of the issues mentioned.

1.4 Academic Research on CRM

Understanding how to effectively manage relationships with customers has become an important topic for academics, requiring a clear and comprehensive set of training programs. Anecdotal evidence and reactions to CRM training would suggest that such training can prevent accidents [6]. Academic questions have arisen as to how much value and efficiency is brought to the company by loyal customers, and whether a long-term relationship with customers can increase profitability. Some researchers have indicated that the loyalty does not improve profitability [23]. Furthermore, there are claims that insufficient satisfaction surveys and loyalty programs are included in the companies' existing business models. The studies show that companies do not perceive a great deal of value in their CRM. It is clear that no adequate methodical research has been done on ascertaining the significance of CRM systems and there is no clear distinction between the customers' perceived value and the customers' satisfaction. In previous studies, the researchers considered perceived value only in terms of satisfaction but did not link this with customer satisfaction and loyalty. Although Harris and Goode [24] demonstrated that the perceived value has an impact on loyalty, they failed to establish the relationship between perceived value, customer satisfaction and loyalty, which is a critical aspect of this thesis. We also study the interaction of companies with the customers and how this has a big impact on loyalty. Companies need to have a dynamic interaction process put in place to communicate with their customers or clients. CRM is not a tool to help business connect to customers in real-time and just-in-time; nor does CRM have any built-in features to respond to and monitor customers via the Internet. Ineffective communication creates dissatisfaction which negatively affects customer satisfaction which in turn generates disloyalty [23]. In order to address these shortcomings, we will present an intelligent advanced framework which includes solutions for customer complaints and an evaluation of the solutions in terms of perceived value, interactivity, customer satisfaction, customer acquisition and loyalty.

Intelligent CRM provides a cutting edge technology and tools which will enable companies to tackle customer-related issues in the competitive marketplace. An examination of the majority of scholarly journal articles reveals that most of them address only the common problems about inefficiency of CRM or CRM in the business processes, whereas, in this thesis, we aim to provide a methodology framework known as intelligent customer relationship management to better address the customer problems and improve CRM systems, so that companies can benefit from the management of customer relationships and address their complaints to achieve better services and produce customer satisfaction and loyalty.

1.5 CRM Challenges

Although CRM as an approach and technology is critically beneficial, its potential to assist in the delivery of future goods and services has been disregarded [2]. Companies today are confronted with thousands of complaints and they do not have the ability to measure or assess them as yet [25]. It has been suggested that up to 80 % of CRM projects and 60 % of web-based CRM implementations, fail [26]. One reason for this is a lack of understanding of CRM [27]. Also, [28] states that of every three CRM implementations, one fails and less than half of the projects are up to standard. It has also been suggested that implementations fail as companies fail to adopt a clear strategy or to make appropriate changes to their business processes. Companies install a CRM application in the belief that it is capable of delivering what they need. The most common fault is that companies focus on the CRM's implementation technology and exclude the people, process and organizational changes required for it to be useful [29]. CRM has been created to address these issues. However, CRM has not yet been broadly designed, implemented or utilized in most small-medium enterprises [29]. Moreover, organizations have inadequate CRM systems, or due to a lack of experts or qualified clients, do not fully exploit the benefits offered by a CRM. Inadequate training in CRM has also been noted as another reason for CRM failure [6]. Only 17 % of companies include customer acquisition, retention and enhancement cost programs in their marketing strategies. In spite of the talk about customer focus, there is little evidence to show that chief executives pay an attention to it. Three quarters of chief management personnel do not have consistent, direct contact with customers [27]. Customer loyalty is not understood by firms; nor do they attempt to secure it [30, 31]. Companies fail to recognize or identify the potential benefits of CRM [32]. Although CRM has emerged as a significant business strategy for electronic commerce, little research has been conducted in evaluating its effectiveness. In addition, companies have not recognized the relationship between CRM and loyalty [27, 33]. Most companies around the world underestimate CRM and take it for granted as part of their daily activities, as they have a strong system to assess all kinds of issues [12]. Companies with customer relationship management systems are often unable to utilize these to establish customer relationships, enhancing productivity or maximizing efficiency. Organizations remain and continue to remain viable because of customers, and without customers, there is no business. Because customers are a company's main asset [34], the relationship between customer and company is crucial to any business. Value has always been the fundamental basis for all business activities and customers' satisfaction and loyalty creates this value [35]. CRM performs as an intermediary between company and customer, enabling communication and thereby leading business forward. However, existing CRM does not provide comprehensive intelligent support to deal with customer concerns, complaints or feedback [12, 36]. Finally, despite great improvements in information technology and marketing science, there is still misunderstanding and vagueness about the concept of CRM and the way companies intend to implement it. This provides the motivation for this research [1, 37]. In this thesis, we intend to address this significant void in CRM systems, analyse customer complaints, and evaluate satisfaction.

1.6 General Understanding of CRM

1.6.1 Customer Knowledge

Customer knowledge is defined as a structured form of information about the customer that will enable companies to have intelligent decision-making procedures by which they can acquire and retain customers [38]. Customer knowledge is one of the key concepts in customer relationship management and is central to ensuring the success of CRM. Interaction can take place when there is customer knowledge and interaction alone can endorse knowledge [37]. To be able to have an effective CRM, companies may need to employ knowledge management systems to intensify the effectiveness of CRM systems and strategies by obtaining knowledge about customers. According to knowledge management rules, knowledge must be generated, distributed and shared by all customers and companies. Without knowledge, the company will not be able to have an effective decision-making process. Moreover, knowledge alone falls into two different categories: the first is the knowledge about the current and potential customers and the second is the knowledge about customers which have been previously processed [38]. In order to acquire enough knowledge for the purposes of this study, we need to utilise various methodologies for and approaches to CRM and customer complaint management systems to acquire related data and transform this into a valid and reliable knowledge base to be utilised by future researchers.

1.6.2 Customer Relationship

First of all, both company and customer need to have a complete view and information about each other. Legitimate information is the basis on which mutual trust is built by both parties and leads to a long-lasting relationship. The information about customers' needs to be updated on a monthly or yearly basis to promote better relationships. Hence, a company must pursue customers via email, telephone and other communication channels. This creates good competition in the marketplace and increases the level of efficiency. Having a good relationship with customers is a matter of breaking down barriers and eliminating disrespect, dishonesty and insecurity. Hence, interaction and communication should be accompanied by trust and honesty [37].

1.6.3 Value Proposition

After getting to know customers and initiating an interaction with them, a company must offer exclusive, flexible and individualized services in a dynamic environment in order to satisfy the customers. The company also needs to expand its relationship with customers using various strategies in order to retain them. In addition, to every customer, the company must provide a mutual commitment [37].

1.6.4 Relationship Management

This is closely linked to knowledge management and the key relationship management procedure includes effective communication to meet customer requirements and create loyalty. To do so, we need to have a contemporary and sophisticated customer service approach followed by appropriate maintenance. Also, knowledge must be central to relationship management as it is a valuable resource [38].

1.7 Benefits of Customer Relationship Management

Customer Relationship Management is a business strategy and has an immense impact on organizational processes. CRM is able to assist business to develop sophisticated client recognition and insight, improving customer fulfilment, and increasing customer faith and customer value by supplying appropriate goods and customized services [39]. Also, it is a strategic way to methodically address and transform pertinent client information into viable information for Decision-making in a company [40]. Likewise, CRM can provide a company with a gratifying level of achievement and increased business, through efficient database management, invention and excogitation of services [41]. Using CRM, companies can segment customers based on their core needs, placing them into different clusters [42]. This can help to identify misunderstandings in an organization, enhancing capabilities to target clients. This then enhances personalised direct marketing messages about goods and services that offer better pricing and value [43]. It is considered as an outstanding business strategy that consolidates internal procedures and externals to generate value for customers and easily eliminates competitive perils by utilizing a roadmap for the company. Furthermore, it may indicate potential pitfalls, thus creating a company triumph [11]. Preparing an effective CRM system may create an environment for client prioritization and for evaluation of market segmentation and it leads the company to new market opportunities and reduces competitive threats [44, 45].

1.8 Research Gaps in the Literature

In previous researches and studies, there have been fascinating improvements in ways to obtain customers and ensure customer satisfaction. Also, it has been shown that in recent years, companies have placed far more emphasis on the customers, making them the central focus and attempting to meet their needs by resolving their problems. Researchers have provided different strategies and systems in customer relationship management to make the businesses more profitable, and keep the customers satisfied and loyal. However, the impact of CRM systems and strategies is not similar and does not bring the same rewards to each company. Also, companies are fielding thousands of complaints because customers have greater expectations, and markets are becoming increasingly competitive. Companies do not have the knowledge and ability to identify and address customer issues. Satisfying the customers and maintaining their loyalty are objectives that companies fail to achieve in the majority of cases. Customers cannot easily trust the companies and can simply change over to competitors who can provide the desired services. This has led us to investigate the problems of customer relationship management, and the intricacies and applicability of CRM, in order to resolve the issues. To this aim, we proposed an intelligent customer relationship management system to acquire more customers, and initiate and establish strong relationships with customers. This will help us better address the complaints and find solutions for them. A robust intelligent customer relationship management system allows customers to be traced and their requirements to be identified, and ascertains whether their feedback is positive or negative. Furthermore, using various approaches, we can identify the issues, and by categorizing them we may be able to prioritise them. Then, we can evaluate customer satisfaction and estimate the loyalty of the customers. Additionally, we can provide solutions to the issues in order to rectify the system and provide control over both customers and the organization.

According to our research, in previous studies researchers have failed to completely identify and gather a comprehensive range of customer issues. Also, they failed to categorise and analyse the issues. To affirm the value of I-CRM and the important role that it can play if implemented by companies, we need hypotheses to test the relationship between its elements and prove their significance. Additionally, although companies are made aware of customer issues, they might not be able to address and evaluate them, or create satisfaction and generate loyalty. The main aim of ensuring customer satisfaction using CRM and complaint management systems is to diminish the level of complaints; however, companies have failed to meet these commitments. Furthermore, companies have not been able to establish approaches that allow them to have positive interactions with customers. To provide satisfaction, companies need to evaluate important elements such as perceived value and loyalty. However, although companies can measure each of these factors, they neglect to provide a clear and methodical way to determine the relationship between these factors in order to create satisfaction, decrease customer complaints and increase loyalty. Finally, none of the previous studies transforms customer complaints into business improvement opportunities for the company. This latter issue is addressed in this thesis.

1.9 Thesis Objectives

This thesis will focus on developing a methodology to carry out analytical and collaborative customer relationship management in the area of customer complaints and customer satisfaction. The case studies used for proof-of-concepts have been defined, namely the Western Australia Department of Transport and Fremantle Port Authority. We obtain the field data by working with the industry partners. Currently, no company in the world has intelligent CRM.

This research aims to describe CRM concepts, offer new opinions, and provide a state-of-the-art conceptual framework. It also proposes to generate recommendations and solutions for customer complaints. Also, utilising intelligent methodology, we address the gaps between CRM and I-CRM systems. The research objectives to achieve this are:

(1) To categorise and analyse customer feedback whether positive or negative and define customer complaints central to CRM.
(2) To identify and address the issues faced by I-CRM and then develop strategies for intelligent customer relationship management.
(3) To test and validate our hypotheses for intelligent customer relationship management and I-CRM evaluators.
(4) To evaluate customer satisfaction in order to decrease the number of complaints.
(5) To evaluate (proof the concept) our proposed solution, i.e. the I-CRM system, using quantitative and qualitative approaches and real-world data from our ARC industry partner: The Fremantle port.
(6) To develop a methodology in the form of a modelling tool to find the relationship between various variables of perceived value and interactivity with customer satisfaction in one way, and customer satisfaction with loyalty and customer acquisition in another way.

1.10 Significance of the Research

To the best of our knowledge, in the current literature review, no methodology exists which can qualitatively and quantitatively evaluate customer satisfaction and loyalty by analysing customer feedback and complaints. In addition, no methodology has been previously proposed to optimise the customer management process. In this study, we present a novel process that simultaneously optimises key customer recognition and multiple customer feedback. We construct a design for the experiments. Then, using several methods, we analyse experimental data and decision-making regarding the prioritization of customers and issues that need to be dealt with. We can also extrapolate customer satisfaction and its development with our proposed framework. Therefore, the significance of this thesis can be summarised as follows:

1. The thesis proposes a new approach to analyse loyalty in relation to interactivity, perceived value and intelligent CRM.
2. This thesis proposes a new Intelligent CRM (I-CRM) framework which focuses on the customers and their feedback and addresses their issues promptly.
3. This thesis proposes a methodology for categorising customer feedback and allows the threshold of complaint probability to be recognised.
4. This thesis proposes a methodology to find key customers and increase potential improvement of the customers. This will intensify the productivity of

the company. This will be determined after reducing the amount of data and categorising customers into groups.
5. This thesis proposes a methodology by which it can assess customer satisfaction and optimise decision-making using linear and non-linear approaches.
6. This thesis proposes a methodology such as linear and non-linear modelling for the analysis of the relationship between customer satisfaction and related variables which have a major influence on customer relationship management.

1.11 Structure of the Thesis

In this thesis, we create a comprehensive methodology for intelligent customer relationship management to achieve customer satisfaction and obtain loyalty. To fulfill this objective, this study comprises nine chapters. In the next section, we will provide a short summary of each chapter.

Chapter 2 provides a review of previous works in the existing literature. This chapter examines current issues in terms of customer relationship management processes. It also focuses on the relationship between the variables in our conceptual framework. Furthermore, it describes the methods that have previously been proposed to solve the problems, and discusses the shortcomings of previous works.

Chapter 3 begins by providing definitions of the concepts applied in this thesis. Also, it defines the problems involved in this study that we will address further throughout the thesis. It also describes the methodologies and tools that will be used in this study.

Chapter 4 provides solutions to the problems identified in Chap. 3. We define terminologies that are used in this thesis, using software to categorise our definitions and perform text mining to find cutting edge definitions. We also provide our conceptual framework in this chapter and separate it into two sections, each of which is discussed separately.

Chapter 5 provides the text-mining process of the data collected in the first round of our data collection process. We introduce a real-world case study as an example and we examine its problems in order to define and discuss the main issues. Then, using the text mining procedure, we categorise and prioritise customer complaints qualitatively. The results of the text mining analysis are presented in this chapter.

Chapter 6 presents our preliminary analysis and hypotheses testing, utilizing two methodologies. In this chapter, we introduce two sets of hypotheses in two separate sections of our conceptual framework. In the first section, we have nine hypotheses and in the second section, we have four hypotheses. We examine the hypotheses to ensure that the relationships between variables are acknowledged.

Chapter 7 presents methodologies to address the issues. This chapter introduces five-step algorithms for analysing customer satisfaction. We need to reduce the data and we intend to identify key customers using data envelopment analysis. We then optimise the recognition of customer complaints and to achieve this, we categorise our customers into various groups using a linear modelling approach.

Chapter 8 describes methodologies to address the relationship between various variables that are used as intelligent CRM evaluators using non-linear modelling and analysis. We determine the linear relationship between customer satisfaction, perceived value, interactivity, loyalty and customer acquisition. Following that, we target our variables using a quantitative method which is ANFIS (network-based fuzzy inference system). ANFIS is a modelling technique that deals with fuzzy uncertainty in the data and displays fuzzy data modelling. Finally, we present a diagram which shows how customer complaints can be converted to business opportunities. This chapter concludes with a discussion of the main contributions of this thesis.

Chapter 9 concludes the thesis and provides insights into and directions for future work. Also, it summarizes the outcomes of the thesis. Additionally, we discuss the contributions of this thesis to the literature.

1.12 Conclusion

The enhancement of customer relationship management is a major concern for many companies and their activities around the world. The realization and adoption of CRM and its components have become central to a great deal of information technology sales and marketing. In this chapter, we provide definitions of CRM and its benefits. Then, we briefly discuss the significance of customer relationship management. We also discuss various CRM systems and the problems with existing CRM solutions. Furthermore, we discuss complaints and how CRM systems and strategies might be able to resolve them and create customer satisfaction. Then, we discuss the motivation for, and the objectives and significance of, this thesis.

In the next chapter, we present an overview of the existing literature on customer relationship management and the outcomes of using it including: customer satisfaction, perceived value, interactivity, loyalty and customer acquisition. The main objective of this thesis is to ensure that the problems that we tackle and address in this study have not been previously addressed like; we accomplish this by conducting this research in an exclusive working environment.

References

1. Faed, A. (2010). A conceptual framework for E-loyalty in digital business environment. In *Digital Ecosystems and Technologies (DEST), 2010 4th IEEE International Conference on*, 2010, pp. 547–552.
2. Ernst, H., Hoyer, W. D., Krafft, M., & Krieger, K. (2011). Customer relationship management and company performance—the mediating role of new product performance. *Journal of the Academy of Marketing Science, 39*, 290–306.

3. A. A., (2003). Classifying and selecting e-CRM applications: an analysis-based proposal. *Management Decision,* 41, 570–577.
4. Becker, J. U., Greve, G., & Albers, S. (2009). The impact of technological and organizational implementation of CRM on customer acquisition, maintenance, and retention. *International Journal of Research in Marketing, 26*, 207–215.
5. Hendler, R. (2004). Revenue management in fabulous Las Vegas: Combining customer relationship management and revenue management to maximise profitability. *Journal of Revenue and Pricing Management, 3*, 73.
6. Krafft, M., Reinartz, W., & Hoyer, W. D. (2004). The customer relationship management process: Its measurement and impact on performance. *Journal of Marketing Research, 41*, 293–305.
7. Hwang, H., Jung, T., & Suh, E. (2004). An LTV model and customer segmentation based on customer value: A case study on the wireless telecommunication industry. *Expert Systems with Applications, 26*, 181–188.
8. Grant, J., & Katsioloudes, M. (2007). Social marketing: Strengthening company-customer bonds. *Journal of Business Strategy, 28*, 56–64.
9. Adamson, W. J. G., & Tapp, A. (2006). From CRM to FRM: Applying CRM in the football industry. *The Journal of Database Marketing and Customer Strategy Management, 13*, 156–172.
10. Sinisalo, J. S., Karjaluoto, J. H., & Leppaniemi, M. (2007). Mobile customer relationship management: Underlying issues and challenges. *Business Process Management, 13*, 771–787.
11. Boulding, R. S., Ehret, W. M., & Johnston, W. J. (2005). A customer relationship management roadmap: What is known, potential pitfalls, and where to go. *Journal of Marketing, 69*, 155–166.
12. Chalmeta, R. (2006). Methodology for customer relationship management. *The Journal of Systems and Software, 79*, 1015–1024.
13. Xu, Y., Duan, Q., Yang, H. (2005). Web-service-oriented customer relationship management system evolution.
14. Harej, K., & Horvat, R. V. (2004). Customer relationship management momentum for business improvement, 1, 107–111.
15. Boland, D., Morrison, D., O'Neill, S. (2002). The future of CRM in the airline industry: A new paradigm for customer management. *IBM Institute for Business Value.*
16. Luck, D., & Lancaster, G. (2003). E-CRM: Customer relationship marketing in the hotel industry. *Managerial Auditing Journal, 18*, 213–231.
17. Stefanou, C. J., Sarmaniotis, C., & Stafyla, A. (2003). CRM and customer-centric knowledge management: An empirical research. *Business Process Management Journal, 9*, 617–634.
18. Yao, X., Li, X., Su, Q. (2005). Study on the customer relationship management and its application in Chinese hospital.
19. Stone, M., Foss, B., & Ekinci, Y. (2008). What makes for CRM system success or failure? *Journal of Database Marketing and Customer Strategy Management, 15*, 68–78.
20. Stevens, E., & Dimitriadis, E. S. (2008). Integrated customer relationship management for service activities an internal/external gap model. *Managing Service Quality, 18*, 496–511.
21. Payne, A., & Frow, P. (2005). A strategic framework for customer relationship management. *Journal of Marketing, 69*, 167–176.
22. Gee, R., Coates, G., & Nicholson, M. (2008). Understanding and profitably managing customer loyalty. *Marketing Intelligence and Planning, 26*, 359–374.
23. Faed, A. (2010). A conceptual framework for E-loyalty in digital business environment, 547–552.
24. Harris, L. C., & Goode, M. M. H. (2004). The four levels of loyalty and the pivotal role of trust: A study of online service dynamics. *Journal of Retailing, 80*, 139–158.
25. Stauss, B. (2002). The dimensions of complaint satisfaction: Process and outcome complaint satisfaction versus cold fact and warm act complaint satisfaction. *Managing Service Quality, 12*, 173–183.

26. Wu, I. L. (2005). A hybrid technology acceptance approach for exploring e-CRM adoption in organizations. *Behaviour and information technology, 24*, 303.
27. Xu, M. (2005). Gaining customer knowledge through analytical CRM. *Industrial Management+Data Systems, 105*, 955.
28. Frow, P., Payne, A., Wilkinson, I. F., & Young, L. (2011). Customer management and CRM: Addressing the dark side. *Journal of Services Marketing, 25*, 79–89.
29. Bolton, M. (2004). Customer centric business processing. *International Journal of Productivity and Performance Management, 53*, 44–51.
30. P. B. Brandtzæg, and J.Heim, "User loyalty and online communities: why members of online communities are not faithful," *The 2nd international Conference on intelligent Technologies For interactive Entertainment*, 1–10, 2008.
31. Ferguson, R. (2008). SegmentTalk: The difference engine: a comparison of loyalty marketing perceptions among specific US consumer segments. *The Journal of Consumer Marketing, 25*, 115.
32. Ryals, L., & Knox, S. (2001). Cross-functional issues in the implementation of relationship marketing through customer relationship management. *European Management Journal, 19*, 534–542.
33. Kim, J. (2003). A model for evaluating the effectiveness of CRM using the balanced scorecard. *Journal of Interactive Marketing, 17*, 5.
34. Rigby, F.R., & Dawson, C. (2003). Winning customer loyalty is the key to a winning CRM strategy. Ivey Business Journal.
35. Eggert, A. (2002). Customer perceived value: A substitute for satisfaction in business markets? *The Journal of Business and Industrial Marketing, 17*, 107.
36. Kim, H. S., & Kim, Y. G. (2007). A study on developing CRM scorecard, 150b–150b.
37. Peelen, E. (2005). *Customer relationship management*. Essex: Prentice Hall.
38. Xu, M., & Walton, J. (2005). Gaining customer knowledge through analytical CRM. *Industrial Management and Data Systems, 105*, 955–971.
39. Sun, Z. H. (2008). Information system and management strategy of customer relationship management. *The 3rd International Conference on Innovative Computing Information and Control (ICICIC'08)*.
40. Alshawi, S., Missi, F., Fitzgerald G., (2005). Why CRM efforts fail? A study of the impact of data quality and data integration Of Data Quality And Data Integration. *Proceedings of the 38th Hawaii International Conference on System Sciences*.
41. Kimiloglu, H., & Zarali, H. (2009). What signifies success in e-CRM? *Marketing Intelligence and Planning, 27*, 246–267.
42. Massey, A. P., Montoya-Weiss, M. M., & Holcom, K. (2001). Re-engineering the customer relationship: leveraging knowledge assets at IBM. *Decision Support Systems, 32*, 155–170.
43. Richards, K. A., & Jones, E. (2008). Customer relationship management: Finding value drivers. *Industrial Marketing Management, 37*, 120–130.
44. Homburg, M. D., & Totzek, C. D. (2008). Customer prioritization: Does it pay off, and how should it be implemented? *Journal of Marketing, 72*, 110–130.
45. Smith, A. (2006). CRM and customer service: Strategic asset or corporate overhead? *Handbook of business strategy, 7*, 87.

Chapter 2
Literature Review

2.1 Introduction

Early in Chap. 1, we introduced and discussed various terms and concepts regarding customer relationship management, its components and the differences between CRM systems and strategies. We also explained the significance of complaints and their importance in any organization. Primarily, customer complaints have the potential to undermine the reputation of an organization. Complaints can even impact on communications, interactions and decision-making processes. In this chapter, we review the current literature on customer relationship management including its definitions, concepts and the solutions provided by CRM systems and strategies. We also examine the underlying components of these systems including customer satisfaction, customer acquisition, customer loyalty, perceived value and interactivity. Furthermore, we investigate the origins of existing problems, what has already been discovered about the existing problems and the various approaches that have attempted to address previous issues. Then, we undertake a critical review and analysis of the literature in order to evaluate, compare, and analyse the various opinions.

In the next section, we discuss various definitions based on the literature review and provide all of the definitions that have previously been formulated by other researchers. Also, we illustrate ways of administering to customers using customer relationship management, the components of CRM, strategies and methods from the past literature reviews which have been discussed. The various definitions of CRM and its main variables are presented in Sect. 2.2. In Sect. 2.3, we discuss various types of CRM. In Sect. 2.4, we examine existing CRM approaches and the relevant current literature. In Sect. 2.4.1, we describe the importance of CRM and the various approaches to it. In Sect. 2.4.2, we debate the impact of CRM on perceived value and its interrelations. In Sect. 2.4.3, we discuss the relationship between CRM and interactivity, the methods of strengthening interactivity, and how these two elements can forge an appropriate bond with each other. The customer-loyalty-focused approach and its importance with respect to customer satisfaction is discussed in Sect. 2.4.4. Section 2.4.5 examines the importance of

A. R. Faed, *An Intelligent Customer Complaint Management System with Application to the Transport and Logistics Industry*, Springer Theses, DOI: 10.1007/978-3-319-00324-5_2,
© Springer International Publishing Switzerland 2013

customer acquisition and its relationship to CRM and other factors such as loyalty and customer satisfaction. In Sect. 2.4.6, we concentrate on customer complaints and current approaches related to this issue. In Sect. 2.4.7, customer satisfaction and various methodologies from extant literature are discussed. In Sect. 2.5, we examine the current methods and tools to measure and evaluate CRM and major factors such as customer satisfaction and loyalty. Section 2.6 presents the existing CRM system together with a discussion of their pros and cons. In 2.7, we critically discuss and evaluate existing literature. In Sect. 2.8, we touch upon open source CRM and cloud and we discuss their importance as a concept. In Sect. 2.9, we provide a summary of the literature and the chapter concludes with Sect. 2.10.

2.2 Concepts and Definitions of CRM

CRM has been defined in numerous ways and with many descriptions. It can be defined as the art of acquiring customers and having a long-lasting relationship with them. Companies must take the initiative to actualize and implement CRM. Also, CRM is a combination of people, processes, and technology in order to understand and obtain customers for the company. It focuses on customer retention and builds up the relationship. To benefit fully from the implementation of CRM, companies must have efficient CRM programs to secure the loyalty of the customers [1]. Furthermore, proper relationships with customers need to be conducted by sophisticated management [2]. In order to compete with business rivals and keep pace with the competition in today's market, businesses need to have more than just a professionally designed Website; they need to engage and involve users with an encyclopaedic system and strategies to support their companies [3].

CRM applications are able to provide an effective connection from front office to back office and touch points with the customers. An organization's touch points include the Internet, E-mail, call centres, face-to-face marketing, fax, pagers and kiosks. CRM can be employed to consolidate these touch points for the benefit of the organization. Using CRM, companies can maximise their interactions with customers and obtain a 360-degree vision of customers [1]. Since the whole study can be accomplished within the context of CRM, it is mandatory to be aware of some important definitions and improvements in this area. In the next section, we highlight and discuss various definitions of customer relationship management from the current literature.

2.2.1 Strategic-Based Definition for Customer Relationship Management

Mendoza et al. define CRM as a process, human factor and technology that produces the best relationship with customers to intensify value, satisfaction and customer loyalty [4]. Also, Ueno [5] consider CRM as an advanced level of marketing strategy to intensify acquisition and retention of customers and creates long-term value and a long-term relationship with customers. Additionally, Özgener, Iraz and Hoots [6, 7], define CRM as a business strategy and a picture of customer requirements to increase profitability and intellectually manage sales, marketing and service procedures. Payne and Frow [8] see CRM as an entity that is a strategic approach that unites marketing activities and information technology to create long-lasting relationship with clients, using human factors, technology and operations. Anderson [9] defines CRM as one of the most robust weapons managers employ to guarantee that customers will remain attached and loyal to the company. Moreover, CRM is defined by Özgener, Iraz and Robinson Jr et al. [6, 10] as a core business strategy and a key driver that creates competition among organizations to provide better goods and services to the customers.

Based on Faed et al. [11], CRM is a strategy used to increase customer retention, customer intimacy and build customer equity. Also, Robinson Jr et al. and Chaudhry [12, 13] maintain that ways of creating and maintaining relationships with customers must be assigned based on customers' value. Furthermore, Lawson Body and Limayem [14] define that CRM is a set of strategies that create an interactive relationship with customers and enable an understanding of their requirements and expectations in order to provide personalised and customized services and products. Limayem [15] also defines CRM as a customer-focused strategy that provides customer satisfaction and a management strategy which utilises a technology to create a long-term relationship with customers. Additionally, Frow et al. [2] claim that customer management is central to customer relationship management and relationship marketing, as customer management provides tactics for dealing with the issues, makes the environment more interactive, and facilitates transactions.

2.2.2 Process-Based Definition in CRM

Customer relationship management is a set of business processes for the purpose of retaining customers and maintaining their loyalty to the company [16–18]. Also, CRM is a systematic management of relationships across all parts of the business, focusing on customers, providing long-term value for them, and increasing customer interaction. It also includes communication channels and offers of different services, thereby producing customer retention and loyalty [13, 19, 20]. It is a concerted effort to acquire customers and gratify their requirements [6, 17]. Furthermore, it is a process-driven strategy that ensures customer retention and face-to-face marketing followed by clarifying customers' preferences [21].

At the same time, it is an inclusive business process which is customer-centric and helps organizations to operate flawlessly and enhance customer interactions [10]. It is also an interactive procedure that creates a balance between a company's success and customer satisfaction to generate profit for them [6]. Other researchers maintain that CRM is a process which identifies current and potential customers, and builds long-term relationships and short-term transactions with them [2, 22]. It is a process of building and managing the mutual relationship between company and the customer to maximize customer relationship and fulfilment [22]. This process customizes services, increases customer focus and individual pricing [23]. While CRM is a comprehensive business process which utilizes information to support business activities, it conceptualizes customer definition and provides a whole view of the customer in order to generate customer intelligence in all stages of a business [20].

On the other hand, CRM can be considered as a process that correlates and systematizes activities which are significant so as to entrench credible, long-lasting, worthwhile experiences for all parties in a business [12]. Moreover, by using CRM approaches and their components, companies can reduce operating costs, have an appropriate interaction with their customers and make the best decisions which in turn are followed by customer support [24, 25]. Finally, it can create a customer-focused data warehouse and express an organization's clear-cut vision [11, 26].

2.2.3 Customer Complaint in CRM

A complaint can be defined as an expression of disagreement between the customer and the company. Solutions must be seen as being fair and impartial, and this is the main criterion when customers select or remain with a company [27]. A complaint mostly results from some disruption to expected services [28]. Complaints represent a comprehensive set of behavioral and non-behavioral responses (non-standard behaviors) made by customers who are engaged in the communication of contradictory understandings, and begins with dissatisfaction with a situation [29]. Additionally, customers may choose various means to express their complaints: verbally, in writing, or face-to-face [30]. Customer complaints reflect real-life situations which must be dealt with by companies using appropriate strategies in order to ultimately satisfy the customers. From the customer's perspective, a complaint should be dealt with promptly and thoroughly by a company, and it should be sufficient to motivate a company to take appropriate action and reach a solution that is satisfactory to the customer [28]. Taken seriously, complaints have the ability to improve the quality of products manufactured by a company and the level of service provided.

According to Stauss [31], a complaint is a verbal or written articulation of discontent that sends a warning to a company or service provider about its behaviors, service or product(s). It is usually created quite deliberately. Customers usually complain if they are confident that there is a chance of compensation and this would be one of the benefits of lodging a complaint.

When a company deals appropriately with a complaint, this will enhance its reputation as customers tend to be favorably disposed towards those companies that listen to them. Complaints have the potential to increase customer loyalty, and promote communication between the company and its customers [32]. Decreasing the incidence of customer complaints often creates customer satisfaction and customer retention. Nevertheless, customers may exhibit different behaviors when expressing complaints. They express their dissatisfactions in different ways which could be positive or negative [30].

Complaint is a significant issue that companies need to deal with in order to increase the level of satisfaction, output generation and profit making [33]. Also, it has been always used as a facilitator between individuals and companies [34]. Each complaint provides an improvement opportunity for a company [35]. The advent of updated complaints could be a starting point for analysis and improvement. From the learning perspective, it can often be a great source of learning, allowing a company to create a favorable situation [36]. Appreciating and giving some credit to complaints as part of a business process promotes a company's long-term relationships with its customers. Complaints must be dealt with in a way that will ensure that the same or similar complaints do not re-occur.

Based on [35], 1–5 % of all complaints are assigned to the headquarters or the management team of the companies. Forty-five percent of all complaints are designated to agents, branches or frontline delegates. Fifty percent of customers who encounter various issues do not make complaints due to feelings of helplessness [35]. The level of complaints varies case by case and when an issue escalates, it becomes more costly. It is shown that those customers whose complaints are dealt with to their satisfaction are 30 % more loyal than those customers who usually do not complain about the issues at all [35].

2.2.4 Complaint Management

Based on Stauss [31], a company needs to have the following factors to be fully capable of openly receiving customer complaints:

People in positions of authority should always be approachable and engage in appropriate customer-centric interactivity. If a customer makes a complaint face-to-face, s/he should be treated in a polite and friendly manner. The one who receives the complaint must have empathy and understand the customer's situation. Also, the receiver must expect and be ready to deal with a variety of customer attitudes and approaches. The receiver of the complaint must take initiatives and be helpful and reliable in order to project trustworthiness to the customer. It is vital that action be taken promptly and appropriately so that the next customer can be dealt with. Finally, the outcomes generated by the system must be fair and legitimate so that customer does not experience discrimination.

According to Stauss [31], there are three types of complaint resolution:

1. Financial solutions: return the money to the complainant; reduce the price of service or product.

2. Tangible solutions: change the product; offer a gift to the customer; propose an alternative, or simply repair the product or re-offer the service.
3. Intangible solutions: offer an apology to the customer, listen to the customer and give convincing reasons for the unsatisfactory situation.

2.2.5 Complaint and Derivations in CRM

Figure 2.1, [37] shows the relationship between complaint and derived concepts—dissatisfaction, defection and retention. Complaints may lead to defection, meaning that a customer no longer wants to have dealings with the company [38]. Some complaints will lead only to dissatisfaction; in this case, the company can remedy those problems in order to retain the customer [28, 36]. The company can strengthen its customer retention because it has responded appropriately. In addition, expectation may have some relationship with extra retention [28, 36].

Defection is a complaint reaction that has been described as departure or leaving attitude (customers start leaving the company) and is often a common reaction following a complaint. A customer may choose to defect to a competitor rather than lodge a complaint. Similarly, customers who express their dissatisfaction by complaining may still choose to leave regardless of the outcome [36]. By defecting, customers signal their desire to no longer use a particular company's product or service, thereby never again buying from or using the same company. Defection is an effective manifestation of a customer's dissatisfaction. Moreover, defection is a customer's decision to completely stop purchasing and supporting a company's services and products. This may occur because of several issues over time and creates a gradual destruction of bonds [30].

Many studies emphasise the benefits of customer retention. Authors in [39] indicate that a 1 % improvement in the customer retention proportion ameliorates and enhances a firm's value by 5 %. Similarly, [38] show that a 5 % increase in customer retention increases a firm's profits in a range of 25–85 %. Companies need to identify previously profitable but currently inactive customers and initiate appropriate activities to reactivate those customers. The focus on retention is established upon the constant and constructive hypotheses that there subsists a durable and robust affiliation between customer retention and profitability:

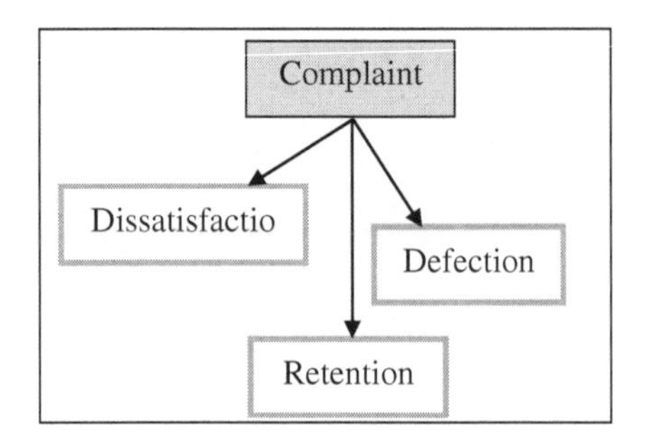

Fig. 2.1 Complaint and derivations in CRM

long-term customers buy more and are less costly to serve, whereas replacing existing customers with 'new' ones is demonstrably a more expensive and risky strategy, since it is feasible to assume that switched customers are more likely to continue their dissatisfied behavior in the near future [40].

Dissatisfaction among the customers has been described as the lack of verification in some service anticipation that has been caused by service breakdown [36]. As shown in Fig. 2.1, the handling of complaints may assist a company to better retain its customers. If the company improves its ability to satisfy its customers and prevent all sorts of defections, it will be able to retain its own customers, as retention is the positive feedback of the complaint. A company should learn from its mistakes. The complaints process is an iterative one, and with each complaint, a company has the opportunity to increase its knowledge about its customers [37].

2.2.6 Definitions of Perceived Value in CRM

One of the antecedents of customer relationship management in the work setting is the new atmosphere produced by even a subtle change in an organization. It generates new experiences for the employees as well as for the customers, and prompts the latter to demand newer experiences [41]. Perceived value is shaped according to the authentic (objective) price and the consumer's reference price. Perceived value and satisfaction are two inter-related elements, but have separate constructs and both are dependent on the consumption model. People have different feelings when buying a product or a service. Customers have pre- purchase and post-purchase feelings. Satisfaction is a post-purchase feeling and perceived value might be either pre- or post-purchase [42]. Perceived value is also defined as social value, emotional value, hedonic value, spiritual value and esteem value [43]. To date, perceived value has been of great interest to many researchers since it may lead to customer satisfaction, customer attitudinal intention and corporate image [44].

Perceived value from the customers' perspective implies the benefits customers acquire which shape their experience based on the planning that has been accomplished [45]. Furthermore, perceived value is created when a customer selects a particular service, is gratified by and acknowledges the service provided by a company and repeatedly seeks the service. In this way, an individual starts evaluating the benefits derived from utilization of the service in terms of incurred costs [46]. Perceived value has a direct link with customer behaviour and is central to the service evaluation. Also, it has a close relationship with satisfaction. The term 'customer value' is interchangeable with 'perceived value' [47].

2.2.7 Definitions of Interactivity in CRM

'Interactivity' is defined as the ability of individuals and companies to communicate personally and straightforwardly with each other irrespective of distance.

The effective interaction and communication of human beings have been always problematic among individuals and organizations. Also, managing interactive and communicative relationships is becoming an issue for organizations regardless of place and region. Effective communication and interpersonal expertise has a significant impact on customer relationships. To date, little research has been done to assess the interaction of companies with their stakeholders. According to [48], managers must have transparency when passing information to customers and in their other interactions. They must also understand the requirements and preferences of a customer [48].

Apart from investigating the source of issues and turning them into new information, customer service personnel must have the ability to learn and remember all the answers and feedbacks. Furthermore, interactivity has been described as a dialogue between the customer and the company. Although previously there has been restricted interactivity, today it is not the same thing and it is totally integrated and interactive [45].

2.2.8 *Definitions of Satisfaction in CRM*

Customer satisfaction is a key factor in the success of any company and is produced when customers' needs have been met and they have derived profit or value from their experience. Also, customer satisfaction brings about new experiences to the customers whose needs have been fulfilled and satisfied.

According to Becker et al., customer satisfaction implies an extended relationship through activities such as selling, increasing revenue of the customers, and generating customer maintenance [38].

Mithas et al. mention that customer satisfaction is a factor that directly or indirectly impacts on a company and society. Companies must perform well, adhere to social contracts and show mutual understanding. Customer satisfaction may have an impact either positive or negative on customer feedback. More satisfaction creates security and decreases loss of clients [49]. Blocker et al. state that customer satisfaction creates positive word of mouth advertising, attracts more customers to the company, and retains existing customers. Moreover, customer satisfaction creates loyalty as a direct effect [50]. Similar to this thesis, Caruana has used 'service quality' as an interchangeable term for 'customer satisfaction' [51].

Satisfaction or service quality has been described as an outcome of customers' expectations based on their comparison and perceptions about goods and services and also how the final result transfers to the customers and to what extent it makes them happier [51]. Using customer satisfaction, the company can create new relationship circles using its customers and its partners. Furthermore, this characteristic will make the company more reliable [52]. Customer satisfaction is defined by Wang and Yang as an important element in creating profitability and building bond and value for customers and it has been greatly utilized by most of the organizations to bring back customers and to promote widespread positive word of mouth [53]. As defined by Coates et al., customer satisfaction is derived from delighting a customer and providing positive surprise which exceeds customers' expectations [54]. Furthermore, Avlonitis and Panagopoulos maintain that customer satisfaction is considered as an information technology surrogate and refers to the way that the company can meet the requirements of the customer and make them happy [41, 55].

2.2.9 Definitions of Loyalty in CRM

'Loyalty' is defined in terms of repeated buying behaviour [56]. Today, the concept of customer loyalty is all-inclusive and it is an important research topic for researchers and companies and one that needs to be studied more in order to achieve maximum effectiveness. Loyalty is considered as a credit in cutting-edge markets of today. Customer loyalty is recognized as the probability of purchase, probability of product repurchase, purchase frequency, repeat purchase behavior and buying order [57, 58]. Based on [43], customer loyalty is a major latent variable, connected to the probability of returning a customer to the system. Then the customer will generate referrals and word of mouth.

At the centre of loyalty programs is the principle of bolstering, whereby it is supposed that behaviours which are rewarded will be repeated [59]. It costs a company more to attract and absorb a new customer than it does to implement a retention strategy. Also, it is said that client acquisition cost is 20–40 % more than finding a new client in the physical marketplace, as mentioned in the study of the Internet clothing market. This produced higher losses in the very first stages of the relationship, but after the 24th to 30th month, the Internet clients probably need to spend twice as much as they used to do in the first six months [60]. A gratified customer has a tendency to be more loyal to a brand or store over time than a customer whose buying occurs due to other causes such as time limitations and information about possible loss or savings [56]. Likewise, loyalty is the outcome of the wise and legitimate relationship that a company establishes with the customer [61].

The main goal of this loyalty is to increase customers' support and their level of faithfulness to the brands and the company and encourage the customers to repeat their purchasing rates [58, 62]. Using an extensive and broad loyalty program, the company will be able to measure, enhance, control and manage customer profitability [63]. The main aim of customer loyalty programs is to increase customers' loyalty to the firm by attempting to increase repeat purchase rates or total purchases [62]. The benefits of customer loyalty to a provider of either services or products include: lower customer price sensitivity, reduced expenditure on attracting new customers; and improved organizational profitability [64]. Loyalty has been measured by the probability of product repurchase [58]. Brand loyalty is the preferential, attitudinal and behavioural response to one or more brands in a product group expressed over a period of time by a series of customers [58].

The notion of enhancing the relationship of a company with its clients goes back many years ago and was examined in earlier days in terms of distribution marketing. Now, it is seen as a bond builder. Nowadays, some organizations regard loyalty as a defensive marketing method and strategy [65]. Defensive marketing strategies concentrate on empowering relationships of the company with its clients for the long term and probably create new connections and businesses with potential clients [66]. However, it focuses on the current and potential customers and it is believed that it should happen in the early stages of the sale [65].

2.2.10 Definitions of Customer Acquisition in CRM

Customer acquisition is a sensitive aspect of business and is one of the outcomes of the customer relationship. Thomas [67] states that customer acquisition is a dependent procedure and in the majority of times impacts on customer retention. It is one of the critical factors in customer relationship and customer acquisition that usually occurs if the system provides for interactivity and if customers are satisfied [67]. Also, based on the M. Lewis framework, there is always uncertainty and doubt in customer acquisition [68]. Becker defines customer acquisition as a first objective of CRM for acquiring new customers and maintains that new customers deserve the same attention as the company gives to potential customers. Based on [38], it is also considered as customer initiation. Customer acquisition has a major impact on CRM and creates effective customer relationships. Also, it may impact on the retention of customers and customer behaviour [69].

2.3 Types of CRM

In order to maintain its viability and success in the market, a company needs to be better than its competitors, focus on its customers, and create value for them. Customer relationship management is an integrated process and business strategy which selects and manages customers to optimise value in the long run. Companies use CRM to identify, attract, satisfy and maintain a close bond with customers and their partners. In terms of typology, CRM is categorised into three groups: analytical, operational or transactional, and collaborative or interactive.

2.3.1 Analytical CRM and its Significance

Analytical CRMs are effective applications which analyse customer data that has been generated by operational tools for the purpose of business performance management. Data gathered by operational CRMs are analysed to classify customers or to identify cross-selling and up-selling potential.

Data collection and analysis are a continuing and iterative process. Ideally, business decisions are refined over time based on feedback from earlier analysis and decisions.

Analytical CRM is a stepwise procedure. Suppose that we have a data source. Using analytical CRM, detailed and also historical data can be generated using operational integration. Using analytical CRM, we obtain all necessary information regarding marketing activities and marketing potentials. We are also able to select a particular market to target which is one of the most important steps in marketing activities and customer relationship processes [15].

In a nutshell, an analytical CRM can reduce customer angst by identifying customers who want to abandon the company and the ties of loyalty. Also, analytical CRM can classify customers and enhance sales by customizing its selling approaches. It controls cost and revenue and improves supply chain management. An analytical CRM investigates the weaknesses throughout the whole system and subsequently turns these into strengths and opportunities.

2.3.2 Operational or Transactional CRM and its Significance

The implementation of operational CRM best practices takes into consideration personnel roles and workplaces. Operational CRM has customer-facing applications that consolidate the front, back and mobile offices, including sales-force automation, enterprise marketing automation, and customer service and support.

Also, operational CRM entails supporting the so-called "front office" business processes which include customer contact (sales, marketing and service). Operational CRM aims at contacting the customers using different touch points such as call centres, web access, e-mail, direct sales and fax. In addition, customers and refined business actions have a direct effect on customer touch points [15]. Operational CRM provides higher customer gratification as the quality of the contacts has been enhanced. It is also a cost-effective approach in customer relationship management due to the consolidation of procedures and process support. It also has stronger consolidation of interaction and communication with customers, using the company's internal procedures. According to [70], the majority of CRM systems have operational functions such as contact management, call centre applications and service support. Approximately 40 % of the CRM systems propose analytical CRM such as knowledge management and about 20 % deal with collaborative CRM. However, about 45 % of the CRM systems offer e-CRM solutions, and Oracle, SAP and other systems provide personalised and customized self-service to the customers to gain trust via the Internet or intranet.

Operational CRM simply delivers customized, personalised and efficient marketing, sales, and service through multichannel collaboration. Furthermore, using operational CRM, sales people and service engineers can access the complete history of all customer interactions with their company, regardless of the touch point. Additionally, it enables a 360-degree view of the customer during the interaction.

2.3.3 Collaborative or Interactive CRM and its Significance

Collaborative services are those that facilitate interactions between customers and businesses (e.g. personalised publishing, e-mail, communities, conferencing, and web-enabled customer interaction centres). Collaborative CRM is used to establish the lifetime value of customers beyond the transaction by creating a partnering relationship. It facilitates interactions with customers through all channels and supports the co-ordination of employee teams and channels. Channels include personal contact, letter, fax, phone, web and email. It is a solution that brings people, processes and data together so companies can better serve and retain their customers. The data and all sorts of information might be structured, unstructured, conversational, and/or transactional in nature [15].

Collaborative CRM enables efficient productive customer interactions across all communication channels. It also enables the Internet participations to reduce customer service costs. It integrates call centers enabling multi-channel personal customer interaction. Collaborative CRM has the ability to integrate a view of the customer while interaction is taking place at the transaction level. It also shows the caller's personal information to the agent before he picks up the call. Collaborative CRM traces a particular customer's past transaction history. One of the very important positive points of all collaborative CRMs is that customers can simply seek assistance from Online FAQs, sales representatives by phone and virtual sales representatives.

2.3.4 Problems with Above CRM Solutions

Much research has been conducted in the area of customer relationship management, customer satisfaction and loyalty. As our focus is on customer complaints, we examine the works in this particular field. It is necessary to state that the research undertaken for this thesis opens up a new horizon for those researchers who want to explore solutions and recommendations to address the issue of customer complaints using customer relationship management processes. To date, no research has attempted to address the issue of complaints in our particular area of interest, that is, transport and logistics. Our case is unique and to the best of our knowledge is the first research to be attempted in this field. Also, we attempt to examine as many complaints as possible in the context of customer relationship management and its components such as customer satisfaction, loyalty, perceived value, interactivity and customer acquisition. We will also investigate the case where a huge organization has failed in its responsibility to satisfy its customers, although the latter, who are drivers and clients, try to maintain their loyalty. In this chapter, we identify in detail the main problems of CRM systems.

2.4 Existing CRM Approaches

In the previous section, we discussed various methods used to analyse the customer relationship and its components. In this section, we discuss different approaches of CRM proposed by previous researchers and since all these approaches are largely from their perspectives of business and marketing, we classify all the existing approaches into five groups for the purpose of analysis. These are: perceived value, interactivity, customer satisfaction, loyalty and customer acquisition. We will then consolidate and evaluate these approaches.

2.4.1 Customer Relationships-Focused CRM Approach

Azvine and Nauck [71] propose an intelligent customer relationship management analytics model to solve customers' business problems. Customers are interviewed to determine their issues and to provide customer satisfaction and achieve the performance of the system. To assess, they introduce a business procedure to optimise the decision-making process. They create a system called Intelligent Universal Service Management System (iUSMS) to manipulate the data. Also, it has the ability to learn and obtain the latest knowledge. When a customer calls to report an issue, an operator will simply find that customer's profile and its details. Meanwhile, the system checks the failure using an automatic evaluation. Someone will be sent to fix the problem and the system will retain the information about the incident. The advantage of a customer focus measure is that it is a tool for creating customized automatic report in order to recognize issues, contacts, and to generate statistics. This system allows the user to customize further information and as the system gives a great deal of freedom to the customers, it engenders trust. Hence, the authors propose a system to be used by customers whereby they are granted permission to use the system to resolve their problem; whereas, the other existing systems do not allow customers to customize their information.

1. The authors consider the optimization and trustworthiness of the system when customers use it. However, the system fails to address the time spent on each customer as companies today have thousands of issues and it is unclear how the system would be able to handle all of these.
2. How many operators must be recruited by the company to solve customer problems? The authors do not show any risk factors of the project and security

problems of the system, a significant issue given that most customers nowadays want to ensure their account security.
3. They do not provide any prediction regarding customers' post-behavior in the event that they are faced with issues in future.

Richards and Jones [23] propose a conceptual model in which they provide customer relationship management's value drivers followed by generating equity for value, brand and relationship which will ultimately lead to customer equity. The rationale for this model is that they believe there is no outcome in terms of benefit to the company. Prior to that, they defined significant CRM core benefits based on various academic papers. Using their proposed model, they want to ascertain whether the management of CRM activities will positively impact on business performance. Furthermore, they adopt customer equity as a measurement tool.

CRM proposed by [23] bridges the gap between distribution pathways. In this way, the information will be used effectively and customers can directly communicate with those in authority within the company. Also, the process of pricing will be improved.

1. The authors clearly discuss the topic, and its importance and definition. However, they neglect to determine the expectations of the customers, and the risk factors cannot be assessed using this model.
2. The authors do not offer an approach to determine the outcome using the conceptual model. While they propose an important hypothesis, they must create a method to produce and evaluate the results. They should have proposed some methodical approach to analyse the data in order to reach a numeric conclusion, rather than adhering to the empirical data. Furthermore, the authors fail to introduce real-time data for analysis.
3. From those proposals and the CRM's core benefits described, the authors do not mention the relationships of those core benefits and ways of improving them. Moreover, the authors omit to discuss the relationship between these factors and their impact on the decision-making process.

Öztayşi [72] proposes a method which greatly enhances the benefits obtained from clients. The authors compare the performance of customer relationship management processes of companies which use multiple criteria decision-making processes. The authors examined the Turkish e-commerce market using an analytical network process and they considered three companies. Company one deals with trade, company two with health and cosmetics products, and company three is a facilitator between other companies. Analysis clarified that company two has the highest standards in CRM performances followed by company one and company three respectively. Finally, the sensitivity analysis showed that ranking is sensitive to shifts in inter-relations and in weights. However, their proposed method has the following drawbacks.

1. They do not propose any means for ensuring the validity and reliability of a given raw data that they need to analyse.

2. They do not propose any method for pictorially illustrating the relationships between major factors.
3. While the analytical network process is an appropriate tool for analysis and decision-making, the authors could have used other methods to determine the significance of each factor involved in the process.
4. When evaluating customer relationship management, the authors do not consider the financial effects of each factor on customer relationship management.
5. They do not consider the effect of customer complaints on customer relationship management.

In further work, Torkzadeh et al. [73] mention the major obstacles to the success of CRM in pharmaceutical companies and propose a customer relationship framework. First, they set up a focus group and initiated discussion in order to generate the problems. Then they conducted a survey and divided the data evenly into two sets. To discuss the main issues and using the first set of data, they used exploratory factor analysis. To verify the factors, and with the second set of data, the authors applied structural equation modelling. The outcomes of the analysis were 7 variables and 21 observed variables that may limit the success of CRM. The issues are as follows:

1. Operating procedure
2. Responsibility and ownership
3. Quality of information created using customer contact
4. Ineffective consolidation of accounting function
5. Difficulty queuing a changing procedure
6. Replacement procedure
7. Time spent in the queue

Based on the current study, these 7 factors may generate attitudinal and behavioral tendencies and in the final stage, the CRM may be successful. Based on this study, successful CRM is comprised of sorting complaints, retaining customers and growth of the customer base.

1. The authors do not specifically discuss customer complaint sorting as a success factor of CRM, and there is no analysis or estimation regarding this issue.
2. Also, while the authors discuss customer retention as a result of CRM success factors, they fail to provide a methodology for this.
3. The authors do discuss the observed variables and sensitivity of those factors in terms of CRM success.
4. They analyse the data and obtain the result using various tools for CRM success. However, they do not provide a complete methodology by which the future success of the system can be predicted.
5. They fail to address how the system will integrate the feedback.
6. They do not explain how customer growth can be ascertained using the approaches in this study; hence, it remains an unclarified aspect of the study.

7. They also fail to provide a means for interaction or effective communication with customers to ensure CRM success.
8. If one wants a successful CRM model, it is necessary to take into account customer satisfaction and evaluate this using current data.

Dimitriadis [74] proffers a consolidated model for designing, evaluating and implementing a CRM system. The author mentions the dearth of research in the area of CRM in service industries. The author defines the subject and components of CRM and then introduces a framework. The author shows that people, strategy, organization and technology are the stepping stones and backbone of CRM; however, all of the main variables may be linked and send their feedbacks to a channel to be integrated. The results derived from this study indicate that there is a great gap between customer expectations and the formulation of a CRM strategy. The second gap is the alignment of management, technology and organization according to CRM strategies which have been neglected. This study has the following shortcomings:

1. The author proposes a methodology to address those factors that cause customer dissatisfaction, but this chapter provides only a descriptive analysis and does not include a mathematical analysis.
2. The author does not determine how to estimate and analyse customer expectations.
3. There is no mention of how customer perception will be assessed and fed back to customer expectations and to the appropriate channels.
4. The author omits to discuss how the system can generate feedback which should be a function of every CRM system.
5. The author does not classify the factors and their importance in terms of customer expectations and perceptions.

Frow et al. [2] identify the major types of vague aspects or negative side attitudes of service providers and offer consolidated methods to overcome the negative side effects. They [2] believe that it is crucial for companies to manage customers appropriately and have a productive relationship with them. It is said that customer relationship management must collaborate tightly with relationship marketing in order to become more effective. The authors argue that poor company behavior can destroy a customer relationship. Also, a poor management system may allow injustice and corruption which might be related to the lack of knowledge. Moreover, deceitful and misleading service providers may categorically misuse customers. What is more, customer complaint is another negative side of the CRM, as the CRM systems in general cannot address customer complaints. The researchers use two data sets and analyse them qualitatively. Their analysis shows that the negative side effects of CRM can create the following negative behaviors.

The first set of data creates information corruption, customer confusion, lack of honesty among customers, and misuse of customer privacy to the company's benefit. Also, the second set of data shows that customer bias in the buying behavior of potential customers is one of the outcomes. Also, customers may be subjected to certain techniques that make it difficult for them to shift to another provider. Furthermore, customers may be neglected at times despite having a good relationship with the company. Companies also might impose unjust monetary penalties on customers in order to increase revenue. The existing study has the following shortcomings:

1. The authors do mention how they are going to estimate CRM consequences, and they also discuss the various dark sides of CRM; however, they do not mention how these factors are going to be analysed.
2. The authors do not discuss how the negative aspects of customer behavior can be determined as they do not have any numerical data for this.
3. Although the authors present a diagram to show the outcomes, they do not generate a broad conceptual framework to depict the relationships.
4. They do not propose any methodology to ascertain the presence and validity of each factor, although all of the factors seem to be explicitly significant in CRM systems and strategies.
5. They also discuss the existence of customer complaints in CRM systems as barriers, but they fail to provide any remedy for this issue.
6. The authors fail to discuss customer satisfaction and how a CRM system can address and improve this.
7. They omit to consider prediction and post-service behavior of customers when challenged with such problems.

King and Burgess [19] propose a novel method by developing a conceptual framework for CRM and changing that simulation model. The authors clearly discuss the successes and failures of CRM systems. They mention that customers complain about the failure of over 50 % of CRM projects and express their lack of trust in the systems. Hence, they created a model as CRM outcomes (operational and development) and divided this into two segments. In the next stage, they discussed the tangible and intangible benefits of CRM. They used a CLD mapping technique for simulation. Results show the difference between the work quality of an experienced CRM user and that of a new CRM user, indicating that for an experienced user, the diagram has an increasing trend. Finally, departmental support given to the users shows a similar increasing trend.

1. The authors use a simulation technique to validate their model; however, this study lacks valid data.
2. The authors do not have a methodology for validation of the main variables.
3. They fail to discuss how they need to address the issues related to cost and quality although these two factors can be considered as the main features of CRM systems.
4. They proposed tangible and intangible benefits for CRM, but they do not mention how to evaluate either of them. For example, they do not mention customer satisfaction evaluation and do not provide any methodical approach to determine it.

Phan and Vogel [75] propose a model for consumer buying arrangements and credit behavior. They initiated their survey ten years ago and have been collecting data ever since. The authors sought to examine the effects of unfair pricing on customer relationships followed by the effect on customer relationships of changing costs or restrictions. They proposed several hypotheses and used online analytical processing (OLAP) for data analysis. They ascertained that unfair pricing will lead to a poor relationship. Also, an appropriate business intelligence system and CRM system will enhance the level of satisfaction and the relationship. Increasing costs do not annoy customers as the customers are paying with their credit cards and in this study the company adapts the products and services based on each customer. The weaknesses of this study are:

1. The authors propose no method by which they estimate future satisfaction and relationships, as the nature of this study is dynamic.
2. They fail to discuss how they are going to address success, as the final outcome of the conceptual framework is success.
3. They do not create any methodology to evaluate each of the factors throughout the model and how each of the factors must be determined.
4. They do not create any prediction for customer satisfaction and for future success.

2.4.2 Customer Perceived Value-Focused CRM Approach

Customer perceived value is an outstanding factor that sometimes can be considered as a substitute for customer satisfaction and customer experience [76]. However, there are differences such as cognitive construct in perceived value and influential construct in satisfaction. Also, perceived value is strategy-centered whereas satisfaction is tactic-centered.

Yang et al. [77] suggest that changes in cost is an issue that may impact on customers' perceived value, satisfaction and loyalty. To support this claim, the authors conducted an online survey focusing on perceived value and satisfaction as a means of obtaining loyalty. Then, they employ a two-step method whereby they use EFA (exploratory factor analysis) and in the second step, CFA (confirmatory factor analysis). According to this study, perceived value is related to both satisfaction and loyalty; however, satisfaction has a bond with loyalty. Meanwhile, moderating factors can be considered only when the satisfaction level is just above average. However:

1. The authors fail to provide a prediction method for future analysis.
2. In their proposed method, the authors do not mention how customer loyalty is going to be measured.
3. There are also other moderating effects available that need to addressed, such as lock-in, managerial issues and time management.

4. Also, there are variables that directly or indirectly affect each of the variables such as customer acquisition, and while the authors estimate loyalty, it would be a good idea to evaluate it as well.

Roig et al. [78] propose a method by which perceived value can be analysed in the banking sector by introducing a GLOVAL scale which measures perceived value in either the purchasing procedure or a customer's experience after the purchase. They conducted a survey for the purpose of data collection followed by CFA and structural equation modelling to validate the proposed scale of perceived value. Based on the analysis of this study, perceived value has various dimensions such as "service's operational value, operational value of establishment, personnel's operational value, price's operational value, social and emotional values". The disadvantages of their proposed method are as follows:

1. This study must be conducted using other sets of data to revalidate the relationships.
2. The authors do not propose any approach and methodology by which the process is optimised.
3. Also, no method is proposed for future prediction, although it is important to regularly estimate the perceived value of the customers in every work setting.

Chang and Wang [79] propose an approach by which the moderating impacts of perceived value are determined. In this study, the authors examine the effect on loyalty of the quality of services, perceived value and satisfaction. They performed studies to validate the self-regulating processes, and evaluated the moderating effects of customer perceived value on satisfaction and loyalty. Their survey questionnaire was based on e-service quality, perceived value, satisfaction and loyalty followed by introducing hypotheses using a conceptual framework. Moreover, structural equation modelling (AMOS 5.0) techniques and linear hierarchical regression models were used to test the causal model.

As depicted by Fig. 2.2, the authors prove that e-service quality influences customer satisfaction, followed by leveraging of customer loyalty. Also, perceived value has a relationship with satisfaction. Finally, those customers who have higher perceived value may experience better satisfaction and provide loyalty.

1. The authors consider the mediating impact of customers with higher perceived value on satisfaction and loyalty; however, they do not mention how they formulate the relationship.
2. The authors do not define perceived value although one of the key factors in this study is perceived value.
3. They do not propose any approach to predict future customer loyalty although customer loyalty and customer satisfaction both have various consequences to be considered.
4. They do not optimise the result by using other tools and approaches. They can optimise results by reducing costs, or providing better services for the customers.

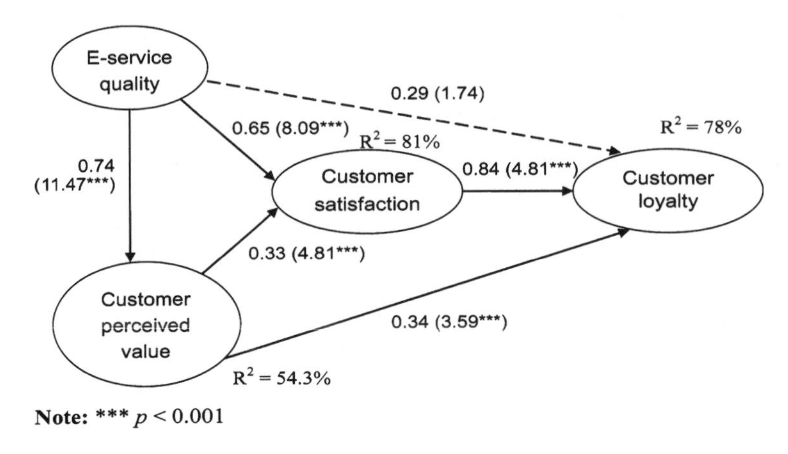

Fig. 2.2 Relationship between variables [79]

Lai et al. [80] propose and test the relationships between service quality, perceived value, satisfaction, and loyalty in a survey conducted to collect data from customers of a mobile company. Also, in addition to distributing a pilot questionnaire to randomly-selected clients, several managers were interviewed. They validate and analyse the conceptual framework using structural equation modelling. It indicated that service quality is positively associated with perceived value and image perceptions. Furthermore, perceived value and image may leverage customer satisfaction. Also, based on this study, perceived value and image are positively associated with loyalty. The disadvantages of their proposed method are as follows:

1. The authors do not provide a methodology by which they evaluate and formulate customer satisfaction and loyalty.
2. They do not mention how loyalty can impact on the service quality. Also, they do not create any prediction method to make recommendations for improvement.
3. Although they use software to establish the hypotheses, they fail to use it to evaluate potential improvement for customers.
4. This study does not define perceived value or how it can be mathematically evaluated.

Ryo et al. [81] discuss the relationship between image, perceived value, behavioral intention and satisfaction. The authors define the main concepts and then develop their hypotheses. They use a focus group, and a questionnaire was administered to evaluate the variables. The research has two sections the first of which deals with semantic items and second measures responses using a 7-point Likert scale. The main variables and the relationship between them pertain to restaurant services. To analyse the data, the authors employed regression analysis to test and validate the relationships.

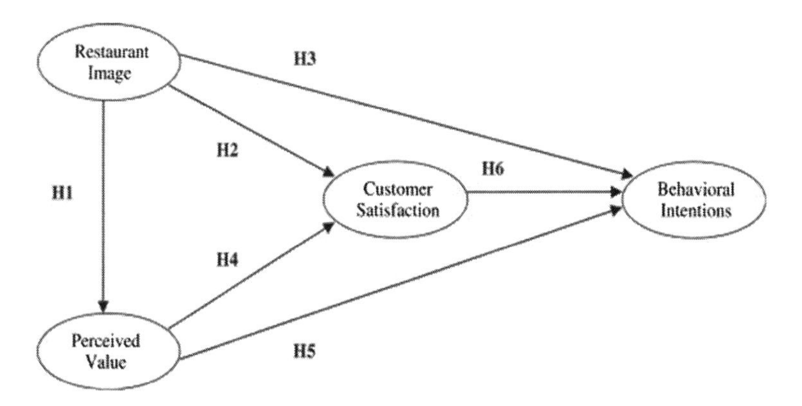

Fig. 2.3 Proposed model showing the relationship among hypotheses [82]

As depicted by Fig. 2.3, image is related to perceived value and satisfaction. Furthermore, perceived value is directly associated with satisfaction. Likewise, restaurant image, perceived value, and satisfaction significantly influence customers' intentions. Figure 2.7 shows that customer satisfaction can perform as a mediator among the rest of the variables. This study has the following drawbacks:

1. Since they have the data, and given the nature of the research (restaurant industry), the authors should have performed optimization analysis, but failed to do so.
2. They cannot deal with customer complaints and get to know their views so that customers' perspectives can be taken into account. The authors do not mention how they predict future customer perceptions or ways by which to maintain customer satisfaction.
3. The authors fail to discuss the varieties of customers they may encounter. Also, they do not take the opportunity to categorise them into various groups and perform the evaluation regarding each of the variables.

According to Chang and Wu [82], a reliable CRM strategy helps the company to obtain revenue, increase customer's perceived value and acquire more customers through its own customer acquisition procedure. In this study, the authors consider two facets. Firstly, customer need is defined as customer perceived value and secondly, CRM is defined as a process of absorbing new customers, providing an interactive relation with them and maintaining them as part of a loyalty program. This study defines the concepts and presents the relations pictorially. Also, this study investigates various dimensions of perceived value in wireless technology which are "business value, customer perceived value and social value" and here in the Internet-based services, customers must choose the amount to be paid for the service. They proposed a value-based model for e-CRM services. They found four levels of customer perceived value: self-actualized value using self-problem solving, social and emotional value, using value generation in a network, added value using e-services, and operational value, using fundamental values. This chapter has following shortcomings:

1. The authors do not introduce any methodology to ascertain the relationship between perceived value and the CRM process.
2. While they discuss emotional values and self-actualization based on Maslow's theory, they need to go beyond that and conduct an analysis based on the sensitivity of the issue.
3. The authors do not offer a methodology to prove the interrelationship between need and customer acquisition.
4. They do not provide a means of examining the impact of motivation as a mediator on the process of customer acquisition.

Minoumi and Valle [83] maintain that, in order to ensure complete customer satisfaction and loyalty, companies must focus on improving customer perceptions. The authors talk about the benefits of loyalty programs and group them into practical benefits, hedonic and symbolic values. They use quantitative studies, including members of loyalty programs. They conduct two surveys to ensure that the benefits achieved are correct. They use exploratory factor analysis in the first study and utilize a 7-point Likert scale for measurement. They use EFA to illustrate the correlation among dimensions. Also, they use confirmatory factor analysis for the second study and use AMOS for evaluation of dimensions which are savings, exploration, entertainment, understanding, and social values. This study establishes that five dimensions have a positive relationship with perceived value and relationship investment, and with these in the work setting, a company may have an appropriate quality of relationship. The disadvantages associated with their proposed approach are as follows:

1. They provide no means by which the given researcher can estimate a loyalty program on the basis of the conceptual framework.
2. They provide no methodical approach by which customer perceived value can be analysed using a mathematical method.
3. They did not take the opportunity to include customer satisfaction in the conceptual framework although the questionnaire contains a question regarding satisfaction.
4. The authors fail to mention how relationship quality is going to be controlled and predicted. This needs to be accomplished in order to examine customer behavior.

2.4.3 Customer Interactivity-Focused CRM Approach

To the best of our knowledge, there has been inadequate research into customer satisfaction in relation to CRM systems.

Blocker et al. [84] presents the essence of proactive customer centralization and tests the related hypotheses. The authors conceptualize active customer centralization as it needs advanced levels of interactivity.

As seen in Fig. 2.4, the authors assess the impact of the proactive customer centralization on generating value using a new method to investigate the "proactive customer orientation → value → satisfaction → loyalty chain", utilizing data from five different countries and applying confirmatory factor analysis. Results suggest that, connected with companies' other abilities, active customer centralization is the most persistent motivation of value. Also, results show the interaction of responsive customer centralization for value creation. There are moderating factors which are influential such as "levels of customer value change, a global relationship scope, and a transnational relationship structure". Overall, findings significantly advance the understanding of the proactive dimension within the market orientation and provide marketers with insights into customer processes [84]. However:

1. The authors omit to include interactivity as a separate layer of the conceptual framework. Also, interactivity must be evaluated separately.
2. They do not provide any means by which loyalty can be assessed. Also, there is no approach for predicting future behavior of customers regarding loyalty.
3. They prove that satisfaction has a direct relationship with loyalty but they fail to formulate this relationship rather than just proving the hypotheses.

Florenthal et al. [85] propose a substitution for interactivity and create four interactivity modes: human, medium, message and product. Then the authors propose a model to integrate all modes. Next, the authors streamline this for various disciplines such as marketing, interaction, psychology, communication and computer science. The model shows the preference of the customer perception for four interactivity modes. Finally, the author concludes that personal and positional

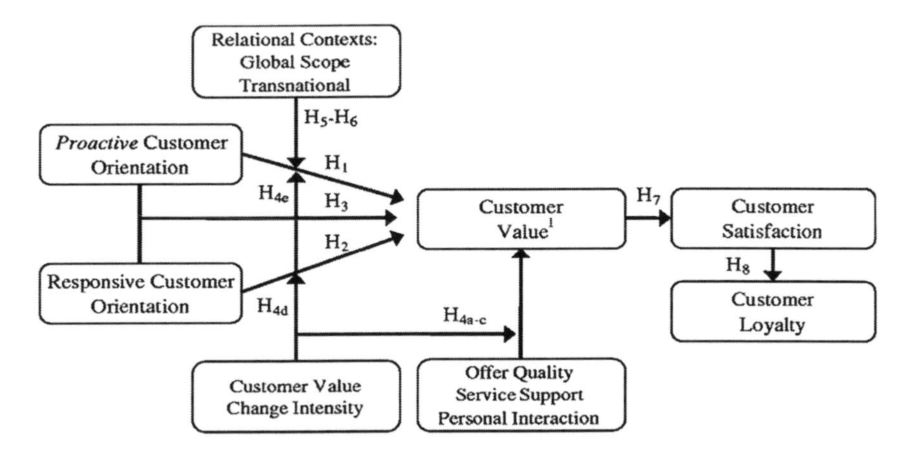

Fig. 2.4 Relationship between hypotheses [1]

characteristics may shape the preferred interaction. Also, when customers observe competition in a market, they prefer to switch to this new market and start their interaction with this market.

Yoo et al. [86] examine the bond between interactivity, perceived value and hedonic values. They conducted a survey comprising five sections and a total of 451 respondents were used for data collection. They define eight hypotheses. In order to analyse the data, they employ structural equation modelling using Amos 6.0 to assess the model.

Figure 2.5 shows how interactivity factors result in perceived value and the acquired perceived value ends in satisfaction. The results illustrate that synchronicity has a big impact on perceived value. Furthermore, bi-directionality has an impact on hedonic values. This study has the following shortcomings:

1. The authors analyse the hypothesis using a descriptive and qualitative method, whereas, they could have conducted this research with higher precision using quantitative data.
2. The authors do not propose any particular methodology by which they can evaluate customer satisfaction.
3. Also, while they define perceived value and analyse this qualitatively, they fail to address it using quantitative analysis.
4. The authors do not provide a means of estimating the impact of interactivity and its mode on perceived value and customer satisfaction.
5. They do not create any prediction for future customer behavior using the current system.

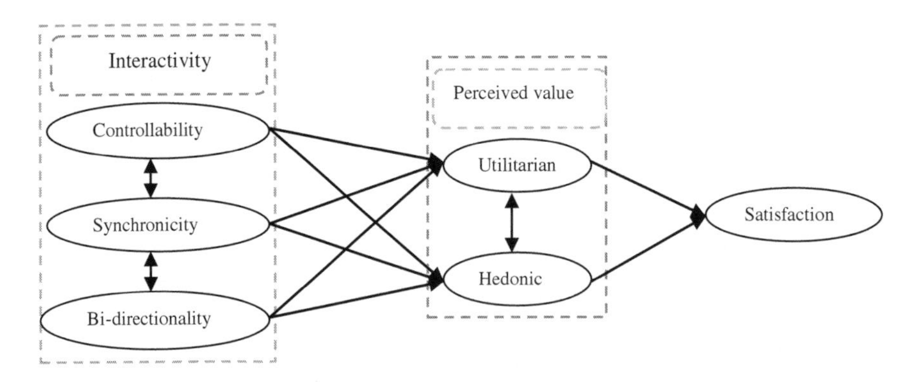

Fig. 2.5 Proposed model of consumer value-creating online interactivity [86]

Bonner [87] examines monitoring mechanisms utilized in new product development (NPD) and their impact on customer interactivity. The author proposes five hypotheses and tests these using a questionnaire derived from various industries. The measurement scale was a 5-point Likert scale. After obtaining the reliability and validity of the measures, statistical regression analysis was conducted to examine the hypotheses. The author concludes that output control and team rewards have a positive relationship with customer interaction and process control has a negative impact on it (Fig. 2.6). This study has the following drawbacks:

1. The author does not formulate customer interactivity using a mathematical method.
2. While output control may increase the degree of customer interactivity, the author fails to propose a methodology to ascertain the consequences of interactivity.
3. The author does not mention the future prediction of customer loyalty.

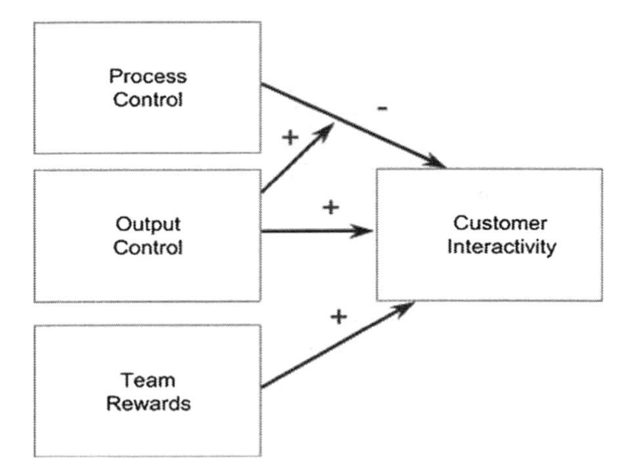

Fig. 2.6 Conceptual framework [87]

Kirk [88] investigated the relationship between interactivity and customer satisfaction. The author stated various hypotheses and conducted a survey. The author also considers age and perceived usefulness as moderators. A 5-point Likert scale is the measurement tool. Then, using descriptive statistics, the author tested the hypotheses.

The analysis illustrates that customers are in favor of the things that they accustomed to, such as a book in preference to an e-book. Also, the age of the customer is significant when interaction takes place.

Lee [89], mentioned that perceiving interactivity in a company creates trust and transaction willingness by customers. Data were gathered from 20 interviews and the measurement tool was 7-point Likert scale. For the analysis, the author used a correlation matrix and Chi square difference test. Based on Fig. 2.7, interactivity is positively associated with satisfaction and adoption intention; however, satisfaction has a positive relation only with adoption intention.

The author ascertains that components of interactivity such as user control, responsiveness, personalisation, connectedness and contextual offer can increase the level of trust in customers and this alone makes them satisfied. Satisfied customers normally have a better interaction with companies. Both of the studies above have these drawbacks:

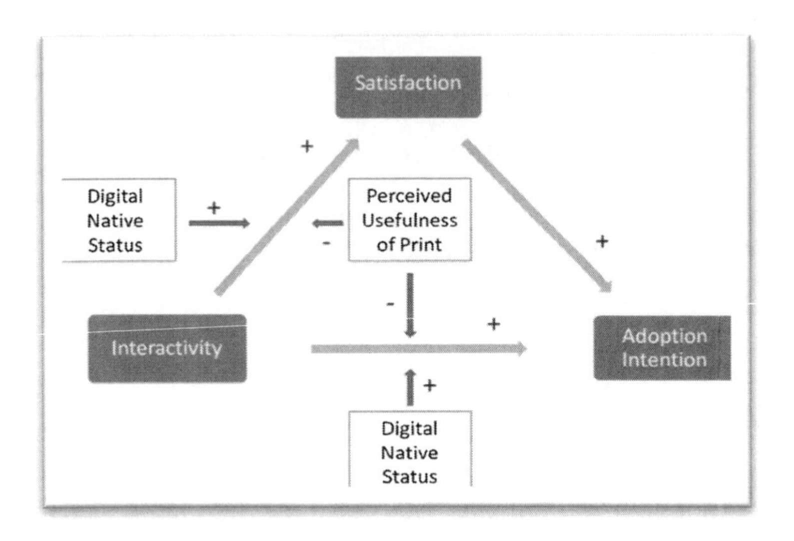

Fig. 2.7 Theoretical model [88]

1. Neither creates an innovative solution to evaluate and measure customer satisfaction.
2. The authors offer no optimization process and they cannot evaluate the sensitivity of their model.
3. None of the studies can offer an effective decision-making process based on their frameworks.

Ballantine [90] presents the facets of on-line shopping which are interactivity and information and their impact on satisfaction of customers. The author proposed two hypotheses which were tested using an Internet-based experiment. Respondents were presented with an on-line store using simulation techniques. The measurement tool was a 7-point Likert scale and respondents were required to state whether they agreed or disagreed with the statements. The data having been collected, the author performed analysis of variance (ANOVA). The results illustrate that interactivity and amount of information have a direct and positive bond with customer satisfaction. The proposed approach has the following disadvantages:

1. Although the author uses an on-line simulated store, a physical store could have produced better outcomes.
2. More data should have been collected in order to effectively verify the hypotheses. In addition, the author did not present a clear definition of interactivity.
3. The author does not provide a robust methodology for measuring interactivity. Also, other tools apart from analysis of variance could have been used to validate the hypotheses.

Liu et al. [91], propose a method for interactivity and discuss the structure and various facets of interactivity. They also turn the definition into theory and practice. Furthermore, they compare various online marketing tools based on interactivity and then establish the dimensions of interactivity including active manipulation, synchronicity and two-way communication. Their results illustrated that interactivity can be measured and cannot be manipulated. Also, the outcomes of the interactivity must be under the control of an authority in the company in order not to be confused with other researchers' results or for the results to be re-used. Based on this study, interactivity dimensions become an interaction process and this alone will produce interaction outcomes which comprise learning, self-efficacy and satisfaction. Below, we state the drawbacks of this study:

1. Although the authors provide a definition of interactivity, convert the definition to a theory and use a model to show the relations, they neglect to apply it to customer relationships, as interactivity can be used as a forceful technique in CRM.
2. Their conceptual framework which is intended to distinguish and clarify relationships is in fact vague.
3. They propose a conceptual framework to ascertain self-efficacy, learning and satisfaction; however, they do not formulate the relationships or discuss the hypotheses in detail.

Roh et al. [92] propose a priority factor framework for the success of CRM. They discuss the intrinsic CRM success which comprises effectiveness and customer satisfaction, and extrinsic CRM success which is profitability. They first conducted in-depth interviews and questionnaire with CRM managers to check the validity. Using a structural equation analysis (AMOS), the authors investigated the CRM system success framework. Based on their analysis, system support and efficiency both result in customer satisfaction; however, system support and quality are not linked with profitability. According to this study, it is unlikely that efficiency converts to profitability, but customer satisfaction can bring about profitability.

1. Due to the exploratory nature of this study, it would have been better if the authors had created an iterative process within the conceptual framework.
2. The authors do not evaluate customer satisfaction and do not discuss customer post-behaviors.
3. While they analyse customer relationship and its components, the authors must include other significant intrinsic and extrinsic factors too; however, it appears that they did not want to use all of the factors in the analysis.
4. There are some rejected hypotheses based on conceptual framework; however, the authors fail to address them and provide recommendations and solutions.
5. The authors must discuss the consequences of customer satisfaction and as one of them is loyalty, this should be included in the conceptual framework.

2.4.4 Customers Loyalty-Focused CRM Approach

Min et al. [93] propose a methodology to control company employee turnover and enhance truck driver satisfaction and loyalty. Likewise, this study emphasises the shortage of drivers and its influence on competitiveness. The authors propose different employment incentives and retention strategies to retain drivers and control driver turnover (customer defection). The companies understand that to monitor the costs, they must stabilize drivers. For purposes of analysis, the authors selected a sample of trucking firms and conducted a survey. Then, the authors selected the most valid questionnaires and used SPSS to evaluate the data. Also, for measurement purposes, they use a 7-point Likert scale. In the very initial stage of analysis, it emerged that one of the reasons that drivers choose to work with a particular company is the facilities and the equipment that the company provides. Other than that, drivers do not like to have operational difficulties, repeated failure, health and safety problems. For job security, drivers hope to have a well-known recruiter. In this study, to better identify important factors, the researchers use exploratory factor analysis and principal component analysis. Using Chi square, they illustrate that some incentives are correlated with others. Using Varimax rotation with Kaiser Normalization, they discovered new factors such as non-driving activities, career enhancement and financial incentives. Also, word of mouth is one of the factors that attracts new clients and creates referral to new members.

According to this study, the factors that discourage individuals from driving and which contribute to lack of satisfaction are: rules and regulations, rigid labor market, drug testing, and lack of driving schools. According to another analysis using Chi square and factor analysis, the authors establish that there are three factors that create obstacles for drivers. These are infrastructure issues and include: working space, parking, hard work, and inadequate facilities for drivers.

Simultaneously, cost would be another issue for drivers according to this study, as 25.8 % of respondents believe that increasing salaries and repayments may exacerbate driver shortage. Furthermore, 70.7 % of the respondents believe that lack of drivers negatively influence their profit. Also, about 84.6 % of the respondents believe that lack of qualified drivers is one of the obstacles and issues preventing companies from initiating loyalty programs [93].

Results confirm that drivers are poorly managed. Drivers maintain that the best motivation for them to remain loyal is financial support and rewards. Furthermore, the behavior of employers directly affects drivers' attitude and positively relates to retaining drivers. Moreover, age, education, family status and owner-operator status should also have an impact on the retention process; however, evidence shows that young (21- to 25-year-old) and experienced drivers (45+) are willing to stay with the same company (become conservative). The authors conclude that employment and retention strategies should be based upon the profiles of individuals [93]. The current study lacks the following:

1. While the authors analyse the relationships between the key variables, they do not provide a conceptual framework to facilitate comprehension.
2. They provide no mathematical framework which effectively allows a researcher or a company to employ the model.
3. Although the authors appropriately addressed the issues of customer retention and loyalty, they do not propose any future direction particularly in relation to any prediction for post- behavior customer analysis.
4. While the authors propose various methods for analyzing the relationship of the hypotheses, they fail to use a fuzzy inference system in their analysis to estimate the sensitivity of each individual variable.

Lin and Wang [94] proposed, developed and validated a loyalty framework. The authors collected the data from mobile commerce systems using a questionnaire and then analysed the data using structural equation modelling. The results of this study illustrated that perceived value is a significant element that impacts on loyalty. Also, level of trust, customer satisfaction and the habitual behaviour of customers have a direct impact on customer loyalty. Furthermore, perceived value and trust are positively associated with customer satisfaction.

1. Although the authors include loyalty in the system, they fail to address recommendations to improve it. Also, they need to provide a methodology to ascertain the level of customer loyalty.
2. They do not propose any means for predicting future loyalty programs and customer behaviour evaluations.
3. While the authors investigate the relationships to prove the hypotheses, they do not mention how they derive the perceived value and trust from the performance.
4. They also fail to discuss prediction in terms of future decision-making procedures regarding customer satisfaction and customer loyalty.

Sweeney and Swait [95] examine the impact of brand credibility on loyalty. To avoid customer attrition, companies must implement CRM and deal with customers in such a way as to create long-term relationships with them. Customer retention is critical and it will not be achieved unless the company improves customer satisfaction and quality of the services. The name of the company must perform as a brand in order to be able to retain its customers as well as its own employees. A survey was conducted using questionnaires and to validate these, the authors use exploratory factor analysis followed by structural equation modelling. Preliminary analysis showed that customer satisfaction and loyalty may lead to positive word of mouth, and brand credibility has a direct and strong influence on satisfaction. However, in the final alternative analysis, it emerged that customer satisfaction and commitment are directly related to credibility. Ultimately, it is credibility which provides word of mouth and intention to stay or leave the company. This study has the following shortcomings:

1. The authors do not propose any means for predicting the future direction for credibility and its impact on word of mouth and inclination of the customers.
2. While their proposed method for making credibility the core of the CRM process is valid, they do not formulate this using a mathematical method.
3. The elements of loyalty, satisfaction and credibility are dynamic in nature and have various levels. Consequently, the authors must consider their antecedents and consequences in the conceptual framework to make it more logical.
4. Using the existing data for this study, the authors could have been able to use fuzzy systems to balance the data and obtain the sensitivity of loyalty as well as credibility.

Ball et al. [96] examine the impact of personalisation of services on loyalty and assess the psychological impact of this process. After defining each of the significant variables in the model, the authors introduce personalisation into loyalty using several hypotheses. They acquire the data by means of a survey administered within banking industry and their purpose is to ascertain whether or not personalisation of services has a relationship with trust, loyalty and satisfaction. The partial least squares (PLS) method is used to evaluate the framework and hypotheses. The results show that personalisation has a major impact on trust, satisfaction and loyalty, respectively. The shortcomings of this study are:

1. The authors do not include customer complaints in their conceptual framework.
2. This study could be conducted using a fuzzy inference system to show the sensitivity of personalisation and its impact on loyalty and satisfaction.
3. The authors do not formulate the relationship to obtain more effective results. Also, the results produced by this study are not applicable to the decision-making process.
4. Using the same conceptual framework, it is best to conduct the performance measurement and optimization process.
5. The authors do not provide a means, based on their analysis and results, of predicting future customer behavior.

Gomwz et al. [97] analyse loyalty in terms of behavior and attraction to build and develop a loyalty program. The data was acquired via a survey of supermarket chains in Spain. Analysis of variance was used to compare two loyalty dimensions. The results illustrate that respondents in loyalty programs are highly behavior-oriented and effectively loyal. Additionally, not all customers are willing to change their purchasing behavior after joining a loyalty program. The final outcome is expected to be the retention of loyal customers. Also, the customer will be linked to the retailer for further interactions. This study suffers from the following drawbacks:

1. The authors do not provide any direction for prediction and decision-making processes.
2. The authors fail to pictorially show the relationship and discuss the relationships.

3. To be able to increase the loyalty of the system and positive attitudes of the customers, the authors need to formulate and analyse the notion of commitment.
4. To have a proper commitment toward customers, the authors must include a strong management system, effective time management, viable infrastructural improvement, and a decrease of customer complaints.
5. The authors analyse behavioral loyalty, but do not provide any methodology to ascertain levels of loyalty and the relationship between loyalty and other factors.
6. The authors fail to show how loyalty programs can be evaluated for the purpose of optimizing the system in future.

Gee et al. [54] discuss customer loyalty and offer managers a broad scope of knowledge in the context of CRM. Based on this study which is analysed both descriptively and using a qualitative method, the authors concluded that companies need to recognize what absorbs customers and makes them content.

A customer-oriented perspective lets a company understand its customers, as different customers have varying needs and wants. Data analysis is central to most accompany procedures and if a company needs to create customer loyalty, it needs to undertake such analysis to improve customer retention. Customer segmentation is a significant thing to do and its operating costs must be controlled to ensure it is spending in a productive way. By profiling potential and current customers, a company can be confident that customer acquisition is likely to occur. A re-gain strategy is advised as previously discussed, as acquiring previous customers is less costly and they are easier to acquire again. Analysis of customer discontent assists the company to search for at-risk customers and retain them before they have a chance to leave. However:

1. The authors do not provide any step-by-step methodology and conceptual framework.
2. They do not provide any means by which formulate the issues and provide solutions for them. Also, they omit to address customer satisfaction.
3. They do not provide any optimization strategy for the raised issues. Also, they fail to provide a prediction approach for future loyalty programs.

Leenheer et al's [98] methodology is intended to strengthen customer loyalty. According to this study, not all loyalty programs are efficient, since customers can be selective when choosing a particular program. The authors define loyalty as a consolidating system whose final goal is to make the members more loyal. However, they need to test the loyalty of the members. Also, they must test whether or not they can increase their 'share-of-wallet' (SOW). They conduct an empirical study and use panel data from a sample of Dutch households. The result shows that a very small proportion of loyalty impacts on share-of-wallet. Also, the SOW of a company is connected with its attraction to the customers. They mention that the SOW results illustrate that creating loyalty membership is a stepping stone to developing SOW and prove that loyalty program membership positively affects company attraction. The shortcomings of their proposed approach are as follows:

1. They do not propose any means for predicting the future loyalty program. What is more, they do not include satisfaction of the customers which is related to loyalty.
2. They do not propose a performance measurement for this study, that the use of which could ascertain the sensitivity of the issues.
3. They do not analyse the effect of feedback of SOW on the whole system. Also, they do not mention how they predict SOW.

Wallenburg [99] suggests that initiatives assist logistics service providers (LSPs) to beat their competitors. However, it is still unclear to what extent companies can implement their strategies to create customer loyalty, and concentration on cost or performance developments is desirable. The data were collected via a web-based questionnaire survey from 298 logistics outsourcing bonds, and structural equation modelling was used to evaluate the impact that cost and performance improvement have on customer loyalty. Simultaneously, they test impact of ''service complexity'' and ''length of contracting period''. The outcomes disclose that proactive cost improvement and proactive performance improvements are motivators of all major dimensions of loyalty which are "retention, extension, and referrals". Also, proactive cost improvement and efficiency are major motivators of loyalty when the period of contracting is relatively quite short. Finally, customer loyalty is basically oriented by an increase in proactive performance and effectiveness. The shortcomings of the researcher's proposed approach are as follows:

1. He does not propose any means by which to analyse the impact of retention on loyalty although these are related to each other.
2. While the SEM is a reliable method, the author could also use fuzzy logic to create sensitivity for each variable.
3. The author does not provide an iterative process whereby all variables can be connected with each other.
4. The author missed the opportunity to work with customer complaints and test the loyalty dimension in relation to customer dissatisfaction.

2.4.5 Customer Acquisition-Focused CRM Approach

Verhoef [100] presents an approach that measures the impact of customer acquisition on satisfaction and loyalty. The author presents the Probit framework which measures the difference between the rates of customer retention and different acquisition pathways. The author proposes an econometric and customer loyalty model using simulation. Verhoef considered customer retention as a binary model. What is more, four groups of insurance were assigned. The Probit model is as follows: "where Yi is the value of the dependent variable for Customer i (retention or cross-selling), and Xi is a vector of explanatory variables, b is a vector of regression parameters, and Φ the cumulative normal distribution function". The Probit model presents the significant of variance in loyalty: $\text{Pro}\beta(\text{Yi} = 1) = \Phi(\text{Xi}\beta)$

In this study, simulation was used for "interpreting the coefficient of the model". This study also investigates the impacts of acquisition channels over a lengthy period. Prior to that, they integrate loyalty and cross-buying to derive effective value from their clients. The mathematical formulation for the average number of products purchased is [100]:

$$\text{Pred}(\#\text{Serv})\text{i}, \text{t} = \text{Prob}(\text{retention})\text{i}, \text{t} \times [1\_\text{Prob}(\text{cross} - \text{buying})\text{i}, \text{t}].$$

Öztayşi et al. [101] proposes that CRM is intended to produce a more effective relationship with customers and customer acquisition would be one of the outcomes. This study uses an analytic network process to analyse the sensitivity and to find the priority of the elements in the network. To create an equation between goals and variables and to create effective decision-making, they used qualitative analysis. The outcomes illustrate that the ranking of the alternatives is sensitive to changes in the parameters. This study accesses customer relationship management performance of organizations using ANP MCDM or multi-criteria decision-making methodology.

Becker et al. [38] maintain that there are many organizations which are dissatisfied with their implemented CRM system. The authors propose that organizations do not have any enhancement in terms of acquisition level. They present a conceptual framework and use it to measure customer retention and customer relationships. They conducted a study of four various industries in ten European countries. They defined four hypotheses that connect comprehensive implementation to CRM performance. They pre-tested the data and distributed questionnaires to project managers. To test the hypotheses, they used the weighted product of the PLS weights and regression models.The results show that CRM has a different impact on each of the variables including customer acquisition [38].

The shortcomings of this study are as follows:

1. A more extensive amount of data was required; their analysis and the results of this study cannot be generalized.
2. The authors do not provide a methodology for evaluation of CRM performance.
3. They do not provide a prediction for enhancement of future CRM performance.
4. As the correlation is high, it is a good idea to use analysis of variance to find the relationship between variables and test them. The authors can use this as the data is normally distributed and they will be able to see whether there is a significant difference in the level of customer acquisition and loyalty.

Schweidel and Bradlow [102] recognize that the two components of customer value are customer retention and acquisition. In this study, they concentrate on the bond between the time that is needed to acquire a customer and the value of the customer. They use a bivariate model of timing to determine the bond. For data collection, they used monthly subscription of data by a telecommunication provider. Finally, they create a bivariate split hazard model to acquire, retain customers, estimate customer discontent and timing procedure.

Arnold et al. [103] provide the managerial insight that every company must have a strategic focus on customer acquisition. They collected their data from 225 business entities, and provided a solution model based upon a conceptual framework. They formulated twelve hypotheses. They conducted six comprehensive interviews and surveyed twenty companies for data collection purposes. To obtain consistency, they built a coefficient of variance to determine the extent to which the five dimensions deviate from the mean. They tested the validity of the hypotheses with LISERAL after applying factor analysis. The outcomes indicate that the influence of customer acquisition and retention on customer knowledge and decision-making and finally on performance, is intensified when a company consistently implements a CRM strategy.

Villanueva et al. [104] state that customers are invaluable assets but they are hard to acquire. They claim that customers vary in terms of creating a bond with them. Hence, they introduced a customer acquisition method and measured its effectiveness. Their method was vector auto regression modelling (VAR) whereby each variable is considered as internal. In order to be estimated, VAR must have an adequate number of time series data. VAR identifies the positive impacts of acquisition on company's performance, cross-effect of acquisition types, feedback impacts, and supporting impacts. For analysis, they used impulse response functions (IRF) that pursue current and future responses from variables. Also, it measures the total impact on a company of unexpected customer acquisition.

The above chapters have the following drawbacks:

1. The authors do not provide a methodology for customer satisfaction and customer acquisition.
2. There is no means by which they can estimate the future behaviour of customers.
3. None of the above studies considers customer complaints and relates them to customer acquisition.
4. In none of the above studies is there a categorisation of customers or a model to optimise the relationships.

2.4.6 Customer Complaints-Focused CRM Approach

Bougie et al. [105] investigate the effect of anger and dissatisfaction on customer behavior. They are basically different qualitatively and both influence customer behavior. Data has been generated based on the previous studies and is experiential. The items for experiential content have been considered as feelings, thoughts, tendencies, actions, and goals. Also, a questionnaire is designed based on these experiential content items. The testing and analysis of the three main hypotheses were done using p value from LISERAL and the measurement tool was a 9-point Likert scale, ranging from 1(not to all) to 9 (very much). The result showed that dissatisfaction with a correlation of 0.934, $p < 0.05$ will lead to customer anger and this anger will lead to customer behavior responses. Also, dissatisfaction can indirectly lead to behavioral responses with correlation of 2.27 which is acceptable $p < 0.05$.

The disadvantages of their proposed method for dealing with customer complaints are as follows:

1. They provide no robust mathematical framework by which an individual can solve the problem.
2. They do not propose any means for predicting customer complaints in the future.
3. The author does not propose any methodical approach by which an individual can estimate when the customer will become angry or dissatisfied with the situation.
4. The author gathers the data from experiential content of anger from various resources in the past; whereas, the author could use new data derived from various channels.
5. The author does not provide a link between dissatisfaction and other factors. Also, behavioural responses could have many antecedents which have been ignored in this study.
6. The author does not propose any approach by which the customer can interact with the company. Moreover, the proposed method lacks a decision-making process.
7. The author could have used another method to categorise the complaints.

Richins [106] proposes the correlation of one response regarding the dissatisfaction and identifies the variables concealed in it. According to this research, customers have various attitudes toward a service or a product when they are not happy. At first, they may not intendto repeat their purchase; in the second place, they may complain and the third possible outcome would be negative word of mouth (WOM). In this chapter, the author has conducted an empirical investigation and proposes several hypotheses. Data collection has been done in two stages using in-depth interviews and an exploratory questionnaire. For analysis, the author has used SPSS. Results showed that as the problem becomes more severe, negative WOM will be increased. Also, when blame is attached to a company, this will have a negative effect on WOM. Also, for severe issues, the customer may incur expenses in response to that discontent. Furthermore, negative customer perception of company's responsiveness may lead to negative WOM.

1. This approach does not take into account the dynamic nature of complaint management.
2. The author does not propose any conceptual framework by which to trace the variables and outcomes.

3. The author does not provide a method to predict when each of the factors will be converted into word of mouth. Also, apart from WOM, dissatisfaction has other outcomes which have been neglected in this research.
4. The author does not categorise the respondents' interviews for purposes of analysis.

Goodman [35] proposes that customer complaints are inevitable and, although unpleasant for the company, every complaint could be turned into a triumph and customer loyalty.

According to this research, complaints differ case by case and the more money is involved, the more vehement and frequent will be the complaints. Issues of mistreatment, level of quality and ineptitude account for only 5–30 % of complaint rates because clients assume that complaining will do them little good. One of the significant weapons that a customer has at his/her disposal is word of mouth. In interviews with the managers of five important service companies, it was ascertained that more than 40 % of new customers and in two cases, more than 50 % of all new clients absorbed by the company were due to the personal referrals from current customers. Goodman's research indicates that complaints are often not made for the following reasons:

1. It does not help them to improve the situation.
2. It is not worth complaining and producing a new problem.
3. They do not know the right channel for complaints.
4. They are afraid of retribution.

For data collection purposes, the author used five surveys.

The shortcomings of the above study can be shown as follows:

1. The author analyses the data using a simple calculation, whereas other approaches and tools could have produced better results.
2. The author does not propose any framework by which an individual can follow the model and solve new problems.
3. This study qualitatively analyses the issues, while the author could have used the data to conduct quantitative analysis as well. Also, the author does not provide any means or tool to analyse the data collected.
4. The author could have used text mining analysis of the interviews.

Jarrar et al. [107] present an ontology-based method for administering online customer complaints. They established a platform and made their e-interactions transparent. They addressed the issues using an ontology approach. They established a conceptual framework intended to obtain the major knowledge about the issue which is customer complaints.

With this model, a company can provide multilingual services since companies sometimes may encounter cultural issues, difficult-to-understand accents and translation issues. This conceptual framework can tackle the issue from various perspectives such as contract, non-contract problems, evidence, privacy problems and purchasing problem. Likewise, they identified each of the issues using a variety of methods such as data collection methods, private data access, permission-based data collection, sales method, content, product problems, contract termination problems, delivery and billing issues. Finally, economic request, symbolic request and information correction request turn into resolution.

The authors' proposed approach suffers from the following drawbacks:

1. The authors do not explain how they intend to establish the reliability of their conceptual framework. While they discuss the lack of trust and confidence in online purchases, they omit to mention the prediction of trust in such transactions.
2. The authors do not propose an approach by which trust and customer confidence can be calculated.
3. The authors fail to propose a method to evaluate customer complaints and provide feedback to those complaints.
4. While customer complaints are related to the time management of the company and customers, the authors fail to consider calculating the time.
5. While it may be true that a multilingual representation of this ontology might be effective in action, the authors fail to formulate this and make it applicable to other work settings.

Hulten [108] presents an exploratory study on managing customer complaints. They used a survey from 57 managers who were in direct contact with customers. A Lindblomian method is employed to analyse and evaluate CRM. To this aim, they provided hypotheses to be examined and a questionnaire was distributed to respondents. They also used the Kolmigorov-Smirnov test of variables and a *t* test was applied on the data. This study forecasted that companies who use computerized CRM systems are different from the companies that do not have a CRM system. Also, the authors propose that in order to handle customer complaints, a company should have a formal policy. The author's approaches have the following shortcomings:

1. The author does not propose a satisfactory resolution for the customer satisfaction process.
2. The author is unable to provide a measurement to predict customer satisfaction and foresee customer complaint initiation.
3. While the author proposes an analysis for CRM and customer complaints, he fails to categorise complaints. Also, complaints should be prioritised.
4. The author does not propose an approach to illustrate the relationship between the factors in CRM; nor is the relationship shown pictorially.

Homburg and Furst [109] illustrate how managing customer complaints may affect justice in the work setting as well as customer satisfaction and loyalty. The authors provided a mechanistic method based on guidelines and an organic approach (similar to the building of a favourable inner ecosystem). They performed a dual analysis based upon a managerial evaluation of complaint management and evaluation of customers who have complaints. They obtained the data from a commercial provider using an interview approach. They selected a 7-point Likert scale as their measurement tool. To analyse the set hypotheses, they selected LISERA and structural equation modelling. Results show that the mechanistic and organic methods both impact on complaining customers' evaluations, but the mechanistic method has a greater effect.

1. The authors could have used another method to collect the data, as the majority of the respondents, especially expert individuals, seem to prefer to answer the questions by phone.
2. The authors could have categorised the complaints rather than just going about finding solutions.
3. To turn the conceptual framework from complaint to justice, we need to have mediators to create a flow in the procedure.
4. The authors do not mention interactivity and its impact on complaint reduction, although it is an important element of customer complaint management.

Stauss [110] proposes an approach for complaint satisfaction. Complaint has two dimensions which are "outcome complaint satisfaction and process complaint satisfaction". Based on the results, these two dimensions have an impact on the satisfaction of customers, increasing the relationship with customers, and increasing customer willingness to repeat the purchase. Stauss also discusses the determinants of customer complaint satisfaction as being customer-oriented and problem-oriented and these might be related to the quality of the customer complaint management system in the company. The researcher conducted an empirical study and selected a random sample for distribution of a questionnaire. For measurement purposes, a 5-point Likert scale was used ranging from "totally satisfied" to "dissatisfied". In the data analysis section, Strauss used the Chi square test to determine whether the variables were dependent or independent; and factor analysis was used to identify the main factors in satisfaction such as "cold fact satisfaction and warm act satisfaction".

1. The author proposes a method to find the dimensions of customer complaint satisfaction, but fails to pictorially show the relationships.
2. While it is true that relationship satisfaction is an important factor, the author does not provide a methodology for it.
3. The author does not propose a method by which customer satisfaction can be estimated and evaluated.
4. No means is proposed for predicting future customer complaints and for calculating future customer satisfaction.

Ro and Wong [111] propose opportunistic resolutions for customer complaints in service encounters. To this aim, they have collected qualitative data by questioning people based on their previous experiences. They then categorised 346 accidents based on the customer complaints and evidence in restaurants and hotels using a critical incident technique. The results of this study indicate that opportunistic customers intend to complain more about tangible products; and people in authority in categorizing and defining complaints, need evidence to ascertain the nature of the complaints. Hotels were found to be more conservative than restaurants for repayments and rectification procedures. Employees in hotels were more skilled in dealing with complainants compared with restaurant employees. Likewise, hotels appear to have more resources than restaurants, enabling them to recognize and follow opportunistic customer complaints.

1. The authors identify and classify the opportunistic complaints; whereas, they could have taken the other less favourable side into account in order to obtain better results.
2. Their proposed method for opportunistic customer complaints just investigates the compensations, but they do not provide any remedial action to prevent complaints.
3. The authors do not propose any means by which a new reader who needs to apply the framework can pictorially model the relationships between the significant factors in order to simplify the representation.
4. The authors could use several tools and techniques for addressing the customer complaints. To this end, they have categorised the complaints using two researchers to read, store and revisit the complaints, in order to provide a solution for each customer. Although they were concerned about their customers and aimed to address their issues, they did not achieve this.
5. The authors do not include the prediction of future customer complaints in their discussion of follow-up procedures.

Karatepe [112] suggests that companies make different responses to similar customer complaints which could be considered as distributive, procedural or communicational justice. Also, the author examined how these could be related to satisfaction and loyalty. Hence, a survey was conducted of Turkish guests in Cyprus hotels. They tested the hypotheses using LISERAL. They discovered that compensation has a positive link with distributive justice and assistantship; agility has a direct link with procedural justice; while apology, explanation and effort were the other important responses that have a direct relationship with communicational justice. The drawbacks of this study are as follows:

1. The author proposed a conceptual framework for dimensions of complaint and its relationship to satisfaction and loyalty, but does not clarify how satisfaction will be addressed.
2. The author proposed a conceptual framework and depicted outcomes of the model as satisfaction and loyalty, but loyalty has not been measured.
3. The author tested the hypotheses correctly; however, the model needs to be formulated to evaluate each of the significant factors.
4. The author does not discuss the level of satisfaction in this study, although it is vital to the discussion.

5. The author considered only the positive points involved in this study; however, negative points could be considered to make the study more comprehensive and challenging.

Davidow [113] proposes a summary for customer complaint sorting and various responses complaints that companies generate. Furthermore, he discusses how companies' responses can impact on future customer attitudes. The author creates a framework by which organizational response which has six dimensions (benefit, punctual, assistance, apology, reliability and cautious) will lead to customer satisfaction and this alone can produce positive post-complaint behaviour which could be positive word of mouth and/or willingness to repeat the purchase.

1. While the author proposes a conceptual framework that illustrates organizational responses to the customer complaints, no approach is provided to analyse customer complaints and address customers' dissatisfaction.
2. The author does not propose any method that deals essentially with complaint sorting and discovering post-complaint customer behaviour.
3. The author does not propose any means for predicting post-complaint customer behaviour and satisfaction.
4. The author does not propose any means by which a reader can understand when the company should expect to receive post customer complaints when they discuss timeliness, attentiveness and facilitation.
5. The author discusses the issues only in a discussion format; there is no methodology available to measure satisfaction.

2.4.7 Customer Satisfaction-Focused CRM Approach

Luo and Homborg [114] claimed that customer satisfaction is a significant factor in profitability. Based on the results of this study, it appears that the issue of customer satisfaction has been neglected. Based on their conceptual framework, customer satisfaction outcomes would be customer-related, effectiveness-related, employee-related and performance-related. Customer-related outcomes will lead to attitudinal willingness and commitment, re-purchasing willingness and intention to pay for an item. Customer behavior derived from this procedure will be turned into loyalty; complaining behavior into defection. They collected data from archival sources and to determine customer satisfaction, they used a survey. For analysis of customer satisfaction, they used data envelopment analysis (DEA). This chapter suffers from the following shortcomings:

1. The authors do not propose a decision-making methodology for customer satisfaction outcomes.
2. Although the authors provide a prediction and estimation methodology for customer satisfaction, they fail to provide a methodology for customer satisfaction and evaluation.

3. They do not have any suggestions for the management system of a company; furthermore, they do not have any prediction regarding time management of each of the procedures.
4. The authors do not present a methodology to prove whether or not the current conceptual framework is applicable in other work settings.

Mithas [49] investigated the impact of CRM on customer satisfaction and customer knowledge. Customer satisfaction is expected to be one of the outcomes of CRM. For this reason, the hypotheses were provided to ascertain the relationships. The author generated the data from a weekly IT publication and used the Probit method. The findings show that CRM applications are related to customer knowledge and customer satisfaction. What is more, customer knowledge acts as an aggregator between the procedure of CRM and customer satisfaction. This chapter suffers from the following drawbacks:

1. The author proposes no method to analyse each of the factors. For example, customer satisfaction must be evaluated and the relationship of satisfaction with its antecedents must be ascertained.
2. The author does not propose any approach to quantitatively analyse the relationships. Furthermore, the author fails to quantify the level of customer knowledge and customer satisfaction.
3. The author did not provide a conceptual framework to depict the relationship between the variables.
4. Although the author mentions that customer knowledge acts as a mediator and creates an impact within the procedure of CRM applications and customer satisfaction, the author did not formulate the relationships or the indirect impact of customer knowledge in that procedure.
5. The author does not provide a prediction model for future customer satisfaction analysis.

Flint et al. [50] propose a customer value anticipation methodology as suppliers do not appear to value the customers. They highlight the importance of customer satisfaction and loyalty. The authors propose a conceptual framework by which the company can assess how customer value anticipation contributes to satisfaction and loyalty. They establish hypotheses to test whether customer value anticipation does link with satisfaction, and loyalty. To achieve this, the authors conducted two surveys and their sampling was randomly selected from mailing lists. They then applied subjective analysis and factor analysis to check the validity in the first survey. Then using LISERAL and structural equation modelling, they analysed the data. For the second survey, they tested the same model with various satisfaction and loyalty measures. Again they conducted the sampling and the analysis based on AMOS. According to the results of the first analysis, customer value anticipation (perceived value) has a direct bond with satisfaction and loyalty. Likewise, satisfaction positively related to loyalty. Based on the second analysis results, perceived value negatively impacts on customer satisfaction. This chapter has the following shortcomings:

1. While the authors have rejected the hypothesis in the second study which is the relationship between customer anticipation value and loyalty, they do not propose any method to revalidate this.
2. They fail to provide any methodology regarding the future prediction of customer satisfaction and customer loyalty.
3. While they discuss the relationship between customer value anticipation with satisfaction and loyalty, they do not discuss the effect of loyalty and satisfaction on customer value anticipation.
4. They do not show how the methodology can predict further satisfaction and loyalty, and how these factors affect customers.

Kwong et al. [115] present a new methodology to ascertain customer satisfaction using a non-linear and neuro-fuzzy inference system. To identify and understand who the customers are and their perceptions, a survey was conducted among laptop users. For measurement purposes, they used a 5-point Likert scale from 1 "very bad" to 5 "very good". Then following a customer satisfaction model, they validated their proposed methodology. The first test indicates that the model established by the proposed method can provide an accurate result of a typical ANFIS using the same datasets convincingly. The results of the second test show that their model surpasses the models developed using statistical regression based on mean errors and variance of errors.

1. The proposed method does not formulate a methodology for customer satisfaction.
2. The proposed methodology does not illustrate the levels of customer satisfaction.
3. While the analysis was completely done based on the fuzzy inference system, it does not illustrate how future customer satisfaction trends will be estimated.
4. They do not show how to determine customer satisfaction and important variables within the process. Hence, they should have used antecedents and descendants of customer satisfaction.

Briggs et al. [116] propose results from an empirical study using an online survey of third party logistics customers to test the impact of these two types of performance on third part logistics (3PL) service satisfaction. When companies accomplish the services, the clients would be able to assess positional and velocity of performance. For analysis, they use Chi square and AMOS with maximum probability of computation.

Based on Fig. 2.8, the results show that the velocity performance is a more effective motivator of satisfaction; however, positional performance positively influences satisfaction. Furthermore, market turbulence and competitive intensity are both moderators in velocity performance. Also, the results propose to implement metrics and use velocity performance during customer negotiation. Below are the shortcomings of this study:

1. The authors fail to include expectation as one of the significant factors in customer satisfaction and a variable which is important in logistics.
2. Although, there is a significant correlation between velocity and positional performance and the model has an acceptable fit to the data, the authors do not formulate satisfaction.
3. While the authors have used an online survey, it is best to conduct interviews and some real-world surveys to ascertain proof of the responses. Also, they could have used other industries from which to collect the data.
4. While the authors prove that satisfaction can be achieved, they do not mention how they address the impact of satisfaction.
5. Also, this study does not discuss the decision-making process and optimization of customer satisfaction.

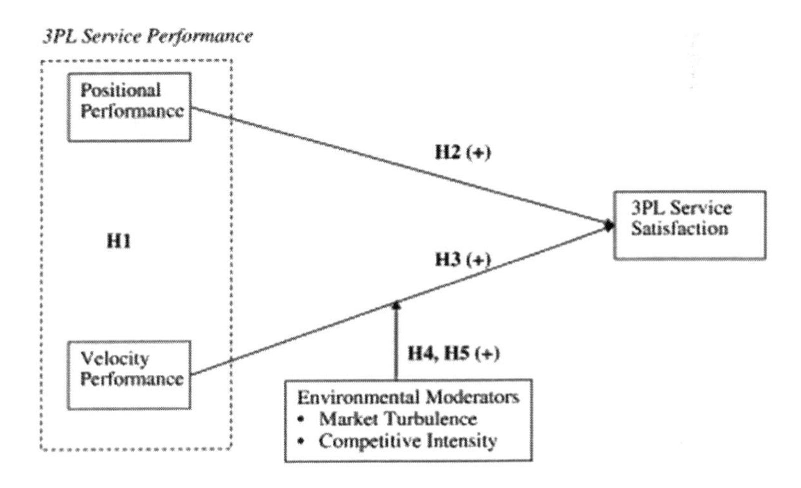

Fig. 2.8 Conceptual model [115]

Sivadas et al. [117] propose a method by which they test and assess the relationship between service quality, customer satisfaction, and loyalty. They defined each of the variables and inter-relationship of those variables. They then propose hypotheses followed by introducing the participants and data collection method which is the computer-assisted telephone interview. Their measurement tool was a 5-point Likert scale. They estimated the model using structural equation modelling.

Figure 2.9 shows that service quality creates satisfaction and attitude. Relative attitude and satisfaction can generate recommendations and encourage the customers to repeat their purchase and these two factors may produce customer loyalty. They assessed that service quality has a direct link with satisfaction. Also, there is a huge probability that customers will produce positive word of mouth and recommend the store to others. Likewise, satisfaction relates to relative attitude. The relative attitude is negatively associated with recommending the store. Finally, satisfaction and attitude are not linked with loyalty. Some of the study's drawbacks are listed below:

1. The authors do not propose a methodology to provide decision-making for satisfaction.
2. They fail to optimise the process as they have an adequate amount of collected data.
3. Although they have introduced the methodology for loyalty, they neglect to discuss the feedback of the loyalty program to the system.
4. The authors do not propose any means for analysing the impact of service quality on loyalty and in turn loyalty on service quality, although the whole process is an iterative one.

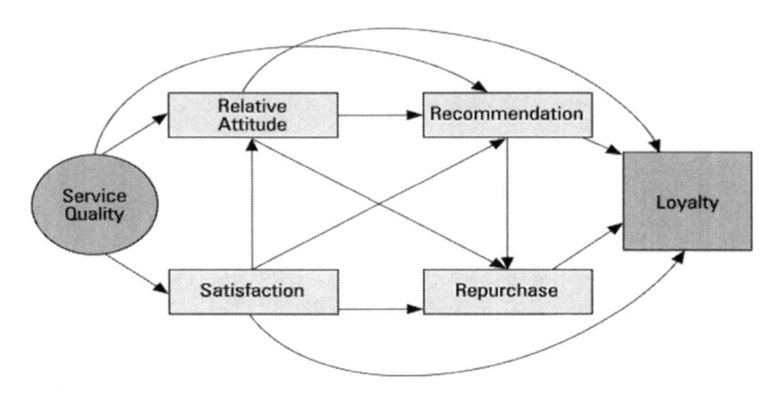

Fig. 2.9 Conceptual framework [116]

Steven et al. [118] propose a methodology by which they find the bonds between customer service, satisfaction, and company performance in airline industry. They introduce hypotheses followed by enhancing the model and data. According to the model, customer service must be linked with satisfaction. Also, customer-centric performances of the company and satisfaction of customers may lead to profitability. To date, airline services have been using previous systems to determine satisfaction. The authors evaluated customer complaints first and then calculated values for these complaints. They collected the data from a panel set. They derived predictions of customer complaints according to the results. According to the assessed customer complaints, all of the customer service variables obtain a 5 % enhancement from their mean values. On-time performance based on customer complaints has the most important impact on satisfaction. "Also, 5 % increase in on-time performance decreases complaints by 2.6 % from 1.077 to 1.049 in each 100,000 passengers. Furthermore, decreasing the number of lost baggage by 5 %,, decreases the number of complaints by 1.9 %, whereas a 5 % decrease in ticket in sales and a 5 % decrease in cancellations of the tickets decline customer complaints by 0.65 % and 1.1 %". Steven et al's [118] study has the following drawbacks:

1. While the authors discuss the prediction about optimization, they do not provide any methodology for it.
2. The authors do not create a complete conceptual framework to show all the relationships. Instead, they depict the outcomes on the model.
3. When analysing the satisfaction and loyalty, they could use other non-linear models to obtain the sensitivity of the relation and sensitivity of the complaints.
4. They do not discuss prediction of customer complaints and customers' post-complaint behavior, after satisfaction.

Bayraktar et al. [119] highlight the concept of satisfaction and loyalty and propose that increasing customer satisfaction and loyalty will result in profitability and market share. They define variables and draw up a conceptual framework. The important factors are corporate image, expectation of customers, perceived quality and perceived value. The authors analysed customer satisfaction and loyalty using DEA. The sample is selected from a Turkish mobile phone brand. A European customer satisfaction index model is used for this study as a base indicator. Then, prior to doing the DEA, they conducted exploratory factor analysis to reduce the amount of data. Based on the comparison and analysis, Nokia performs as the most effective brand followed by LG and Sony-Ericsson for effectiveness in customer satisfaction and loyalty, while Motorola, Samsung and Panasonic are ranked as the bottom line brands. Below, we provide the drawbacks of this study:

1. Although, the data is transformed to a different scale to become normal, it is more effective to use principal component analysis to reduce the amount of data.
2. They do not propose any means for predicting the future satisfaction and loyalty value and to make future decisions.
3. They do not provide any future direction for customer satisfaction and loyalty regarding post-complaint behaviour of the customers.
4. They only ranked the brands but did not suggest ways by which brands could improve and become top brands in future.
5. They do not illustrate whether or not customer satisfaction and loyalty are fed back to the system.
6. The authors do not mention whether or not each of the main variables has a direct impact on customer satisfaction and loyalty.

Edvardsson and Gustafsson [120]'s method is intended to enhance and control operational relationships with customers. They use an exploratory, qualitative research method and for data collection they use in-depth interviews with personnel from the Volvo Auto Company. They analyse the impact of enhanced internal and external quality on productivity and customer satisfaction when the company sells trucks. Productivity must lead the system toward cost reduction and customer satisfaction will be turned into loyalty. If the company meets these important criteria, they will increase profitability. Results illustrate the major impediments to improvements in the Volvo car manufacturing company. The company cannot always focus on customers due to a problem in the CRM system. People rely on product quality but there is a plethora of customer complaints regarding the problems they have with their trucks. The company aims to offer the best ever service experience to its customers, but there are times when the company is not able to provide the desired services due to problems with the CRM system. Company employees cannot meet their commitments, and this may reduce competence. Based on Fig. 2.10, good quality of services and products may result in improved internal and external quality, whereby the internal one may improve productivity and the external produces satisfaction. Simultaneously, improving the quality leads to cost reduction and satisfaction results in loyalty; together, they create profitability and growth. Information technology can bring about much advanced communication and interactivity between company and employees at the Volvo Company and may enable the employees as well as customers to become more loyal. However, the information technology processes are inadequate and give rise to unfavourable issues. Customer satisfaction, which is one of the key factors of CRM, must be monitored on a daily basis by expert operators.

1. This study analyses the variables using qualitative analysis, whereas having data for variables, the authors may have more precise results.
2. The authors do not propose a methodology by which they can estimate customer satisfaction as well as customer loyalty.
3. The authors fail to provide a direction for future prediction of customer satisfaction.

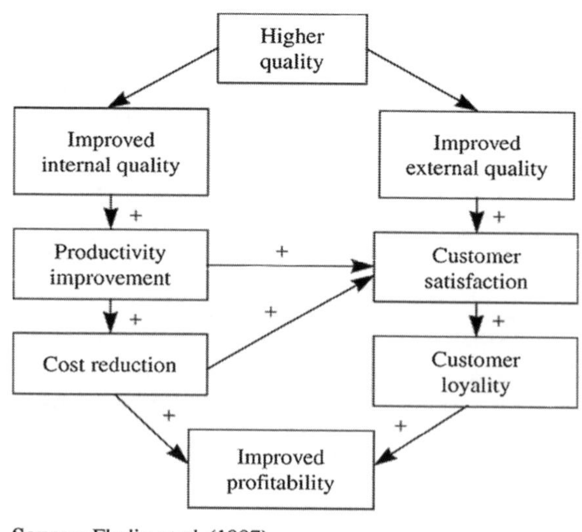

Fig. 2.10 Relationships between main variables [119]

4. Also, the authors fail to mention the sensitivity of each of the variables. Also, the reader cannot determine which variable in the conceptual framework has the most impact.

2.5 Models and Methods Used in CRM

2.5.1 AHP Method

Since the creation of the analytic hierarchy process, it has been used as a decision-making tool for multiple criteria. Based on [121], AHP can be integrated with linear programming to analyse tangible and intangible elements. Also, using AHP, we can determine the priority of the major factors.

Javadi [122] proposed an analytical hierarchy method to analyse data and Expert Choice software is used as the analysis tool. The Analytic Hierarchy Process (AHP) is a structured technique for dealing with complicated decisions. An advantage of the hierarchical analysis procedure is manipulation of decision compatibility. Also, using AHP, researchers are able to obtain the weight of variables, model the issues, prioritise them and make appropriate decisions and judgement regarding the issues.

The AHP is designed to solve complex multi-criteria problems. It is based on the innate human ability to make sound judgments about small problems. It also creates a flow in the decision-making process by means of including recognitions, emotions and judgement in a model [123].

2.5.2 ANP Method

"ANP is a generalization of analytic hierarchy process" or simply, it is an abstraction of AHP. According to Öztayşi [72], the analytical network process is a decision-making approach and can be used for multiple criteria. "Analytical network process (ANP) is a multi-criteria decision making approach in which, it can consider the inner and outer dependencies among multiple criteria". In this study, Öztayşi [72] compared CRM performance of companies who deal with e-commerce. ANP finds the priority of factors in the network and determines the best decision. The main challenge of ANP is to discover the priorities of the factors in the network and the alternatives of the decision. They also conducted sensitivity analysis to determine the strength of the priorities and ranking among alternative websites against changes in weights and interdependency situations.

2.5.3 Clustering Method

Clustering or the linguistic data sequences are proposed by fuzzy data sequences and a fuzzy compatible relation. In fact, individuals utilise various attributes when choosing a service or a product and these attributes are highlighted by clustering or linguistic data sequence. It was established to ascertain the binary relationship between two data sequences. Then, a fuzzy equivalence relation is deducted by conclusion using maximum and minimum transitive conclusions from the fuzzy compatible relationships. According to the fuzzy equivalence relationship, the linguistic data sequences for that particular product or service can be simply categorised into clusters. The clusters presenting various clients' preferences for a product or service are considered as the basis for enhancing customer relationship management [124]. As an example of clustering and based on [124], a bank wanted to employ CRM in relation to credit cards. To achieve this aim, the managers surveyed clients regarding fees, interest, loans, line of credit and other criteria. Then, they grouped all criteria into five clusters and the linguistic preferences are pinpointed as "very low" to "very high" based on a 7-point Likert scale. Next, they evaluated the application of credit cards.

2.5.4 An Object-Oriented Analysis Method

This method is used to create a use-case diagram that characterizes the view of the customers and their attitudes and preferences when using the system and it also creates an activity diagram that illustrates the enhancement of that specific behaviour to satisfy the expectations. Furthermore, it produces a model that shows internal objects and entities that cooperate to enrich these behaviours. As mentioned above, an object-oriented analysis method can be applied for the analysis of a CRM [125].

2.5.5 Content Analysis

Content analysis is an approach used for the text mining of text-based data and to find the frequency of the words and important phrases within a text. As the information about CRM is widespread and ongoing, using that, researchers can streamline the literature in this area [126]. Also, Bowen and Sparks [127] used content analysis with a comprehensive overview of literature and provided a summary of most of the analyses done by researchers in the marketing area. McDonald [128] introduced content analysis as a method of observation to categorise and summarize the interviews which could be either verbal or written. In this study, the author used content analysis to qualitatively analyse quantitative data.

After content analysis, the researcher did the clustering and then developed new strategies based on the results. Gebauer [129] proposed content analysis for the factors that users consider significant. Then, using structural equation modelling, their performance and usability was found. McAlister [130] proposed a content analysis approach to monitor, categorise and code customer complaints. Using this method, the author analysed the words and concepts and found relationships throughout the text.

2.5.6 Single-Case Approach

A single-case approach is used because of the complex nature of relationship management and its generic implementation [22]. When using this approach, we need to have many variables to determine their relationships and the way in which variables are being managed. Although, many researchers think that this method produces only superficial results, the use of real surveys and secondary data in this approach will provide appropriate results.

Single-case creates discussion regarding the significance of the case and clarifies the main variables in the system that may create the issues. It also formulates research questions that clearly describe the problematic areas. Using this approach, we can describe and utilise quality control measurements, provide an acceptable conclusion, and verify the results. This approach can be trusted as we are able to ensure the validity of results [131].

2.5.7 A Simulation Model Approach

This is applied to create the conceptual model and convert it into a simulation framework. However, the process of simulation must be manipulated and arranged according to the criteria and perception of the benefits. Simulation may encourage more explorations of possible results utilising the framework of a realistic situation. When obtaining various results, the usefulness of various performances in a system can be assessed [19]. Also, based on Lemon et al. [132], simulation was used to investigate the leverage of customer mental analysis concerning the effects of satisfaction.

2.5.8 Web-Based Questionnaire Survey

Nambisan [133] collected data utilising a web-based questionnaire which is a very cost-effective approach. It is the best hassle-free option which allows more questionnaires to be distributed to a wider range of respondents outside a region.

However, the data collection would be random. In another web-based survey conducted by Limayem [134], an email invitation was sent to the respondents, and to encourage them to participate in the survey questionnaire, a voucher was given. Respondents then answered the questions to assess perceived usefulness and customer satisfaction.

2.5.9 Survey, Questionnaire, Interview and Hypothesis Testing

Mithas [49] surveyed the top IT managers in more than 300 large U.S. firms. Using a quantitative approach, the author distributed a survey to collect the data and performed statistical analysis on it. The survey method is an approach to collect data from a particular selected sample. It has two subsets which are questionnaire and sampling. For example, [6, 135] structured a questionnaire to gather information about the managerial areas of CRM.

Hypotheses testing is another way of finding the relationship between the variables [12, 136]. Based on Stein [137], an in-depth interview is proposed for data collection in a qualitative research and to identify every small characteristic of an individual, product, and a service. The interviewees are given the opportunity to express themselves.

2.5.10 Data Mining

According to Rygielski [138], the data mining technique can be used for the analysis of data in CRM. Data mining implies a comprehensive search for data using statistical algorithms to reveal the correlations between the data. Data mining can be described as a complicated data search to find patterns and correlations. It can collect the data and build a realistic model for interpretation. The model is useful for prediction, discovery of the data and various data analysis. In this study, Rygielski adopted two case studies and used two different data mining approaches, namely Chi square automated interaction detection (CHAID) and neural network. Regarding the first case, they utilised a neural network to optimise the profit and provide a solution, and in the second case, they used CHAID to increase the effectiveness of targeting extant customers and to entrench predictive models.

2.5.11 Case-Based Reasoning (CBR)

According to Choy et al. [139], case-based reasoning is one of the sub-divisions of knowledge-based systems. It explains activities in case form and thus creates useful information to guide evaluation. CBR is a general model of an intelligent science-based approach to solving problems. Using previous information and reusing it, CBR defines how individuals use previous information to deal with a situation. CBR builds a conceptual framework in which to store operator experience and makes that experience available to various users to create a flow in the situation assessment. Its process includes retrieval of identical cases which it uses to find a solution to the problem, adjust the proposed solution if necessary, and keep the solution as a new case.

2.6 Existing CRM Systems

Since companies face the challenge of winning customers from their competitors in the existing market, they need to have a stringent system with high management capability and prediction of issues prior to their occurrence. Moreover, companies are challenged to implement the most effective system to optimise and streamline customer-oriented procedures across all functional areas to enhance effectiveness and productivity. Each system creates a new solution and each system has its own characteristics and each of the systems meets various customer and business requirements [140].

2.6.1 SAP CRM and its Challenges

Based on the information provided by [140], SAP is a giant software organization which was founded in Germany in 1972. It concentrates mainly on enterprises and mid-size companies. It also helps companies like AMD and Colgate-Palmolive. SAP proposes various CRM systems for a variety of reasons. The SAP business model is on-premise and hosted. SAP proposes on-demand subscription-based CRM resolutions that are Internet-based. SAP CRM was improved to address the unique, end-to-end requirements of major industries including: automotive, chemical, client products, retail, telecommunications, professional services, public sector, media, utilities, oil and gas, and wholesale distribution [141].

SAP has about 3,000 CRM clients and more than 90,000 customers around the globe. About a third of these have been catered for completely. Several clients obtain mySAP CRM licenses but some have progressed to full implementations. Various SAP customers required mySAP CRM quickly but did not have the appropriate facilities to accommodate its applications. Many SMEs may not be able to acquire SAP's impressive on demand CRM software, as SAP's framework prevents customers from crossing over to the SMEs. Still, SAP has not revealed what is the estimated capacity and space each customer may achieve with an individual usage time framework. Even a well-known company such as SAP cannot fully deal with customer complaints.

2.6.2 Salesforce.com and its Challenges

According to [140], salesforce.com, provides customized CRM applications and could be utilized by all sorts of companies regardless of their size. It was founded in California, USA in 1999. The business model of salesforce.com is hosted. This firm offers a considerable number of CRM and business application services [141]. It assists businesses to monitor customer accounts, follow sales leads, assess marketing campaigns, and prepare after-sales services. It concentrates on sales force automation, marketing automation, and customer service and support automation. In addition, it creates one of the best security systems for a company. Moreover, in recent years big companies like salesforce.com are becoming cloud-oriented; their services are becoming accessible by everyone with free investment.

Offline client customization is not endorsed by salesforce.com. There is a shortage of efficient marketing automation. The sales procedure ability is restricted. There is no allowance made for users to efficiently work with groups. Data is often incorrect and irrelevant. The mail merge facility is deficient. Salesforce.com provides no simple channel for conveying data. It does not have adequate tools to guarantee data quality. In addition, its mobile and offline abilities are restricted. It is expensive to customize and personalise and relies on the Sandbox environment. There are restricted enterprise intelligence resolutions and it needs capable information technology experts to make important changes to the solutions [142]. Its inadequacy in satisfying customers is another important issue. Also, the high rate of customer angst, customers' website accountability, and heavy limitations are the other significant problems that the company must resolve. Ultimately, there is a lack of capability to ensure best practices [142].

2.6.3 Oracle-Siebel and its Challenges

According to business-software.com [140], Oracle is one of the biggest application software firms in the world catering for virtually all of a company's requirements to control its business. It was founded in California in 1977; the business model is hosted and is on-premise and serves companies such as Dell, Visa and AT&T. Oracle has created PeopleSoft and Siebel in the last two years. It has various functionalities such as sales automation, marketing capabilities, and implemented intelligent reporting system. Clients can easily prepare CRM software from either Oracle, Siebel, or PeopleSoft's Siebel CRM on Demand clients [141]. Oracle does not provide appropriate and successful training [143]. It cannot provide feedback and recommendations following customer complaints.

2.6.4 *Maximizer and its Challenges*

Maximizer is said [140] to be designed for larger companies and was founded in 1995 in Canada. The business model is on-premise and serves various companies like Cathay Pacific and Oxford University. Maximizer CRM prepares a full-highlighted CRM encompassing marketing computerization and customer service and support. Maximizer CRM provides different connection options such as desktop, Web and PDA. By facilitating mobile tools with full-featured CRM, Maximizer enhances cooperation and customer engagement. Maximizer Software Inc. generates an affirmative, manageable CRM and contact management resolution that may assist SMEs to increase sales, simplify marketing, and improve customer service and support [141]. With the advance to Maximizer 10, Crystal Reports which are intelligent reporting tools, no longer play an effective role. This proves to be a problem with linking to the information source area [144].

2.6.5 *Sugar CRM and its Challenges*

Sugar CRM [140] was founded in California in 2004. It is an open source CRM application which is an on-premise and hosted business model. Creating an enriched set of business procedures, it improves marketing productivity and effectiveness, improves sales, amends and increases customer satisfaction and increases a company's level of business performance. Sugar CRM is implemented by companies such as Avis, Coca Cola and Athena Health [140]. It is a preferred choice for all kinds of customers across a broad range of industries due to its coordination and administration abilities, adjusting to the performance of every firm [145]. Sugar Suite is the best choice because of its improved features. It suits all sizes of companies and industries [141]. This software is easy to install and trouble-free, with an administration facility that provides a variety of selections and tools. Sugar CRM is open-source and as it has a flexible delivery model, there are no arrangement limitations. Also, it gives personalised service and workflow guarantee adjustment to the user.

Based on the evidence, Sugar CRM loads slower than V-Tiger CRM and is troublesome to the user. Issues arise when the user does not lock the installation upon completing it. Unlike V-Tiger CRM, many add-ons are not free to install and are supplementary.

2.6.6 *Sage and its Challenges*

According to the evidence [140], Sage was founded in Irvine, California in 1981 for mid-size and small enterprises. Sage is user friendly, prompt-to-adjust, on-premise and on-demand CRM software resolution with out-of-the-box constructible business procedure automation. It serves companies such as Avent, Legoland and Panasonic.

It is a CRM solution that uses the convenience of the web to create a company's marketing, sales, and customer care teams with the required devices to market more efficiently [141]. Customer service and support encompass the following aspects: centralized client information, consolidated service and support, strong prediction system and reporting, improved opportunity management, comprehensive mobile CRM, prompt access to data, and effective resolution of client problems [146].

However, it provides no customer feedback facility. Customers are often dissatisfied with the product because they do not understand what they required it for at all [147].

2.6.7 *Microsoft Dynamics and its Challenges*

Microsoft Dynamics [140] was founded in Redmond, WA in 1975, with on-premise and hosted business model and its focus is on SMEs. Microsoft Dynamics for customer relationship management encourages employees to increase sales, satisfaction, and service with an automated CRM that is easy to use, customize, and maintain. Microsoft Dynamics business software offers a wide range of enriched, easy to access CRM resolutions to assist firms fulfill their needs. Using Microsoft Dynamics CRM customer service solutions, users can convert customer service into a strategic advantage. With a 360-degree view of the client, representatives can resolve problems promptly and decrease adjustment times with improved client service software. In addition, with computerized procedures, companies can decrease expenses and ensure that customer service is carried out across all touch points. The client service solution encompasses the following: accounts, contracts, knowledge base, programming, workflows in a setting, and analytics [141]. Business solutions from Microsoft Dynamics CRM are: (1) flexible, with selections for arrangements, purchase, and approach; (2) familiar and easy to use because it works like other Microsoft products; (3) designed to suit a business through broad customization and partner proposal [148].

Customer feedback indicates that customers do not have an adequate communication CRM server for Microsoft Dynamics. Also, the toolbar in the software tends to disappear. What is more, sometimes accounts which need to be separated from each other are combined. Also, users may receive many unwanted emails continuously. Microsoft Explorer may close abruptly without warning [149].

2.7 Critical Evaluation of the Existing Literature

In this section, four major areas of research work related to CRM are considered: definition, approach, methodology and software. We identify and define all relevant variables within our conceptual framework. For the approaches, we consider those methods which have been useful and essential for our thesis, particularly those approaches which clarify the link between our variables. For methodology, we choose those approaches that are the most efficient and that will ensure consistency. Finally, we need to choose the most efficient and applicable software to analyse the data.

2.7.1 Interpretations and Definitions of CRM

Different researchers have proposed various definitions of CRM. For example, [6, 7, 14] define CRM as a set of strategies that creates an interactive relationship with customers to provide customized services and products and increase profitability, and manages sales, marketing and services. Also, [6, 8–10] define CRM as an entity that is a core strategic approach that unites marketing activities and information technology to establish a long-term bond with clients and creates competition among organizations. While these definitions are acceptable, we create a new definition for CRM which covers all definitions in the literature intelligently and will be fully defined in Chap. 4.

Our main focus in this thesis is on customer satisfaction which is a growing concern that is strongly connected with its antecedents and expected consequences such as loyalty and customer acquisition. The studies that were discussed previously had relevant and new ideas, but contained no particular features that were found useful for our research; however, each is unique with distinctive features.

Customer satisfaction refers to the ways that the company can meet the requirements of the customer and make them happier than they used to be and provide services beyond customer's expectations [41, 54]. However, we use a much broader definition in this thesis. Customer satisfaction has been previously analysed by many researchers. For instance, satisfaction has been evaluated using an empirical study of third party logistics customers [116], using a quantitative approach. Also, [117] propose a quantitative approach by which customer satisfaction can be examined. Likewise, [115] use a non-linear and neuro-fuzzy approach to analyse customer satisfaction. However, none of the approaches provides a complete methodology to meet the requirements of CRM systems. By a comprehensive methodology, we mean an approach to the issue of customer satisfaction that covers a complaint management system and creates thorough interactivity. However, none of the above approaches ascertain customer satisfaction.

2.7.2 CRM Approaches Which Focus on a Specific Area

There are plenty of CRM approaches and we have considered those which are relevant to our thesis. For example, [71] propose an analytical CRM which deals with quantitative data and is intended to solve customer problems. However, the authors do not consider optimization and trust of the system and fail to address the time spent on each customer. In another study by [23], a conceptual model is proposed in which they provide value drivers, brand and relationship, but neglect to determine the expectation of the customers, and the risk factors cannot be assessed using this model. [72] proposes an approach whereby they increased the benefits achieved from clients using quantitative analysis, but they do not determine the validity and reliability of the data. Furthermore, [73] propose an approach to find the impediments to the success of CRM; however, they do not mention customer complaint sorting as a success factor of CRM.

To analyse customer satisfaction, [117], propose hypotheses followed by computer-assisted telephone interviewing. Their measurement tool is a 5-point Likert scale. [118] proposes a methodology to find links between customer service, satisfaction, and company performance in the airline industry. They introduce hypotheses followed by enhancing the model and data and collecting the data from a panel set. [49] provided hypotheses to ascertain the relationships and generate the data from weekly IT publications and then used the Probit method. In another study [119], the author highlights the concept of satisfaction by drawing a conceptual framework and the sample is selected from Turkish mobile phone brands. A European customer satisfaction index model is used for this study as a base indicator. While all of the methodologies are valid, they do not fully address the issue of customer satisfaction as they cannot deal with customer complaints and their categorization. Simultaneously, they are not adequately interactive and do not give the customers the best experience.

2.7.3 Different Models and Methods Used to Analyse CRM

To analyse CRM, [71] use an analytics model to solve customers' business problems to optimise the decision-making process. However, they do not provide any prediction or optimization regarding customers' post-complaint behavior. Also, [23] consider that managing CRM activities will positively impact on business performance, but they do not have a performance measurement. [72] propose an approach in which they increased the benefits achieved from clients, but they do not consider the effect of customer complaints on customer relationship management. [73] mention major obstacles to the success of CRM in pharmaceutical companies, but they do not discuss customer complaint sorting as a success factor of CRM, and there is no analysis and estimation of it. [74] proposes an integrated approach for designing, evaluating and implementing a CRM system, but it does not determine how to analyse customer expectation and satisfaction. [2] provide consolidated methods to address the shortcomings of CRM, but the authors do not mention how they are going to estimate CRM consequences, such as customer satisfaction and loyalty. Among the other approaches which measure CRM, [19] proposes CRM operational and development consequences, but this study lacks valid data and does not have a methodology for validation of word of mouth and customer satisfaction.

As seen in the customer complaint section, the literature reveals various approaches to defining and analysing customer complaints. Some are quite unique with distinctive features that will be examined in the evaluation section. For example, [105] examine the impact of anger and dissatisfaction on customer behavior, but they do not propose a mathematical framework to solve the problem. Also, [106] proposes the correlation of one response regarding the dissatisfaction, but does not take into account the dynamic nature of complaint management. Also [35] maintains that customer complaints are inevitable, but the author does not provide any tool or means of analysing the data collected, and fails to estimate the level of customer satisfaction.

Customer perceived value is a significant factor that sometimes can be considered as synonymous with customer satisfaction and customer experience [76]. However, there are differences such as cognitive construct in perceived value and influential construct in satisfaction. Also, perceived value is strategy-centered whereas satisfaction is tactics-centered.

As discussed earlier, loyalty is one of the consequences of CRM and customer satisfaction, and it pinpoints the strength of customer retention. Min et al. [93] propose an approach to control turnover of the company and enhance truck driver satisfaction and loyalty. However, they do not propose any future direction and prediction for post-complaint customer behavior analysis. [94] developed a method for establishing and validating a loyalty framework, but they do not mention how they evaluate the perceived value and trust from the performance. [96] examine the impact of personalisation of services on loyalty and evaluate the psychological impact on loyalty, but fail to include the customer complaints can be dealt with using a fuzzy inference system to show the sensitivity of personalisation.

2.7.4 Different Systems and Tools Used for CRM

Based on the methodologies and the software that have been examined through the literature review, we select the most significant of these but in no particular order. For example, [71] use a customer focus measure as a tool to create a customized automatic report in order to recognize issues and contacts, and to generate statistics. To control the process, they use a dashboard. Also, they use business analytics software, but this method does not allow for the prediction of customer satisfaction and behavior. In another study, [72] uses an analytical network process, but this study does not consider the effect of customer complaints on customer relationship management. Also, [73] use a focus group and the survey method to collect data and identify problems. The collected data was divided equally. To discuss the main issues and using the first set of data, exploratory factor analysis was used. To verify the factors and with the second set of data, the authors applied structural equation modelling. They do not clearly discuss customer complaint sorting as a success factor of CRM. Nor do they provide a complete methodology for the analysis and estimation of this success factor.

To analyse perceived value, [77] propose a two-step method in that they first use EFA (exploratory factor analysis) and in the second step, confirmatory factor analysis (CFA), but they fail to evaluate customer loyalty. Roig et al. [78] use a survey for data collection followed by CFA and structural equation modelling to validate the propose scale of perceived value. However, this study must be conducted using other sets of data to revalidate the relationships. Chang [79] devised a questionnaire based on e-service quality, perceived value, satisfaction and loyalty, and then introduced several hypotheses using a conceptual framework. Moreover, structural equation modelling (AMOS 5.0) techniques and linear hierarchical regression models were used to test the causal model, the formulation of the relationship was not explained; nor was the process for evaluating customer satisfaction and loyalty.

To analyse interactivity, [84] presents confirmatory factor analysis but does not provide any means by which loyalty can be assessed. Also, there is no approach for predicting future behavior of customers regarding loyalty. Also, [85] proposes a qualitative analysis approach, but it is unlikely to obtain a precise result on interactivity. Additionally, [86] introduce hypotheses, and to analyse the data, structural equation modelling is used (Amos 6.0) to assess the model, but this model cannot evaluate interactivity and customer satisfaction.

As seen in the customer satisfaction section, researchers use various techniques to evaluate customer satisfaction. In this section, we will summarize those that are most effective and applicable. [114] use a survey for data collection, and for analysis of customer satisfaction, they use data envelopment analysis (DEA), but they do not propose a decision-making methodology for customer satisfaction outcomes. Nor do they make recommendations to ensure satisfaction. [49] generate the data from weekly IT publications and use the Probit method, but they do not evaluate customer satisfaction and its antecedents. Furthermore, [50] provide two surveys and the sampling is randomly selected from mailing lists. They then applied subjective analysis and factor analysis to check the validity in the first survey. Then using LISERAL and structural equation modelling, they analysed the data. For the second survey, they tested the same model with slightly different satisfaction and loyalty measures. Again they conducted the sampling and the analysis based on AMOS. While this methodology and its analysis are valid, they cannot evaluate and formulate customer satisfaction.

This section builds upon the previous critical evaluation of various methodologies. It contributes to a consolidated effort to determine the major issues that emerge from the literature and that must be addressed. To the best of our knowledge, insufficient attempt has been made to design a comprehensive methodology for customer relationship management and its elements. We acknowledge that it is quite difficult to estimate and evaluate each of the factors included in customer relationship management. However, in Chap. 3 we discuss the issues and in Chap. 4 we will pictorially illustrate our conceptual framework and discuss our methodology in detail. Firstly, we need to have a complete methodology which pinpoints and presents different facets that ought to be analysed and may lead to better decision-making processes.

The literature demonstrates the various approaches proposed by different researchers to analyse customer relationship management related to complaints or other variables such as perceived value, interactivity, customer satisfaction, loyalty and customer acquisition. However, although numerous, none of these approaches presents a complete methodology that illustrates the interaction process of the company with a customer.

When we talk about a complete or comprehensive methodology, we really mean that although we are determining the level of perceived value, for example, we need to take into account the possible outcomes that will disadvantage a company. This alone has a negative relationship with customer satisfaction.

2.8 Open Source CRM on the Cloud

Open-source is a software the initial codes and source code of which can be published and made accessible to everyone. They are becoming very popular because of their low cost. There are many open-source software packages, which are competing to obtain a greater share of the market. In this thesis, due to the time constraint, we discuss only the major open-source CRM software including their functionality and their pros and cons.

2.8.1 Advantages of Adopting CRM Open-Source Software

Sales activities are better organised and facilitated using customer relationship management software. The software can easily track and identify business opportunities and weaknesses based on a company's requirements and administer current relationships with the customers. Apart from that, open-source software has been created to provide better interaction within companies. Using the software, a company can better address customers' feedback and generate recommendations and solutions for it. Moreover, the software in CRM is implemented to avoid wasting both company and customer time. This allows the company to accurately track and search the issues and find appropriate solutions for them. Below, we will discuss some of the most important, useful and commonly-used open sources in the market.

2.8.1.1 Vtiger

According to [140], "V-tiger CRM which is an open source can be utilized to administer all activities. It supports customers, services, marketing automation, and fulfillment effectively". V-tiger provides year-long 24/7 customer-service-oriented activities with technical and email support, appropriate and timely scheduled backup and recovery of customer data [150].

Vtiger was first introduced in 2004 and is constantly being updated. It is utilised by more than one hundred thousand companies around the world and is accessible in fifteen languages. It is one of the best-known CRM packages that can be utilized by both small and medium-sized enterprises. Anyone is able to use, copy or disseminate it without requiring permission or payment. It is constructed from SugarCRM with additional sales force automation. So far, there have been over two million downloads from the website due to its effectiveness and applicability. In 2010, Vtiger provided a cloud solution in addition to its other services.

It has the ability to cover fifteen different aspects of business including marketing, customer service and support, sales and managerial functions. It is an ideal solution for companies and can be simply customised using different features. It creates customer report, establishes a flow in the work setting; controls user permission, can be accessed from mobile phones, consolidates with emails and selects a third party extension.

As it provides on-demand services and solutions, it has various advantages. For example, it does not need any installation, thereby saving company costs. There is no need to pay extra for system upgrade. Also, the system regularly upgrades and scans itself and it is accessible from anywhere. It is also absolutely secure as it is hosted on reputable servers such as Amazon EC2.

On the website, there is a preference page that will allow the user to configure personal settings such as time zone, language and personal information. One can also change the logo of the company using the website and update the company's details. Vtiger also allows the users and customer to submit, track trouble tickets, access the knowledge and other important documents which will be provided in further detail. Customers are able to create new contact details and search through the lists. One of the powerful features of Vtiger is that Microsoft Outlook or Word can be completely installed as add-ons.

2.8.1.2 Workflow Stages in Vtiger

Users must supply their names in order to obtain an account and login information. Then, based on their issues, they will be able to submit their trouble ticket. After the customers get through to the web site, they encounter the knowledge base part of the web page. If customers can find a solution or recommendation for their problem, they will leave the page; otherwise, they must fill in a trouble ticket in a ticket submission section. Then, the ticket will transfer to the server and will be presented to the help desk module. In the next stage, the administrator in charge of customer support finds the ticket and solves the problem. However, prior to that, the system automatically may send an email regarding the status of the customer's ticket. Then the solution is shown in the customer portal page. This process is interactive and every now and then the customer can make changes to the ticket and obtain more feedback from the system. Although it has various advantages, it also has disadvantages like any other software. To be completely installed, it may need different file changes as the software cannot supply and process the database while being installed.

2.8.1.3 Sugar CRM

Sugar CRM is accessible in open-source and commercial source applications. It has different functions such as sales force automation, marketing automation including campaign management and email marketing, collaboration, report and customer service and support includes case management and pursuing an accident. To meet customers' needs and for security, Sugar CRM proposes on-demand, on-premise CRM solutions.

Using in-house resources, organizations can customize and extend CRM applications. Sugar Professional and Sugar Enterprise are two commercial products of Sugar CRM which make this brand distinctive and increases it competency and the commercial versions would be a better source for most of the organizations because of their variety of features.

Although Sugar CRM is affordable, has proper documentation, is collaborative and customizable, it has some disadvantages. Open-source software has been introduced to science, but it is still not completely accepted by businesses. Sugar CRM provides solutions only for small businesses. Overall, its pipelines, predictions and reporting are very weak in comparison with major open-source systems.

2.8.1.4 OpenCRX

OpenCRX is an open-source customer relationship management system that provides solutions and is able to meet many company requirements in terms of sales, coordination, marketing and service activities to most of the stakeholders of the company, including mediators, suppliers, customers and partners. This software is customizable based on XML, fast, flexible and can always be measured. Its security is high and as the software is a professional one, it has various features regarding sales, call centre, management problems and marketing issues. It also performs on different platforms such as MySQL, SQL, MS, Oracle and Apache Tomcat. It is also a commercial open-source CRM.

2.8.1.5 Daffodil CRM

This CRM package again is a commercial open-source CRM and enables companies to create solutions and administer their relationships effectively. In general, it covers all facets of interaction that a company is likely to have with its clients. It develops sustainability within the company and facilitates interactions with customers. It also consolidates all facets of customer dealings from identification, customer acquisition to maintaining the customer and giving the customer lifetime value in order to secure their loyalty to the company. It is an automatic way of performing functions in the company as the system is completely time-dependent and critically analyses the data to obtain the best possible solution.

What must be remembered is that Daffodil CRM does forecasting, manages performance and attends to customer satisfaction which simply makes the open source distinct among other software. Daffodil CRM includes effective customer support to create loyalty and allow management to acquire new customers. It is user-friendly, being easy and simple to learn and use, and affordable for companies [151].

2.8.1.6 Hypergate CRM

This is an open-source Java customer relationship management system that provides customer support, marketing, virus tracker, project management, content management, intranet, webmail and sharing. It has various modules such as, for example, the ability to include rich media such as video and flash. It has a quick form-based query wizard and its reports are online and in HTML format. One of the very good features of the software is that the company provides training courses for anyone who wishes to use it.

2.8.1.7 Salesforce CRM

Salesforce is one of the well-known CRM open source packages. It has already covered the majority of customer touch-points in dealing with customers. Customers can customize and consolidate CRM based on their requirements. It assists companies to manage customer accounts, track sales leads, assess marketing, and provide post sales services. It is focused on sales force automation, marketing automation, and customer service and support automation. It also provides one of the best security systems for an organization.

One of the failings of this software is that offline customers are not offered customization. There is a shortage of effective marketing automation and sales process capability. There is no provision for customers to effectively work with groups. Sometimes, data is irrelevant and not genuine. There are limited business intelligence solutions and qualified information technology expertise is required to make significant changes to the solution [142]. In terms of customer satisfaction, it is inadequate.

2.9 Major Weaknesses of Current CRM

Galitsky [152] maintains that in order to handle customer complaints effectively, companies must be equipped with a complaint management system. Also, complaints should be considered as an opportunity to improve the quality of the services. However, although Ro [111] examined complaints in the hotel and restaurant industry as opportunities for improvement, the issues were not addressed methodically and no recommendations were provided.

According to Stauss [110], customer complaint satisfaction has a big influence on the post-behavior of the customers and companies need to know the structure of each individual complaint. Companies mostly receive the complaints and archive them in their repositories and forget to address and investigate them as they think they can always retain their own customers. However, as a consequence of ignoring complaints, the company may encounter negative word of mouth and lose its reputation. If the company cannot provide a proper response to a received complaint and meet the expectations of the complainant, the customer remains dissatisfied.

Atalik et al. [153] mentioned that although customer complaints are on the increase, studies are restricted in this field. Also, big companies cannot manage complaints effectively and the failures exacerbate the situations. Also, according to [153], once the complaint arrives at the system, companies simply create service recovery which in the majority of the cases does not resolve the issues.

Fergusen et al. [154] mentioned that companies are deficient in providing feedback to the customers and satisfying them. They analysed negative customer voices within the company and third party recipient of complaints. They also investigated the complaint recipients and analysed their behaviors.

Based on [155], customer complaints systems have a common low level of listening and they only accumulate the complaints, while failing to methodically address customer complaints and propose a valid solution for the issues. According to [156], the majority of companies using CRM systems have service failures and encounter a huge number of complaints. Their inability to sort out and attend to complaints has led to customer defection and negative word of mouth. According to [35], customers do not make their complaints known because they believe that they will either be ignored by the company or any action taken by the company will be unsatisfactory for the customer. Also, customers do not go looking for disasters and they do not know how to go about lodging a complaint and to whom. Finally, most of the complaints addressed by companies satisfy 30–70 % of the customers, and the rest of them remain dissatisfied.

Ro and Wong [150] examined opportunistic customer (dysfunctional customers) issues and they identified and categorised the complaints. Through their results, they were able to establish the percentage of customers who are unhappy about product, service or both product and service failures, and the source of the complaints. However, they fail to provide a complete methodology for the approach. Predicated upon [150], when a company alone cannot easily handle complaints, they call on a third party which is usually a government agency to act as mediator in resolving the issue. Here, the researchers used content analysis to identify and address the complaints.

Cossument [157] provided a methodology to address the complaints using automatic email classification which differentiates between various customer feedbacks. While they thoroughly analyse the issues, they fail to provide a solution and recommendations. In a similar study conducted by Galitsky et al. [158], the customer complaints have been modeled and categorised and then analysed for validity. Also, they mention that complaints can turn into knowledge about the customers and add to their profile. While all of the statements are valid and they successfully processed the complaints using pictorial representation of the communicative actions and attack relations, they fail to provide a complete methodology to illustrate the quantitative relationship between the factors. Also, they did not provide any solution to address the customer complaints. In none of the above scholarly works have researchers transformed the dissatisfaction to satisfaction and provided a complete methodology for customer satisfaction and loyalty. Likewise, to the best of our knowledge, none of the academic papers to date has proposed a methodology to create business opportunity from customer complaints which is one aim of our research.

From the literature review and to the best of our knowledge, no previous researches in the CRM area have tackled the issue of customer complaints in order to effectively address them and provide strategic recommendations. Likewise, this is the first thesis to address customer issues from the perspective of converting them into opportunities for customers as well as employees at the Fremantle port. Additionally, in this thesis, we convert customer dissatisfaction to customer satisfaction. The model that we propose has the potential to decrease customer complaints which follow service failures.

2.10 Summary of the Literature Review and the Need for an Intelligent CRM

According to our understanding of extant literature, all the above challenges give rise to our thesis which is different in terms of definitions and methodology. The most salient feature in this thesis is that our respondents are exclusive, given our area of interest. Hence, we use an optimization approach to search for the key customers. To the best of our knowledge, no researchers have previously conducted a study in the area of logistics which focuses on customer relationship management and customer satisfaction.

Intelligent customer relationship management has not been previously defined from the perspective of categorizing and analyzing customer complaints in a work setting which deals with various customers. Furthermore, sometimes the customers can be the employees of that organization. The major inadequacy of the current approaches in the literature review conducted to identify the gaps and to propose a comprehensive methodology can be summarized thus:

1. There is no methodical approach for converting perceived value to customer satisfaction in individual cases, especially when key customers of that company must be evaluated. Also, in this thesis the feedback of I-CRM will be converted to perceived value.
2. There is no methodology to illustrate the relationship between interactivity and customer satisfaction. Using feedback from I-CRM, we must evaluate customer satisfaction.
3. There is no methodology for converting customer satisfaction into loyalty using key customers.
4. There is no methodology to convert customer satisfaction into customer acquisition using key customers.
5. None of the current methodologies are adaptive and learner, but using methods, the approaches and models can be learners too.

The current literature on CRM focuses mainly on the business-to-customer relationship, which is the traditional CRM according to our research. Our aim is to integrate B2C and B2B processes which are central to all sorts of transaction involving purchasing and business processes. In the current literature, as discussed in this chapter, none of the researches addresses the issues in a B2B environment. For instance, a stevedore company which is a giant entity and does business with more than 200 road transport logistics companies may deal with this particular situation and work with plenty of other enterprises using CRM. The CRM that we introduce is quite different from the traditional CRM. In this thesis, we address this exclusive situation that has not been investigated before.

As discussed in the previous section, the current literature does not propose an appropriate methodology which allows customers to voice their opinion. The company must provide a safe and trustworthy environment on which customers and employees can rely, so that if they need to complain about a matter, they can lodge their complaint without fear of recrimination. Individuals also expect to obtain appropriate and prompt feedback. Every CRM system must have this ability to estimate the threshold of customer complaints as complaints sorting is a pivotal part of the service evaluation procedure. Only after the implementation of such a system might a company be able to generate customer satisfaction and customer loyalty. To the best of our knowledge, no existing CRM system can generate feedback regarding customer satisfaction and loyalty, and provide a means for customer acquisition. The visual modelling of I-CRM will be extremely beneficial. In Chap. 3, we will define this problem precisely from the perspective of the current literature. In Chap. 4, we propose the solution overview for this problem. We also illustrate that our CRM is intelligent because it can be adapted to specific cases using various approaches. Using intelligent CRM, employees as well as customers will be able to lodge their complaints whenever they wish. Moreover, they have a chance to benefit from the feedback given to other individuals. Intelligent CRMs try to prevent customers from leaving a company by providing guidelines, incentives and steps for improvement that are visible and responsive to customer needs and requirements.

2.11 Conclusion

In this chapter, we conducted a survey of the current literature relevant to our subject. After providing various definitions for customer relationship management and its major components for the purposes of this thesis, we categorised various approaches based upon this research. Ultimately, we critically assessed the current literature in order to categorise customer complaints in terms of intelligent customer relationship management. We then evaluated customer satisfaction in relation to its antecedents and its outcomes. In Chap. 3, we define the problem that we attempt to address in this thesis.

References

1. Chen, I. J., & Popovich, K. (2003). Understanding customer relationship management (CRM). *Business Process Management Journal, 9*, 672–688.
2. Frow, P., Payne, A., Wilkinson, I. F., & Young, L. (2011). Customer management and CRM: Addressing the dark side. *Journal of Services Marketing, 25*, 79–89.
3. Chen, K., & Sockel, H. (2004). The impact of interactivity on business website visibility. *International Journal of Web Engineering and Technology, 1*, 202–217.
4. Mendoza, L. E., Marius, A., Pérez, M., & Grimán, A. C. (2007). Critical success factors for a customer relationship management strategy. *Information and Software Technology, 49*, 913–945.
5. Ueno, S. (2006). The impact of Customer Relationship Management. *USJP Occasional Paper,* pp. 6–13.
6. Özgener, S., & Iraz, R. (2006). Customer relationship management in small-medium enterprises: The case of Turkish tourism industry. *Tourism Management, 27*, 1356–1363.
7. Hoots, M. (2005). Customer relationship management for facility managers. *Journal of Facilities Management, 3*, 346–361.
8. Payne, A., & Frow, P. (2005). A strategic framework for customer relationship management. *Journal of Marketing, 69*, 167–176.
9. Kerr, C., & Anderson. K. (2002, 01-may-2011). *Customer Relationship Management.*New York: McGraw-Hill.
10. Kim, H.-S., & Kim, Y.-G. (2009). A CRM performance measurement framework: Its development process and application. *Industrial Marketing Management, 38*, 477–489.
11. Faed, A., Ashouri, A., & Wu, C. (2010). The efficient bond among mobile commerce, CRM and E-loyalty to maximise the productivity of companies. In *Information sciences and interaction sciences (ICIS), 2010 3rd international conference on*, pp. 312–317.
12. Robinson, L., Jr, Neeley, S. E., & Williamson, K. (2011). Implementing service recovery through customer relationship management: Identifying the antecedents. *Journal of Services Marketing, 25*, 90–100.
13. Chaudhry, P. E. (2007). Developing a process to enhance customer relationship management for small entrepreneurial businesses in the service sector. *Journal of Research in Marketing and Entrepreneurship, 9*, 4–23.
14. Lawson Body, A., & Limayem, M. (2004). The impact of customer relationship management on customer loyalty: The moderating role of web site characteristics. *Journal of Computer Mediated Communication, 9*, 00–00.
15. Limayem, M. (2006). *Customer relationship management: Aims and objectives.* Tehran: Tarbiat Modares University of Tehran.

16. Ngai, E. W. T., Xiu, L., & Chau, D. C. K. (2009). Application of data mining techniques in customer relationship management: A literature review and classification. *Expert Systems with Applications, 36*, 2592–2602.
17. Bose, R. (2002). Customer relationship management: Key components for IT success. *Industrial Management & Data Systems, 102*, 89–97.
18. Furuholt, B., & Skutle, N. (2007). Strategic use of customer relationship management (CRM) in Sports: The Rosenborg case. *Advances in information systems development*, 123–133.
19. King, S. F., & Burgess, T. F. (2008). Understanding success and failure in customer relationship management. *Industrial Marketing Management, 37*, 421–431.
20. Reinartz, W., Krafft, M., & Hoyer, W. D. (2004). The customer relationship management process: Its measurement and impact on performance. *Journal of Marketing Research, 41*, 293–305.
21. Goldsmith, R. E. (2010). The Goals of Customer Relationship Management. *International Journal of Customer Relationship Marketing and Management (IJCRMM), 1*, 16.
22. Lindgreen, A., Palmer, R., Vanhamme, J., & Wouters, J. (2006). A relationship-management assessment tool: Questioning, identifying, and prioritizing critical aspects of customer relationships. *Industrial Marketing Management, 35*, 57–71.
23. Richards, K. A., & Jones, E. (2008). Customer relationship management: Finding value drivers. *Industrial Marketing Management, 37*, 120–130.
24. Iriana, R., & Buttle, F. (2007). Strategic, operational, and analytical customer relationship management. *Journal of Relationship Marketing, 5*, 23–42.
25. Foss, B., Stone, M., & Ekinci, Y. (2008). What makes for CRM system success—Or failure? *Journal of Database Marketing & Customer Strategy Management, 15*, 68–78.
26. Chan, J. O. (2005). Toward a unified view of customer relationship management. *Journal of American Academy of Business, 6*, 32–38.
27. Cho, Y., Im, I., & Hiltz, R. (2003). The impact of e-services failures and customer complaints on electronic commerce customer relationship management. *Journal of Consumer Satisfaction Dissatisfaction and Complaining Behavior, 16*, 106–118.
28. Reinartz, W., Thomas, J. S., & Kumar, V. (2005). Balancing acquisition and retention resources to maximize customer profitability. *Journal of Marketing, 69*, 63–79.
29. Heung, V., & Lam, T. (2003). Customer complaint behaviour towards hotel restaurant services. *International Journal of Contemporary Hospitality Management, 15*, 283–289.
30. Ndubisi, N. O., & Ling, T. Y. (2006). Complaint behaviour of Malaysian consumers. *Management Research News, 29*, 65–76.
31. Stauss, B., & Seidel, W. (2010). *Complaint management*. Wiley Online Library.
32. Cho, Y., Im, I., Hiltz, R., & Fjermestad, J. (2001). Causes and outcomes of online customer complaining behavior: Implications for customer relationship management (CRM). In *Proceedings of the 7th americas conference on information systems*, pp. 900–907.
33. Florenthal, B. & Shoham, A. Four-mode channel interactivity concept and channel preferences. *Journal of Services Marketing, 24*, 29–41.
34. Liu, Y., & Shrum, L. (2002). What is interactivity and is it always such a good thing? Implications of definition, person, and situation for the influence of interactivity on advertising effectiveness. *Journal of Advertising, 31*, 53–64.
35. Goodman, J. (2006). Manage complaints to enhance loyalty. *Quality Control and Applied Statistics, 51*, 535.
36. Vos, J. F. J., Huitema, G. B., & de Lange-Ros, E. (2008). How organisations can learn from complaints. *TQM Journal, 20*, 8.
37. Faed, A. (2010). A conceptual model for interactivity, complaint and expectation for CRM," in Computer Information Systems and Industrial Management Applications (CISIM). International conference on 2010, pp. 314–318.
38. Becker, J. U., Greve, G., & Albers, S. (2009). The impact of technological and organizational implementation of CRM on customer acquisition, maintenance, and retention. *International Journal of Research in Marketing, 26*, 207–215.

39. Hidalgo, P., Manzur, E., Olavarrieta, S., & Farías, P. (2008). Customer retention and price matching: The AFPs case. *Journal of Business Research, 61*, 691–696.
40. Larivière, B., & Van den Poel, D. (2005). Predicting customer retention and profitability by using random forests and regression forests techniques. *Expert Systems with Applications, 29*, 472–484.
41. Avlonitis, G. J., & Panagopoulos, N. G. (2005). Antecedents and consequences of CRM technology acceptance in the sales force. *Industrial Marketing Management, 34*, 355–368.
42. Sánchez-Fernández, R., & Iniesta-Bonillo, M. Á. (2009). Efficiency and quality as economic dimensions of perceived value: Conceptualization, measurement, and effect on satisfaction. *Journal of Retailing and Consumer Services, 16*, 425–433.
43. Chen, P.-T., & Hu, H.-H. (2010). The effect of relational benefits on perceived value in relation to customer loyalty: An empirical study in the Australian coffee outlets industry. *International Journal of Hospitality Management, 29*, 405–412.
44. Hua, H. H., Kandampullyb, J., & Juwaheer T. D. (2009). Relationships and impacts of service quality, perceived value, customer satisfaction, and image: An empirical study. *The Service Industries Journal, 29*(2), 111–125.
45. Faed, A. (2011). Maximizing productivity using CRM within the context of M-Commerce. *International Journal of Information Processing and Management, 2*, 1–9.
46. Korda, A. P., & Snoj, B. (2010). Development, validity and reliability of perceived service quality in retail banking and its relationship with perceived value and customer satisfaction. *Managing Global Transitions, 8*, 187–205.
47. Hu, H–. H., Kandampully, J., & Juwaheer, T. D. (2009). Relationships and impacts of service quality, perceived value, customer satisfaction, and image: An empirical study. *The Service Industries Journal, 29*, 111–125.
48. Georges, L., Eggert, A., & Goala, G. (2010). The impact of Key Account Managers' Communication on Customer-Perceived Value and Satisfaction. *URL:* http://www.cr2m.net/membres/ngoala/travaux/pdfs/D-Gilles%20NGoala-Recherche-Article%20AMS%20EMAC%20KAM-KAMcommunication%20emac.pdf. Quoted, vol. 26.
49. Mithas, S., Krishnan, M. S., & Fornell, C. (2005). Why do customer relationship management applications affect customer satisfaction? *Journal of Marketing, 69*, 201–209.
50. Flint, D. J., Blocker, C. P., & Boutin, P. J., Jr. (2011). Customer value anticipation, customer satisfaction and loyalty: An empirical examination. *Industrial Marketing Management, 40*, 219–230.
51. Caruana, A. (2002). Service loyalty: The effects of service quality and the mediating role of customer satisfaction. *European Journal of Marketing, 36*, 811–828.
52. Minami, C., & Dawson, J. (2008). The CRM process in retail and service sector firms in Japan: Loyalty development and financial return. *Journal of Retailing and Consumer Services, 15*, 375–385.
53. Wang, M. L., & Yang, F. F. (2010). How does CRM create better customer outcomes for small educational institutions? *African Journal of Business Management, 4*, 3541–3549.
54. Gee, R., Coates, G., & Nicholson, M. (2008). Understanding and profitably managing customer loyalty. *Marketing Intelligence & Planning, 26*, 359–374.
55. Faed, A., & Chang, E. (2012) Adaptive Neuro-Fuzzy inference system based approach to examine customer complaint issues. *Presented at the second world conference on soft computing*, Baku, Azerbaijan.
56. Gommans, M., Krishnan, K. S., & Scheffold, K. B. (2001). From brand loyalty to e-loyalty: A conceptual framework. *Journal of Economic and Social research, 3*, 43–58.
57. Kumar, V., & Shah, D. (2004). Building and sustaining profitable customer loyalty for the 21st century. *Journal of Retailing, 80*, 317–329.
58. Anderson, R. E., & Srinivasan, S. S. (2003). E-satisfaction and E-loyalty: A contingency framework. *Psychology and Marketing, 20*, 123–138.
59. Bridson, K., Evans, J., & Hickman, M. (2008). Assessing the relationship between loyalty program attributes, store satisfaction and store loyalty. *Journal of Retailing and Consumer Services, 15*, 364–374.

60. Lee-Kelley, L., Gilbert, D., & Mannicom, R. (2003). How e-CRM can enhance customer loyalty. *Marketing Intelligence & Planning, 21*, 239–248.
61. Reynolds, K. E., & Beatty, S. E. (1999). Customer benefits and company consequences of customer-salesperson relationships in retailing. *Journal of Retailing, 75*, 11–32.
62. Cortiñas, M., Elorz, M., & Múgica, J. M. (2008). The use of loyalty-cards databases: Differences in regular price and discount sensitivity in the brand choice decision between card and non-card holders. *Journal of Retailing and Consumer Services, 15*, 52–62.
63. Reinartz, W., Thomas, J. S., & Kumar, V. (2005). Balancing acquisition and retention resources to maximize customer profitability. *Journal of Marketing, 69*, 63–79.
64. Rowley, J. (2005). The four Cs of customer loyalty. *Marketing Intelligence & Planning, 23*, 574–581.
65. Sharp, B., & Sharp, A. (1997). Loyalty programs and their impact on repeat-purchase loyalty patterns. *International Journal of Research in Marketing, 14*, 473–486.
66. Bridges, E., & Freytag, P. V. (2009). When do firms invest in offensive and/or defensive marketing? *Journal of Business Research, 62*, 745–749.
67. Thomas, J. S. (2001) A methodology for linking customer acquisition to customer retention. *Journal of Marketing Research,* 262–268.
68. Lewis, M. (2006). Customer acquisition promotions and customer asset value. *Journal of Marketing Research, 43*, 195–203.
69. Verhoef, P. C., & Donkers, B. (2005). The effect of acquisition channels on customer loyalty and cross-buying. *Journal of Interactive Marketing, 19*, 31–43.
70. Xu, M., & Walton, J. (2005). Gaining customer knowledge through analytical CRM. *Industrial Management & Data Systems, 105*, 955–971.
71. Azvine, B., Nauck, D., Ho, C., Broszat, K., & Lim, J. (2006). Intelligent process analytics for CRM. *BT technology journal, 24*, 60–69.
72. Öztaysi, B., Kaya, T., & Kahraman, C. (2011). Performance comparison based on customer relationship management using analytic network process. *Expert Systems with Applications, 38*, 9788–9798.
73. Torkzadeh, G., Chang, J. C.-J., & Hansen, G. W. (2006). Identifying issues in customer relationship management at Merck-Medco. *Decision Support Systems, 42*, 1116–1130.
74. Dimitriadis, S., & Stevens, E. (2008). Integrated customer relationship management for service activities: An internal/external gap model. *Managing Service Quality, 18*, 496–511.
75. Phan, D. D., & Vogel, D. R. (2010). A model of customer relationship management and business intelligence systems for catalogue and online retailers. *Information & Management, 47*, 69–77.
76. Eggert, A., & Ulaga, W. (2002). Customer perceived value: A substitute for satisfaction in business markets? *Journal of Business & Industrial Marketing, 17*, 107–118.
77. Yang, Z., & Peterson, R. T. (2004). Customer perceived value, satisfaction, and loyalty: The role of switching costs. *Psychology and Marketing, 21*, 799–822.
78. Roig, J. C. F., Garcia, J. S., Tena, M. A. M., & Monzonis, J. L. (2006). Customer perceived value in banking services. *International Journal of Bank Marketing, 24*, 266–283.
79. Chang, H. H., & Wang, H. W. (2011). The moderating effect of customer perceived value on online shopping behaviour. *Online Information Review, 35*, 333–359.
80. Lai, F., Griffin, M., & Babin, B. J. (2009). How quality, value, image, and satisfaction create loyalty at a Chinese telecom. *Journal of Business Research, 62*, 980–986.
81. Ryu, K., Han, H., & Kim, T. H. (2008). The relationships among overall quick-casual restaurant image, perceived value, customer satisfaction, and behavioral intentions. *International Journal of Hospitality Management, 27*, 459–469.
82. Chang, W. L., & Wu, Y. X. (2011). A framework for CRM E-services: From customer value perspective. *Exploring the Grand Challenges for Next Generation E-Business, 52*, 235–242.
83. Mimouni-Chaabane, A., & Volle, P. (2010). Perceived benefits of loyalty programs: Scale development and implications for relational strategies. *Journal of Business Research, 63*, 32–37.

84. Blocker, C. P., Flint, D. J., Myers, M. B., & Slater, S. F. (2011). Proactive customer orientation and its role for creating customer value in global markets. *Journal of the Academy of Marketing Science, 39*, 216–233.
85. Florenthal, B., & Shoham, A. (2010). Four-mode channel interactivity concept and channel preferences. *Journal of Services Marketing, 24*, 29–41.
86. Yoo, W.-S., Lee, Y., & Park, J. (2010). The role of interactivity in e-tailing: Creating value and increasing satisfaction. *Journal of Retailing and Consumer Services, 17*, 89–96.
87. Bonner, J. M. (2005). The influence of formal controls on customer interactivity in new product development. *Industrial Marketing Management, 34*, 63–69.
88. Kirk, C. P., Chiagouris, L., & Gopalakrishna, P. (2012). Some people just want to read: The roles of age, interactivity, and perceived usefulness of print in the consumption of digital information products. *Journal of Retailing and Consumer Services, 19*, 168–178.
89. Lee, T. M. (2005). The impact of perceptions of interactivity on customer trust and transaction intentions in mobile commerce. *Journal of Electronic Commerce Research, 6*(3), 165–180.
90. Ballantine, P. W. (2005). Effects of interactivity and product information on consumer satisfaction in an online retail setting. *International Journal of Retail & Distribution Management, 33*, 461–471.
91. Liu, Y., & Shrum, L. (2002). What is interactivity and is it always such a good thing? Implications of definition, person, and situation for the influence of interactivity on advertising effectiveness. *Journal of Advertising*, 53–64.
92. Roh, T. H., Ahn, C. K., & Han, I. (2005). The priority factor model for customer relationship management system success. *Expert Systems with Applications, 28*, 641–654.
93. Min, H., & Lambert, T. (2002). Truck driver shortage revisited. *Transportation journal, 42*, 5–16.
94. Lin, H. H., & Wang, Y. S. (2006). An examination of the determinants of customer loyalty in mobile commerce contexts. *Information & Management, 43*, 271–282.
95. Sweeney, J., & Swait, J. (2008). The effects of brand credibility on customer loyalty. *Journal of Retailing and Consumer Services, 15*, 179–193.
96. Ball, D., Coelho, P. S., & Vilares, M. J. (2006). Service personalization and loyalty. *Journal of Services Marketing, 20*, 391–403.
97. Gómez, B. G., Arranz, A. G., & Cillan, J. G. (2006). The role of loyalty programs in behavioral and affective loyalty. *Journal of Consumer Marketing, 23*, 387–396.
98. Leenheer, J., Van Heerde, H. J., Bijmolt, T. H. A., & Smidts, A. (2007). Do loyalty programs really enhance behavioral loyalty? An empirical analysis accounting for self-selecting members. *International Journal of Research in Marketing, 24*, 31–47.
99. Wallenburg, C. (2009). Innovation in logistics outsourcing relationships: Proactive improvement by logistics service providers as a driver of customer loyalty. *Journal of Supply Chain Management, 45*, 75–93.
100. Verhoef, P. C., & Donkers, B. (2005). The effect of acquisition channels on customer loyalty and cross-buying. *Journal of Interactive Marketing, 19*, 31–43.
101. Oztaysi, B., Kaya, T., & Kahraman, C. (2011). Performance comparison based on customer relationship management using analytic network process. *Expert Systems with Applications, 38*, 9788–9798.
102. Schweidel, D. A., Fader, P. S., & Bradlow, E. T. (2008). A bivariate timing model of customer acquisition and retention. *Marketing Science, 27*, 829–843.
103. Arnold, T. J., Fang, E., & Palmatier, R. W. (2011). The effects of customer acquisition and retention orientations on a firm's radical and incremental innovation performance. *Journal of the Academy of Marketing Science, 39*, 234–251.
104. Villanueva, J., Yoo, S., & Hanssens, D. M. (2008). The impact of marketing-induced versus word-of-mouth customer acquisition on customer equity growth. *Journal of Marketing Research, 45*, 48–59.

105. Bougie, R., Pieters, R., & Zeelenberg, M. (2003). Angry customers don't come back, they get back: The experience and behavioral implications of anger and dissatisfaction in services. *Journal of the Academy of Marketing Science, 31*, 377–393.
106. Richins, M. L. (1983). Negative word-of-mouth by dissatisfied consumers: A pilot study. *The Journal of Marketing*, 68–78.
107. M. Jarrar, R. Verlinden, and R. Meersman (2003). Ontology-based customer complaint management. In *On the move to meaningful internet systems 2003: OTM 2003 workshops* (pp. 594–606). Berlin Heidelberg, Springer.
108. Hulten, P. (2011). A Lindblomian perspective on customer complaint management policies. *Journal of Business Research, 65*, 788–793.
109. Homburg, C., & Fürst, A. (2005). How organizational complaint handling drives customer loyalty: An analysis of the mechanistic and the organic approach. *Journal of Marketing, 69*, 95–114.
110. Stauss, B. (2002). The dimensions of complaint satisfaction: Process and outcome complaint satisfaction versus cold fact and warm act complaint satisfaction. *Managing Service Quality, 12*, 173–183.
111. Ro, H., & Wong, J. (2012). Customer opportunistic complaints management: A critical incident approach. *International Journal of Hospitality Management, 31*, 419–427.
112. Karatepe, O. M. (2006). Customer complaints and organizational responses: The effects of complainants' perceptions of justice on satisfaction and loyalty. *International Journal of Hospitality Management, 25*, 69–90.
113. Davidow, M. (2003). Organizational responses to customer complaints: What works and what doesn't. *Journal of Service Research, 5*, 225–250.
114. Luo, X., & Homburg, C. (2007). Neglected outcomes of customer satisfaction. *Journal of Marketing, 71*, 133–149.
115. Kwong, C. K., Wong, T. C., & Chan, K. Y. (2009). A methodology of generating customer satisfaction models for new product development using a neuro-fuzzy approach. *Expert Systems with Applications, 36*, 11262–11270.
116. Briggs, E., Landry, T. D., & Daugherty, P. J. (2010). Investigating the influence of velocity performance on satisfaction with third party logistics service. *Industrial Marketing Management, 39*, 640–649.
117. Sivadas, E., & Baker-Prewitt, J. L. (2000). An examination of the relationship between service quality, customer satisfaction, and store loyalty. *International Journal of Retail & Distribution Management, 28*, 73–82.
118. Steven, A. B., Dong, Y., & Dresner, M. (2012). Linkages between customer service, customer satisfaction and performance in the airline industry: Investigation of non-linearities and moderating effects. *Transportation Research Part E: Logistics and Transportation Review, 48*, 743–754.
119. Bayraktar, E., Tatoglu, E., Turkyilmaz, A., Delen, D., & Zaim, S. (2012). Measuring the efficiency of customer satisfaction and loyalty for mobile phone brands with DEA. *Expert Systems with Applications, 39*, 99–106.
120. Edvardsson, B., Gustafsson, A., & Roos, L. U. (2010). Improving the prerequisites for customer satisfaction and performance: A study of policy deployment in a global truck company. *International Journal of Quality and Service Sciences, 2*, 239–258.
121. Tahriri, F., Osman, M. R., Ali, A., Yusuff, R. M., & Esfandiary, A. (2008). AHP approach for supplier evaluation and selection in a steel manufacturing company. *Journal of Industrial Engineering and Management, 1*, 54–76.
122. Moshref Javadi, M. H., & Azmoon, Z. (2011). Ranking branches of system group company in Terms of acceptance preparation of electronic customer relationship management using AHP method. *Procedia Computer Science, 3*, 1243–1248.
123. Bayazit, O. (2005). Use of AHP in decision-making for flexible manufacturing systems. *Journal of Manufacturing Technology Management, 16*, 808–819.
124. Wang, Y.-J. (2010). A clustering method based on fuzzy equivalence relation for customer relationship management. *Expert Systems with Applications, 37*, 6421–6428.

125. Lin, J., & Lee, M.-C. (2004). An object-oriented analysis method for customer relationship management information systems. *Information and Software Technology, 46*, 433–443.
126. Anderson, J. L., Jolly, L. D., & Fairhurst, A. E. (2007). Customer relationship management in retailing: A content analysis of retail trade journals. *Journal of Retailing and Consumer Services, 14*, 394–399.
127. Bowen, J. T., & Sparks, B. A. (1998). Hospitality marketing research: A content analysis and implications for future research. *International Journal of Hospitality Management, 17*, 125–144.
128. McDonald, W. J. (1994). Developing international direct marketing strategies with a consumer decision-making content analysis. *Journal of Direct Marketing, 8*, 18–27.
129. Gebauer, J., Tang, Y., & Baimai, C. (2007). *User requirements of mobile technology—Results from a content analysis of user reviews*, Champaign: University of Illinois at Urbana-Champaign.
130. McAlister, D. T., & Erffmeyer, R. C. (2003). A content analysis of outcomes and responsibilities for consumer complaints to third-party organizations. *Journal of Business Research, 56*, 341–351.
131. Atkins, C., & Sampson, J. (2002). Critical appraisal guidelines for single case study research. In *Proceedings of the European Conference on Information Systems, June* (pp. 6–8).
132. Lemon, K. N., White, T. B., & Winer, R. S. (2002). Dynamic customer relationship management: Incorporating future considerations into the service retention decision. *The Journal of Marketing,* 1–14.
133. Nambisan, S., & Baron, R. A. (2007). Interactions in virtual customer environments: Implications for product support and customer relationship management. *Journal of Interactive Marketing, 21*, 42–62.
134. Limayem, M., & Cheung, C. M. K. (2008). Understanding information systems continuance: The case of Internet-based learning technologies. *Information & Management, 45*, 227–232.
135. Ko, E., Kim, S. H., Kim, M., & Woo, J. Y. (2008). Organizational characteristics and the CRM adoption process. *Journal of Business Research, 61*, 65–74.
136. Krasnikov, A., Jayachandran, S., & Kumar, V. (2009). The impact of customer relationship management implementation on cost and profit efficiencies: Evidence from the US commercial banking industry. *Journal of Marketing, 73*, 61–76.
137. Stein, A., & Smith, M. (2009). CRM systems and organizational learning: An exploration of the relationship between CRM effectiveness and the customer information orientation of the firm in industrial markets. *Industrial Marketing Management, 38*, 198–206.
138. Rygielski, C., Wang, J. C., & Yen, D. C. (2002). Data mining techniques for customer relationship management. *Technology in Society, 24*, 483–502.
139. Choy, K. L., Lee, W. B., & Lo, V. (2002). Development of a case based intelligent customer–supplier relationship management system. *Expert Systems with Applications, 23*, 281–297.
140. Business-Software. (2012). TOP 40 CRM Software Vendors. Available: http://www.business-software.com/crm/crm.php.
141. Business-Software. (2010). TOP 40 CRM Software vendors revealed. 100. Available: http://www.business-software.com/top-40-crm-vendors.php.
142. Integrity, Q. (2009). Sage SalesLogix vs. SalesForce.com. Available: http://www.qualityintegrity.com/compare_saleslogix_vs_salesforce.asp.
143. N. R. Inc. (2007). Guidebook, Oracle's Siebel *CRM* on demand. Available: http://crmondemand.oracle.com/ocom/groups/public/@crmondemand/documents/webcontent/6071_en.pdf.
144. M. Software. (2009). Maximizer CRM Central. Available: http://www.maximizercrmcentral.com/forums/p/183/490.aspx.
145. SiteGround. (2009). What is Vtiger. Available: http://www.siteground.com/tutorials/vtiger/.
146. S. Software. (2007). SageCRM Customer Care. Available: http://www.sagecrmsolutions.com/assets/Collateral/SageCRMCustSupprtLo.pdf.

147. U. S. Inc. (2007). Sage CRM—What it Can and Cannot Do—Part 1. Available: http://blog.unisoft.net/category/sage-crm/.
148. M. Dynamics. (2009). Microsoft Dynamics CRM. Available: http://crm.dynamics.com/solutions/crm-solutions-overview.aspx.
149. Faed, A., Wu, C., & Chang, E. (2010) Intelligent CRM on the Cloud. In *Network-Based Information Systems (NBiS), 2010 13th International Conference on 2010*, pp. 216–223.
150. H. V. LLC. (2009). VTiger Summary. Available: http://www.seekdotnet.com/vtigerhosting.aspx.
151. D. CRM. (2011). Daffodil. Available: http://crm.daffodilsw.com/.
152. Galitsky, B. A., González, M. P., & Chesηevar, C. I. (2009). A novel approach for classifying customer complaints through graphs similarities in argumentative dialogues. *Decision Support Systems, 46*, 717–729.
153. Atalik, Ö. (2007). Customer complaints about airline service: A preliminary study of Turkish frequent flyers. *Management Research News, 30*, 409–419.
154. Ferguson, J. L., & Johnston, W. J. (2011). Customer response to dissatisfaction: A synthesis of literature and conceptual framework. *Industrial Marketing Management, 40*, 118–127.
155. Bennett, R., & Savani, S. (2011). Complaints-handling procedures of human services charities: Prevalence, antecedents, and outcomes of strategic approaches. *Managing Service Quality, 21*, 484–510.
156. Kaltcheva, V. D., Winsor, R. D., & Parasuraman, A. (2013). Do customer relationships mitigate or amplify failure responses? *Journal of Business Research, 66*, 525–532.
157. Coussement, K., & Van den Poel, D. (2008). Improving customer complaint management by automatic email classification using linguistic style features as predictors. *Decision Support Systems, 44*, 870–882.
158. Galitsky, B. A., González, M. P., & Chesñevar, C. I. (2009). A novel approach for classifying customer complaints through graphs similarities in argumentative dialogues. *Decision Support Systems, 46*, 717–729.

Chapter 3
Problem Definition

3.1 Introduction

In the first chapter, the significance of customer relationship management was discussed and in Chap. 2, the existing literature on the topic was reviewed and evaluated. It was noted that quite a few improvements have been made to different CRM areas including customer satisfaction and complaint management systems. Researchers around the world have made various contributions to this body of knowledge and proposed different models, strategies and approaches.

In Chap. 2, it was stated that none of the current research on customer relationship management proposes a methodical way to analyse complaints in a conjoint qualitative and quantitative manner. Furthermore, none of the existing research presents a comprehensive and precise methodology for categorizing and analyzing customer complaints. Also, no complete methodology exists for categorizing customers.

Given the customer satisfaction methodologies discussed in Chap. 2, there is the need for a method to find the relationship between customer satisfaction and its related variables.

In Chap. 2, the major deficiencies and weaknesses in the existing literature were discussed. To propose a new methodology, all the issues identified need to be addressed. In this chapter, the issues which this thesis is intended to address are formally presented. In Sect. 3.2, various concepts are defined so as to provide a precise understanding of how they are used in the thesis. In Sect. 3.3, the problem is overviewed and discussed. In Sect. 3.4, the need for a new CRM solution proposal is presented. In Sect. 3.5, research issues are presented and Sect. 3.6 contains the choice of research method that will be utilised in addressing the research issues identified in Sects. 3.6.1 and 3.6.2. The chapter concludes with Sect. 3.7.

A. R. Faed, *An Intelligent Customer Complaint Management System with Application to the Transport and Logistics Industry*, Springer Theses, DOI: 10.1007/978-3-319-00324-5_3,

© Springer International Publishing Switzerland 2013

3.2 Key Concepts Used in this Thesis

In this section, the concepts that will be used in the rest of the thesis are defined. Whilst structural definitions of these terms have been proposed previously, the definitions presented here are intended to clarify the way these concepts and terms will be utilised throughout the thesis. Below is a list of the terms which will be used in this chapter to formulate the issues and subsequently offer solutions to address them.

3.2.1 Basic Concepts

Product: A product is a material or a finished good that an individual intends to purchase to meet a specific need.

Service: A service is a non-material item that is purchased from the company to meet a specific need, expectation and requirements of individuals or entities.

Technology: Technology is an approach, system or tool that an organization utilizes to resolve issues and provide optimum services.

Reputation: Reputation is defined as an outcome of customers' and a community's assessment and evaluation of a company's or entity's set of variables in a specific context. Reputation certifies the trust of the customers in a particular brand, service or product. The reputation of a company reflects the level and quality of that company's communication and interactivity with its customers, partners and competitors [1]. Also, reputation can be achieved by maintaining confidentiality and a consistent attitude throughout transactions, and by not allocating a profit to newcomers [2]. Likewise, reputation is what needs to be believed about the attributes and characteristics of an individual, service or a product. Reputation can be deemed as an aggregate metric of trustworthiness in accordance with third party of individuals in society and has the power to link to a group or to an individual [3].

Intelligent systems: Intelligent systems are a complex combination of hardware and software used by an organization to perform complicated functions such as decision-making and business activities.

Business activity: Business activity is comprised of all marketing, management and distribution procedures and strategies that are involved in purchasing and selling a product or service in a service-driven ecosystem.

Customer: The customer is an individual who currently purchases and uses a particular product or a service or potentially may do so. Also, a customer is the entity for whom the company aims to create value. Likewise, a customer is an entity who relies on a company to provide a reliable service or product followed by updated instructions or after-sales service.

Relationship: Relationship implies a bond between two or more parties for a particular purpose.

Management: Management refers to the planning, organization and manipulation of an organization using groups of individuals to accomplish aims and objectives.

Complaints: According to [4–7], there is a plethora of definitions for customer complaints. However, we define a complaint as negative feedback regarding a poor product or service provided by a service provider or a manufacturer.

3.2.2 Additional Concepts Introduced in CRM

Customer retention is a process by which the company prevents customer loss and keeps the customer attached to the business using customer relationship management strategies. A company must provide strategies to prevent customer dissatisfaction before the customer thinks about leaving or shifting to another company. To achieve this aim, the company must recognize the service or product drop cycle and evaluate the placement of the person in that cycle. Customer attrition or customer discontent may arise when the company cannot provide any value for the customers. The company should not take customers for granted and must listen to their voice.

Interactivity: Based on the literature review, there are various definitions for interactivity such as those given by [8, 9]; however, here another definition is provided: Interactivity is the awareness and ability of an individual or a company to create and generate active communication with customers.

Customer acquisition is one of the significant terms and factors that identify the customers and absorb, locate and maintain them in the long term. Following customer acquisition, the company must provide a secure environment. To acquire a customer, a company may need to take various actions such as advertising, identifying the customers in different markets, listening to the customers' feedback, answering their inquiries and providing a prompt response to them.

Customer satisfaction does have various definitions as in [10–12]. However, for the purposes of this research, the following definition is given: Customer satisfaction is a post judgement and assessment of individuals based on their willingness to buy a service or a product.

Loyalty is defined and characterised in terms of repeat purchasing behaviour.

Perceived Value: Based on the literature review in Chap. 2 and definitions in [11, 13], there are many definitions for perceived value, but, the most significant ones have been selected and interpreted as follows: Perceived value is a direct antecedent of customer satisfaction but it may not instantly turn into loyalty or e-loyalty.

Customer centric or customer focused concept is one of the basic premises of CRM and suggests that the customer comes first.

Cloud Service is a framework for facilitating favorable, on-demand network admission to a shared blend of constructible computing resources that can be promptly supplied and delivered with minimal managerial effort or communication with a service provider [14]. Cloud computing includes both the applications released as services in the Internet and the software and hardware in the database that generate those services [15]. There are also, many other definitions regarding cloud computing, however, we modify them as the following definition: Cloud computing is a shared model for creating a favourable, on-demand and agreeable approach to shared knowledge and information resources.

3.2.3 Advanced Definition of CRM

In this section, a definition of CRM as it applies to this research is given as follows: CRM is the art of acquiring customers and establishing a long-lasting relationship with them so that both parties can derive mutual benefits from the association. Customer relationship management is a proven and methodical approach to identifying and acquiring potential customers. It also includes strategies for identifying, satisfying, retaining and maximizing the value of a company's best customers.

3.3 Problem Overview

As discussed in Chap. 2, customer relationship management, interactivity models, perceived value, customer satisfaction and loyalty programs greatly assist organizations to perform effectively in all business processes. Highlighted are the main elements that have a direct and significant influence on customer relationship management. Complaints and the complaints management system and the interrelationship of the two are discussed. Moreover, an analysis is provided of perceived value, interactivity, customer acquisition, customer loyalty and customer satisfaction and their relationships with CRM. Likewise, many reasons have been provided to confirm that there are numerous emerging issues related to customer dissatisfaction with service providers.

In the logistics industries, entities which are considered as customers enter the system hoping to encounter a sophisticated and friendly atmosphere and acquire new knowledge regarding logistics, behaviour, trust and loyalty. Simultaneously, they expect the company to assist them to enhance their operational effectiveness and improve their experiences. Moreover, they want to be satisfied with the work they intend to do. Additionally, these entities must try hard to achieve experience in that specific job or to obtain some degree of achievement in those jobs. Customers and organizations play a vital role in this cycle. They need to provide an eco-friendly and collaborative environment for the employees and make them feel happy and satisfied in that work setting. When this mutual collaboration exists, an organization is heading toward improvement and growth.

When a company is recruiting genuine and savvy entities, it is responsible for providing specific services to the customers (clients) by establishing a rule-based, safe and healthy environment with the necessary infrastructure to meet the requirements of the individuals. The company must recognize that the customers/clients need to have a proper management system to monitor matters in the work setting. Managers and personnel in authority must treat customers equally in order to avoid discrimination.

In a nutshell, if a company does not listen to the customer's voice, whether positive or negative, it will inevitably produce customer dissatisfaction and mistrust. Consequently, the customers might withdraw and gradually, the company will face lack of loyalty and subsequent loss of customers.

In this thesis, an investigation will be undertaken on how to categorise, analyse and respond to customer feedback in an efficient and optimised manner. Organizations normally respond to the customers' feedbacks in different ways to meet their needs and to create value for them.

Not only must an organization respond to the customer feedback in a way that meets the need(s) of the customer, it also needs to create an appropriate message to be able to communicate with the customers and society to increase the productivity. Effective communication and interaction with customers builds and develops relationships with them and with society. This may lead to positive and memorable experiences for the customers.

Our case is unique because it concerns customers at Fremantle port, and as they cannot go to an alternative service provider, they need to have a chance to openly and freely discuss any issues. This will improve the system and prevent service shortcomings and extra problems; it may also be a means of acquiring new customers. For this reason, an organization in accordance with certain rules and regulations must deal with different issues and respond to the requirements of the customers in a timely fashion. The company must ascertain the reasons for customer complaints. The company must address each customer's complaint separately to meet the expectations of individual customers in a reasonable way. The company needs to enhance customer satisfaction by analysing variables such as perceived value and interactivity as antecedents of satisfaction and customer acquisition and loyalty as its consequences. The evaluation of these critical factors improves productivity and reputation of the company and decreases the level of customer complaints.

As discussed in the previous chapter, many researchers have investigated customer relationship management and made numerous contributions to this area. While significantly influencing and improving the body of knowledge in the existing literature, they have nevertheless failed to evaluate and comprehensively categorise the customer complaints, group the customers based on their issues, decrease the level of customer complaints and increase the customer base—issues that are addressed in this research. This thesis will be concerned with: customer complaints, identification of key customers, the introduction of various ways to evaluate the relationship between the main variables, and an evaluation of customer satisfaction and its relevant relationships in the context of customer relationship management strategies. Firstly, it is necessary to obtain and measure interactivity and perceived value from the perspective of the customers. Perceived value and interactivity must be evaluated prior to conducting an analysis of customer satisfaction. Firstly, hypotheses are proposed to test the relationship between these variables. Then, using real data, the relationships are analysed. This is followed by proposing recommendations and rules using mathematical formulations and modelling in Chaps. 6, 7 and 8 respectively. That will enable an analysis and evaluation of customer acquisition and enhance the level of customer loyalty. With this approach, the organizational vulnerability will be decreased and the company will enhance its reputation which in turn increases its profitability. In the final phase, customer loyalty and customer acquisition are optimised.

In concert with this, the main elements of the approach should improve the productivity and effectiveness of the system. However, the application of each element such as perceived value, interactivity, customer satisfaction, loyalty and customer acquisition which were defined earlier in this chapter and throughout the framework, could create confusion and must be clearly addressed to mitigate the issues and produce acceptable outcomes. As an example, in the past, there were quite a lot of researches that dealt with perceived value, interactivity, dealing with complaints, customer satisfaction and customer loyalty and the researchers proposed different approaches to analyse and determine the result and successfully increase the positive effects of these factors.

In this thesis, only one of the domains in customer relationship management (CRM) is discussed, which is intelligent customer relationship management and its application in the port and logistics industry. There are multiple service providers and they are always competing to win customers. While, there are a variety of service providers, it is clear that there is a disparity in services, products and procedures; in fact, there may be many stakeholders. The domain selected for the purposes of this research is unique as the customers are truck drivers who operate in the port precinct; this limits the range of the customer base and they have very specific requirements that must be addressed. Hence, the previous conceptual frameworks and approaches are no longer applicable in terms of increasing customer satisfaction since, in this research, the drivers are the customers.

Previously, researchers have assessed factors such as customer satisfaction, perceived value, interactivity, customer acquisition and customer loyalty, using either qualitative or quantitative researches which have been used in either of those approaches and they did not utilise them all in a conceptual framework. Using either of these two approaches, researchers will be able to determine the likely elements which may leverage the customer satisfaction process. In the majority of the research, only one of either method, namely qualitative or quantitative, has been considered. This research will use both qualitative and qualitative approaches to conjointly produce more stringent analysis and results, followed by the recommendations and decision-making process.

In summary, in this research, the focus is on customer feedback, mainly negative feedback and an attempt will be made to address this issue using categorization and analysis. This will be followed by an estimation of customer satisfaction. A methodical approach is needed whereby the operator or the person in charge can act to prevent a problem from occurring. The problem which will be the focus of this discussion is customer complaint. The system and the company must undertake procedures in order to pre-empt complaints or respond to them in a timely manner before they escalate. In order to investigate the issue of customer complaint management which is highly interactive and deals with customer satisfaction with provision of services rather than product, the Fremantle port including stevedores and road transport operators will be used as the basis of this study. It is also important to emphasise, however, that while this study has been confined to a specific location and service provision, given the complexity of the types of services provided and interactivity, the results obtained will be able to be generalized to many other situations which have complex service provision and high interactivity.

3.4 The Need for New CRM Solutions

Based on the above discussion and the facts, the major goal of this thesis is:

> To investigate the factors which influence customer relationships and customer satisfaction with existing facilities and services at Fremantle port to formulate potential changes to practices and facilities and provide recommendations for any company that needs to enhance the level of customer satisfaction at its work setting. We will qualitatively and quantitatively analyse customer satisfaction and take into account the existing complaints to resolve the issues and provide recommendations.

In order to achieve this goal, it will be necessary to develop a general methodology to qualitatively and quantitatively measure and analyse customer satisfaction utilizing existing customer feedback and customer complaints. While CRM has many cutting edge facets, the scope of this study will be limited to focusing attention on customer complaint management and customer satisfaction due to a limitation of time. In the next part, we present an overview of the previous problems and the research issues that must be addressed to resolve the problems that arise.

Based on the introduction and literature review, the following problems must be addressed and a remedial solution must be sought. In this section, the research issues will be discussed in detail to resolve the problem of customer complaints.

1. In the context of customer relationship management, the following have been defined: complaint, complement, customer relationship management, customer, relationship, management, perceived value, interactivity, customer satisfaction, customer acquisition, customer loyalty and intelligent CRM.
2. A methodical approach needs to be proposed which enables a company to effectively categorise and analyse customer complaints for an intelligent customer relationship management system; therefore, this will be taken into account in the planning stage for the Fremantle port.

3. An explanation must be found for the increasing dissatisfaction produced by intensifying issues at the Fremantle port. Recommendations are needed in order to address the main concerns of this thesis.
4. The main factors contributing to the increasing dissatisfaction with the current Fremantle port booking system must be identified, and remedial actions taken to improve the situation.
5. The main factors contributing to the increasing dissatisfaction with the current Fremantle port costs must be determined, and remedial action must be taken to address this issue.
6. Solutions and recommendations are needed to mitigate the dissatisfaction caused by poor management at the Fremantle port.
7. Recommendations are needed to improve the system of container sorting and labeling as this is another source of dissatisfaction at the Fremantle port.
8. Recommendations and solutions are needed to resolve infrastructural issues and limitations at the Fremantle port that generate complaints and dissatisfaction.
9. Inadequate health and safety measures are significant concerns at the Fremantle port and need to be addressed.
10. A solution is required for the dissatisfaction arising from the shortage of jurisdiction and the lack of a professional and updated policy at the Fremantle port.
11. Weak stevedore performance at the Fremantle port needs to be resolved with the aid of recommendations.
12. Recommendations are required in order to improve the time management as there has been an issue regarding weaknesses in administering the time in different sections of the Fremantle port.
13. The proposed methodologies for addressing research issues 2, 3, 4, 5, 6, 7, 8, 9, 10 and 11, will be validated by using survey and experimental tools.
14. A methodology is proposed to define the relationship between intelligent customer relationship management evaluators which are perceived value, interactivity, loyalty and customer acquisition. Furthermore, the related hypotheses must be tested and validated.
15. This thesis presents a complete methodology for modelling customer satisfaction in relation with other variables to decrease customer dissatisfaction.

3.5 Research Issues

In this section, we present and discuss the research issues that need to be addressed in detail to solve the aforementioned problems of customer complaints and customer dissatisfaction with a service-oriented environment, specifically the port environment.

3.5.1 Research Issue 1: Conceptual Definitions

In this section, we define intelligent customer relationship management for services and the importance of its implementation in the port as an exclusive case study that has not yet been investigated. As discussed in the previous chapter, I-CRM is a system and strategy that needs to be considered by contemporary organizations. Given the pivotal influence of CRM, it has been interpreted and described by many researchers in accordance with the objectives of their studies. Essentially, considering all existing definitions, CRM is a disciplined business strategy intended to establish a long-term relationship with the customers. It should be noted that although there are various definitions of CRM, none of them has made an attempt to focus on the solution of customer complaints and categorisation of the customers.

In this thesis, we intend to examine the significance of customer complaints and investigate the importance of various customers; this is followed by an evaluation of customer satisfaction. In Chap. 2, we presented an extensive discussion of this area of study as reported in the literature. For example, researchers have proposed various definitions and frameworks for CRM to evaluate customer satisfaction, customer loyalty and customer acquisition. However, essentially, none of the approaches articulated in Chap. 2 defines CRM as a consolidated strategy to categorise customer complaints and develop customer satisfaction using perceived value and interactivity. We also examined other consequences of customer satisfaction, including the fact that it may help to establish customer loyalty. To achieve the objective of this thesis, our main focus must be on defining I-CRM and its various components in order to comprehensively address the issues which are the concerns of this research. We also introduce our conceptual framework to show the relationships in two different segments. Further, as we intend to implement our model in the domain of the logistics industry, we define each of the elements based on our need to address the issues here. Additionally, the definitions should be capable of generalization so that they can be adopted by other industries.

3.5.2 Research Issue 2: Propose a Methodology for Complaint Analysis and Management

There are numerous and various complaints from different customers and companies; however, there is no methodical approach to differentiating, categorizing and analysing customer complaints. The question is, how to categorise, prioritise and analyse customer complaints in a way that allows a proper categorization and analysis of them.

To do this, a real-world case is needed in order to generate real data. This thesis uses both qualitative and quantitative methodologies conjointly. Chapter 5 focuses on the qualitative approach and deals with the customer complaints. As we conducted our research in the domain of logistics, we acquired our data by targeting truck drivers who are considered as customers. To this end, an interview survey was distributed among the survey participants. Using the qualitative approach, the major priorities were identified. And to achieve that, a method based on qualitative research is needed in order to find and categorise the main complaints.

There is a need to understand why customers complain about the work environment and the work itself, and why they are sometimes reluctant to provide feedback. Additionally, there are enormous amounts of customer feedback whether negative, positive or neutral, but no-one wants to investigate and analyse this information. If a company had a department devoted to analysing the customer complaints, they could find solutions to improve its work flow systems and processes. By establishing a section that deals with customer complaints, a company can determine customers' concerns and complaints regarding specific issues early on and this alone may create effective services for customers. By analysing customer complaints, a company can obtain a complete view of customer complaint issues, enabling them to provide new strategies to decrease the incidence of complaints.

Apart from analysing the issues qualitatively, the impact of each issue on the total set of interviews should be considered, and solutions should be found to address them. This should be methodical and according to the facts and analysis. Since a qualitative analysis is being carried out, the main words related to each significant issue should be weighted according to their level of importance. It is also clear that a logistics company considers the main customer complaints as a platform and uses these as a basis to formulate recommendations and solutions. This platform and recommendations can be changed to some extent according to the nature of the company. In future, if the company is faced with new issues, it can refer to the previous methodology and recommendations and use these to solve other emerging complaints and issues.

3.5.3 Research Issue 3: Propose a Methodology to Formulate Hypotheses and Provide Recommendations

In Chap. 4, the conceptual framework is designed and the relationships between the main variables that have a positive influence on I-CRM are illustrated. In Chap. 5, the customer issues are prioritised and their importance is determined based on the weight of the issues. In Chap. 6, the relationship between variables is examined by proposing hypotheses. Additionally, as the main focus is on customer complaints, the relationship between customer dissatisfaction and the issues that were identified in Chap. 5 is explored. Then, using the new methodology, the link between hypotheses can be ascertained. Based on Chap. 4, as the conceptual framework has two separate parts that are in conjunction with each other and create an iterative process, different sets of hypotheses are needed for the purposes of analysis and to provide recommendations to address the issues. In this section, we intend to quantitatively analyse the relationships and propose recommendations to address the issues.

To analyse and test the hypotheses, data is required from our industry logistics partner which is the Fremantle port. To this end, a questionnaire survey will be developed and this will be subjected to validity and reliability testing which will be further discussed in detail in Chap. 9. To the best of our knowledge, the approaches in the literature employed one methodology or a software to analyse the hypotheses and to validate the relationship between variables and hypotheses; whereas, in this research, integrated methods are used to test the hypotheses. In this thesis, not only do we need to test the relationship among latent variables; we also need to ascertain the relationship between each of the main latent variables and our observed variables. Observed variables are the ones that have a close interaction with the main variables.

3.5.4 Research Issue 4: Scientific Models for Customer Feedback, and Customer Satisfaction and Preliminary Hypotheses Validation

This issue deals with determining and validating customer satisfaction and finding the key stakeholders using a complete methodology. Also, we need to find the key customers and determine the shortest route to reaching complete customer satisfaction. In fact, we must optimise cost and time to improve customer satisfaction. To this end, we need to have an algorithm to screen the customers first and to find the key customers. Likewise, as the data is huge, we need to reduce the size of the data. As discussed earlier in Chap. 2, companies are dealing with a huge amount of negative feedback and no-one is taking responsibility for addressing the issues and providing a solution for them. For example, in the case of Fremantle port, customers are challenged by a variety of issues such as poor management which dominates some other issues like time management and stevedore performance. Also, this is a lack of adequate health and safety measures, and infrastructure problems are other problematic areas that make the work difficult at the port. For example, there are some drivers coming from eastern states that stay overnight and need accommodation.

Likewise, in the port there is not a proper container sorting mechanism to help drivers load and unload their trucks. Additionally, there are drivers who have to put up with an inefficient booking system and this needs to be addressed. Also, cost is another important issue for the majority of drivers, as they expect not to incur too much cost when they work at the port. Also, there are issues with regulations and jurisdictions that are a source of discontent. As an example, drivers are not happy with road regulations and extra costs and they claim that this is a reason why there are so few drivers. Also, in terms of jurisdiction, another issue that the drivers may have is accreditation. The drivers drive all around Australia and WA's accreditation needs to be recognized nationally and by all states, whereas, currently Western Australia's licences are not recognized inter-state. This HVA, or Heavy Vehicle Accreditation licence, has become an issue for the drivers.

To the best of our knowledge, no other researchers in their previous works have formulated a mathematical methodology for measuring customer satisfaction and customer complaints using a quantitative approach. We must use a methodology to find and analyse key customer recognition optimization and to find key customer complaint recognition.

3.5.5 Research Issue 5: Propose a Methodology for the Relationship Between Customer Complaints with Variables of Perceived Value, Interactivity, Loyalty and Customer Acquisition

After we ascertain and formulate the relationships, we provide recommendations to address the issues. Here, it is essential to evaluate the antecedents and consequences of customers' satisfaction regarding major customer complaints. A solution to this is presented using a complete methodology that has not been provided previously in the literature.

What has emerged in Chap. 6 is that, based on all the data collected, the majority of the hypotheses tested were supported; however, the relationship between customer satisfaction and loyalty was refuted. Using formulation, these hypotheses are confirmed in Chap. 8. Also, there is the need to calculate the total degree of arguments for each of the customers for each of the variables including perceived value, interactivity, customer loyalty and customer acquisition. Rules need to be generated to confirm the relationship between the main variables, as discussed above, and the customer complaints.

3.5.6 Research Issue 6: Validate the Proposed Methodologies

The methodologies and recommendations provided to address research issues 2–5 need to be validated. The process of validation enables the researcher to clearly establish that the conceptual framework and methodology is solid and applicable in the work setting. Also, by validation, the methodology is established, and the relationship between all the latent variables and the bond between latent variables and observed variables will be clarified. This verifies and strengthens confidence that the proffered methodology for each of the chapters is relevant to customer satisfaction and customer relationship management and leverage customer complaint to evaluate and decrease customer complaints. In Chap. 10, an overview of the results is given, followed by a discussion of the contribution that this study makes to the existing body of knowledge.

3.6 Choice of Research Methodologies

In addressing the stated problem, this thesis focuses on the development and subsequent testing and validation of a methodology for risk-based decision-making in business interactions in collaborative environments. In order to propose a solution for the research issues listed in the previous section, a systematic scientific approach must be adopted to ensure the methodology development is scientifically-based. Therefore, in this section, an overview is given of the existing scientifically-based research methods together with the reasons for choosing a particular research method.

3.6.1 Research Methods

There are two broad categories of research in information systems, namely:

a. Science and engineering approach, and
b. Social science approach.

3.6.2 Science and Engineering-Based Research

Science and engineering-based research is concerned with confirming theoretical predictions. Gallier [16] states that in the engineering field, the spirit of 'making something work' is essential and has three levels: conceptual level, perceptual level and the practical level, as explained below:

- Conceptual level (level one): creating new ideas and new concepts through analysis.
- Perceptual level (level two): formulating a new method and a new approach through design and building the tools or environment or system through implementation.
- Practical level (level three): carrying out testing and validation through experimentation with real-world examples, using laboratory or field testing.

Science and engineering research may lead to new techniques, new architectures, new methodologies, new devices or a set of new concepts which together form a new theoretical framework. Frequently, it not only addresses the issue of what problems need to be addressed, but also proposes a solution.

3.6.3 Social Science Research

Social science research can be either quantitative or qualitative research. It is often carried out through survey or interview processes. Quantitative research involves extensive data gathering usually using methods such as surveys, and statistical analysis of the gathered data in order to prove or disprove various hypotheses that have been formulated. Qualitative research frequently involves in-depth structured or semi-structured interviews that allow one to pursue particular issues of interest that may arise during the interview. It does not normally involve a large sample of data and the information gathered may not be in a form that readily allows statistical analysis. A typical social science research approach is to use survey forms to identify problems which are subsequently formulated as hypotheses. The goal of social science research is to obtain evidence to support or refute a formulated hypothesis [17–19]. It assists the researcher to understand people and social issues such as culture, within the area of research. Kaplan and Maxwell [20] argue that the ability to understand a phenomenon within its social and cultural context is forfeited when textual data results are quantified. This kind of research can indicate the extent to which the methodology is or is not accepted and sometimes may be able to give the reason for this. However, unlike engineering-based research, this type of research does not explain what a methodology should be and how to produce a new methodology for problem solving. This research only tests or evaluates a method that has already been produced by science and engineering research.

This thesis uses a survey approach which includes a social science and science and engineering methodology. Basically, social science is applied to identify the issues and the science and engineering approach is employed to synthesize the data and issues.

3.6.4 Choice of Multi-Disciplinary Approach to Problem Solving

In this thesis, a science and engineering-based research approach was chosen as the research method for the proposed solution development. An overview of this research method is depicted in Fig. 3.1.

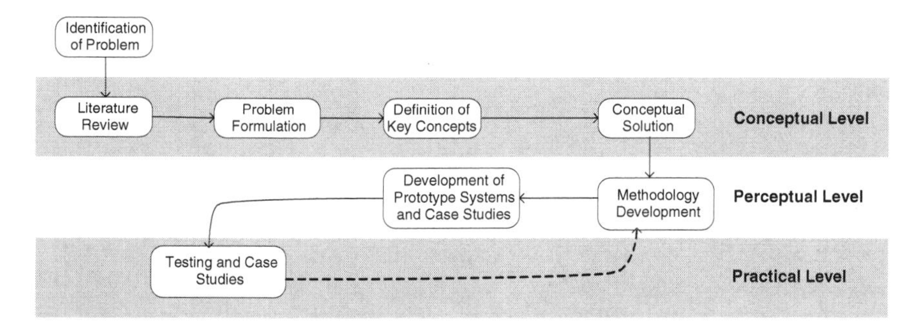

Fig. 3.1 Overview of science and engineering-based research method

The recognized issues were progressively broken down into the major research objectives that must be achieved in order to address the problem outlined earlier in this thesis. Each research objective was broadly defined in respect to the current literature. Furthermore, various research methods were investigated. For the purposes of this study, it was concluded that the most appropriate approach is the science and engineering-based research methodology together with a social science approach that is suitable for information systems research.

Firstly, the research problems were identified together with the major issues, using social science methodology. A comprehensive and detailed range of relevant literature was collected and analysed. Following an in-depth review of the existing literature, the issues to be addressed were formulated. Subsequently, definitions were provided of significant concepts of customer relationship management, complaint, loyalty, customer satisfaction, services, perceived value, interaction and interactivity, taking into account their context- specific and dynamic nature. These definitions will be utilized in developing the conceptual solution. Then, the conceptual solution to the issue being addressed in this thesis was delineated. All processes ranging from literature review to conceptual solution shape the conceptual level. At the perceptual level, an improved methodology is proposed for handling complaints and categorising them using customer relationship management tools. Also, a comprehensive and real-world case study was developed which would be used later to test the proposed methodology. The processes of methodology development and the case study are located in the perceptual level.

Subsequently, prototype systems were engineered and several case studies were developed for later use when testing the proposed methodology. The processes of methodology development and development of prototype systems and case studies constitute the perceptual level of this work. Once the prototype systems have been engineered, they are used together with the developed case studies to validate the proposed methodology and the hypotheses utilizing methodical approaches. At the practical level, based on the results obtained, the proposed methodology was then evaluated and validated. Based on the outcome of the evaluation and validation process, the proposed methodology was then fine-tuned. The evaluation and validation process of the developed methodology constituted the practical level of this work.

With regard to the practical aspects of research output evaluation and validation, Nunamaker et al. [19] argue that typical research follows a pattern of problem definition, hypothesis, analysis and argument. In such a scenario, problems are encountered and analysis is performed in the form of proofs and developed solutions. The results of the analysis and development form the basis for the evaluation of research outcomes. The methodology proposed by Nunamaker et al. [19] consists of the problem definition, conceptual solution and system prototypes processes. This research method proposed by Nunamaker et al. [19] is adopted for validation and verification of this study's research output, through proof of concept.

3.7 Conclusion

This chapter began with an identification of the research problems. An extensive amount of literature on topics related to the study was collected and analysed. Based on an extensive review of the existing literature, the problem that needs to be addressed was formulated. Subsequently, several key concepts that will be used in addressing the problem were defined, taking into account the characteristics of the interaction. These definitions will be used when developing the conceptual solution. Subsequently, the conceptual solution for the problem being addressed in this thesis was formulated. All processes from the literature review to the conceptual solution are included in the conceptual level.

References

1. Huynh, T. D., Jennings, N. R., & Shadbolt, N. R. (2006). An integrated trust and reputation model for open multi-agent systems. *Autonomous Agents and Multi-Agent Systems, 13*, 119–154.
2. Kamvar, S.D., Schlosser, M.T., & Garcia-Molina, H. (2003) The eigentrust algorithm for reputation management in p2p networks, pp. 640–651.
3. Jøsang, A., Ismail, R., & Boyd, C. (2007). A survey of trust and reputation systems for online service provision. *Decision Support Systems, 43*, 618–644.
4. Hulten, P. (2011) A Lindblomian perspective on customer complaint management policies. *Journal of Business Research.*
5. Stauss, B. & Seidel, W. (2010) *Complaint management*: Wiley Online Library.
6. Reinartz, W., Thomas, J. S., & Kumar, V. (2005). Balancing acquisition and retention resources to maximize customer profitability. *Journal of Marketing, 69*, 63–79.
7. Vos, J. F. J., Huitema, G. B., & de Lange-Ros, E. (2008). How organisations can learn from complaints. *TQM Journal, 20*, 8.
8. Faed, A. (2011) Maximizing productivity using CRM within the context of M-commerce.

9. Georges, L., Eggert, A., & Goala, G. (2010) The impact of key account managers' communication on customer-perceived value and satisfaction. *URL:* http://www.cr2m.net/membres/ngoala/travaux/pdfs/D-Gilles%20NGoala-Recherche-Article%20AMS%20EMAC%20KAM-KAMcommunication%20emac.pdf. Quoted, vol 26.
10. Becker, J. U., Greve, G., & Albers, S. (2009). The impact of technological and organizational implementation of CRM on customer acquisition, maintenance, and retention. *International Journal of Research in Marketing, 26*, 207–215.
11. Avlonitis, G. J., & Panagopoulos, N. G. (2005). Antecedents and consequences of CRM technology acceptance in the sales force. *Industrial Marketing Management, 34*, 355–368.
12. Flint, D. J., Blocker, C. P., & Boutin, P. J, Jr. (2011). Customer value anticipation, customer satisfaction and loyalty: An empirical examination. *Industrial Marketing Management, 40*, 219–230.
13. Sánchez-Fernández, R., & Iniesta-Bonillo, M. Á. (2009). Efficiency and quality as economic dimensions of perceived value: Conceptualization, measurement, and effect on satisfaction. *Journal of Retailing and Consumer Services, 16*, 425–433.
14. Dillon, T., Wu, C., & Chang, E. (2010) Cloud computing: Issues and challenges, pp. 27–33.
15. Armbrust, M., Fox, A., Griffith, R., Joseph, A. D., Katz, R., Konwinski, A., et al. (2010). A view of cloud computing. *Communications of the ACM, 53*, 50–58.
16. Galliers, R. D. (1992). *Information systems research: Issues, methods and practical guidelines*. Oxford: Blackwell Scientific Publications.
17. McTavish, D.G. & Loether, H.J. (1999) Social research: Allyn & Bacon.
18. Burstein, F., & Gregor, S. (1999) The systems development or engineering approach to research in information systems: An action research perspective. *Proceedings of the 10th Australasian Conference on Information Systems (ACIS'99)*, Wellington, New Zealand, pp. 1222–1234.
19. Nunamaker, J. F., Chen, M., & Purdin, T. D. M. (1991). Systems development in information systems research. *Journal of Management Information Systems, 7*, 89–106.
20. Kaplan, B., & Maxwell, J.A. (1994) Qualitative research methods for evaluating computer information systems. *Evaluating Health Care Information Systems Methods and Applications,* 45–68.

Chapter 4
Solution Overview

4.1 Introduction

As discussed in Chap. 2, considerable research has been undertaken to enhance and resolve the issues of customer relationship management regarding negative customer feedback or what are termed 'customer complaints' in both the real and virtual worlds. Company managers dread receiving and dealing with customer complaints, and attempt to avoid these. They may feel that complaints are unfair and they do not wish to be blamed. Additionally, the management team does not have adequate time to allocate complaints, only to handle them and they need to provide strategies for doing so. In this way, complaints may be received by other divisions or departments in a company and further investigated. However, neither the departments nor the clients may be willing to deal with complaints, thereby ignoring them rather than investigating the reasons for their occurrence. In fact, a company has the option of utilising the complaints to its advantage by converting the information into a strength and opportunity; however, they shy away from analysing the complaints. Even customers or clients might not lodge their complaints as they do not want to be recognised as a complainant.

Also, they think that by making a complaint, they jeopardize their working conditions and they feel insecure. In this chapter, based on the literature and problem definition stated in Chaps. 2 and 3, a new model for I-CRM is proposed which is intended to remove the barriers facing dissatisfied customers and encourage them to openly lodge their complaints and manage their complaints by providing recommendations and solutions. Also provided in this chapter is an overview of the solution for the research problems. Based on the proposed conceptual framework which will be discussed further, definitions will be provided of the terms and concepts related to this framework. These concepts will be utilised throughout the rest of the thesis. As part of the conceptual framework, definitions will be given for all concepts relating to intelligent customer relationship management (I-CRM). This is followed by a discussion of the conceptual framework including the concepts and the solution framework for each of the identified problems.

A. R. Faed, *An Intelligent Customer Complaint Management System with Application to the Transport and Logistics Industry*, Springer Theses, DOI: 10.1007/978-3-319-00324-5_4,

© Springer International Publishing Switzerland 2013

4.2 The Proposal of an Intelligent CRM (I-CRM) and its Components

In this research, intelligent customer relationship management is defined as an art and a methodology for addressing the issues of customer acquisition, satisfaction and loyalty through people, process and technology, to achieve customer retention, and to create customer life-time value and business profitability (Fig. 4.1).

While customer relationship management literature addresses real-world issues such as improving customer relations, absorbing and retaining more customers, creating value propositions for the customers, and customer satisfaction, it also suggests advanced services and supports for customers. However, there are several areas where there is an obvious lack of research. For example, little if any research has been conducted on ways to decrease the incidence of customer complaints, and CRM alone is not an adequate tool for handling all sorts of qualitative complaints. Sometimes, people may lose interest in purchasing a service or a product from a service provider or a manufacturer due to delays which must be rectified. These weaknesses and gaps have been discussed in detail in the literature review in Chap. 2. Hence, in this thesis, the focus is on the real-world problems that must be addressed in order to maintain customers and increase customer loyalty and profitability. Failure by companies to address problems may induce customers to leave the company and seek to meet their requirements elsewhere.

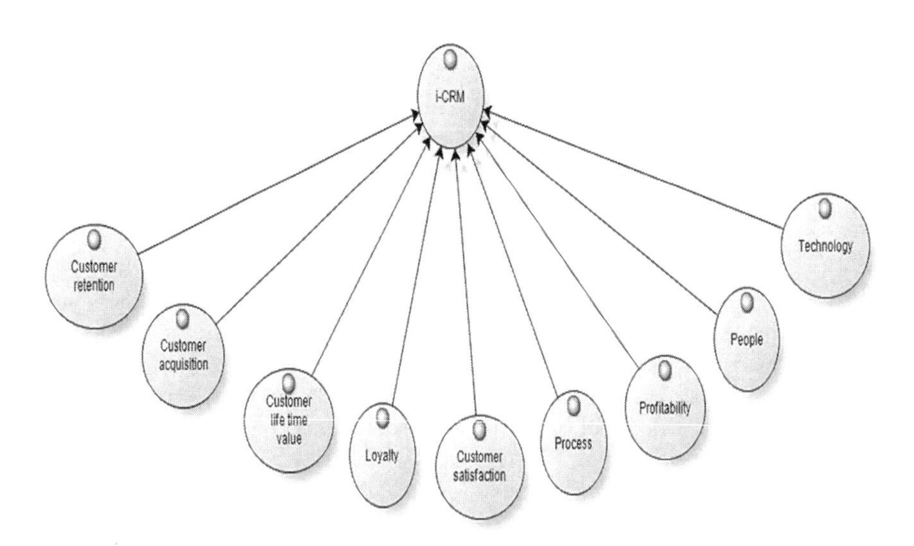

Fig. 4.1 I-CRM model, I-CRM components

4.2.1 The Foundation for the Proposed I-CRM

In Chap. 2, from the literature review, fifty-three different definitions of customer relationship management were identified. While there is considerable discussion regarding customer relationship management, its definitions, its importance in developing company performance, improving relationships with other competitors, retaining customers and satisfying them, gaps still exist. Hence, for this research, another definition is provided that is purely customer-oriented and focuses on creating long-term value for customers. The text-mining approach was used to analyse the definitions. As the results show, various words were encountered that were used as 'code' for different matters in this thesis. These codes have key meanings in the customer relationship management sphere, and each of these highlighted those aspects of customer relationship management definitions that have higher weight and importance. Based on the weights of the important words, a word frequency tag cloud was created. Synonyms were ignored. Based on the importance and relevance, the codes are: interactive, acquisition, loyalty, relationship, retention, satisfaction, business, process, technology and value. According to the following cloud, the essential codes for I-CRM centre on customer, business, relationship and process, and some of the important variables correspond to each other. Based on statistics, this cloud shows the importance of the codes in CRM definitions (Fig. 4.2).

The following table clearly illustrates the weight of each concept involved in intelligent customer relationship management based on its significance. The table is divided into columns showing length of the word used, frequency of the word in the text that was analysed, weight of each word and the importance of the word based on a formula created in the following part.

Weight of each word/Total weight = Importance of the word.

Fig. 4.2 I-CRM tag cloud

According to the Table 4.1, customer, business, relationship, process, technology and value have the greatest weight and are therefore of greatest importance. This demonstrates that I-CRM is a process and technology whereby businesses want to establish a long-lasting relationship with customers to create value for them. Moreover, our statistics illustrate that I-CRM has the powerful ability to interact with customers and this differentiates I-CRM from other tools and methods, thereby making it secure and trustworthy (Fig. 4.3).

The term 'art' of 'customer acquisition', 'customer satisfaction', 'customer loyalty', 'people, technology and process integration', 'customer retention',

Table 4.1 The importance of each code

Word	Length	Count	Weighted percentage (%)	Importance (%)	Similar words
Customer	8	152	13.24	0.32	Customer
Relationship	12	66	5.75	0.14	Relationship, relationships
Business	8	36	3.14	0.08	Business, businesses
Process	7	35	3.05	0.07	Process
Technology	10	34	2.96	0.07	Technological, technologies, technology
Value	5	25	2.18	0.05	Value
Interactive	11	18	1.57	0.04	Interact, interaction, interactions, interactive, interactivity
Profitable	10	15	1.31	0.03	Profit, profitability, profitable, profitably, profits
Loyalty	7	14	1.22	0.03	Loyalty
Integration	11	12	1.05	0.03	Integrate, integrated, integrates, integration, integrative
Retention	9	11	0.96	0.02	Retention
Satisfaction	12	9	0.78	0.02	Satisfaction
Communication	13	7	0.61	0.01	Communication, communications, communicative
Identify	8	7	0.61	0.01	Identifies, identify, identifying
Acquisition	11	5	0.44	0.01	Acquisition
Maximize	8	5	0.44	0.01	Maximize, maximizing
People	6	4	0.35	0.01	People
Revenue	7	4	0.35	0.01	Revenue, revenues
Expectations	12	3	0.26	0.01	Expectation, expectations
Intelligence	12	3	0.26	0.01	Intelligence, intelligently
Lifetime	8	3	0.26	0.01	Lifetime
Fulfilment	10	2	0.17	0.00	Fulfilment
Personalised	12	2	0.17	0.00	Personalised
Preferences	11	1	0.09	0.00	Preferences
Profitable	9	1	0.09	0.00	Profitable
Secure	6	1	0.09	0.00	Secure
Trust	5	1	0.09	0.00	Trust
Total			41.49	1.00	

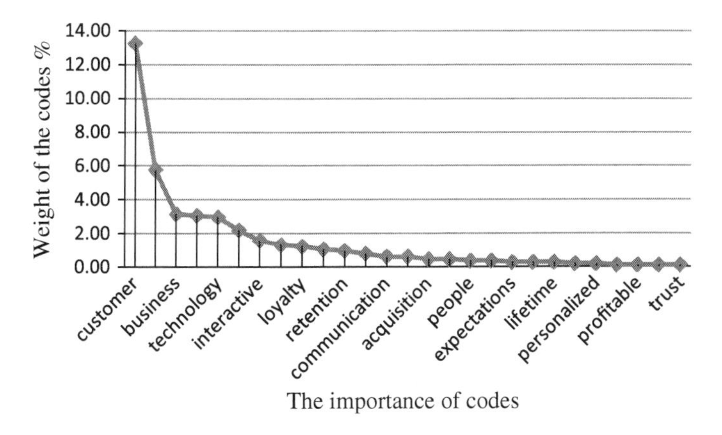

Fig. 4.3 The importance of CRM codes

'profitability', 'customer life-time value' are critical in intelligent customer relationship management. These terms are the crucial stepping stones of CRM and each is explained in the following section.

4.2.2 'Art' Defined in Terms of I-CRM

Art is defined as a procedure or a material that is intentionally or unintentionally absorbed and/or leveraged on people. It is found in a variety of facets in products or services. Customer relationship management is among the systems and strategies that could manage to turn art into science. Just like art, CRM can be either structured or unstructured. While the provider of the service or product provides these intuitively, the person in charge may be considered creative and an artist. However, the semantics of this word could be considered differently in different domains. The author chose this word as he considers customer relationship management to be a state-of-the-art system and procedure.

4.2.3 'Customer Acquisition' as Defined for I-CRM

Acquisition implies ways of acquiring customers for a business and steps that a company takes to purchase or integrate with another company. Following customer acquisition, the company will be able to manage customer views and address different questions posed by them, using CRM techniques. Customers are a company's greatest asset, so the company should do whatever it can to retain them. At the same time, retention can be costly. These days, customers are better-educated and have quick access to the Internet. They can easily shift from one store to another if they are not satisfied. This means that companies are often using different approaches and strategies in their struggle to maintain their customer base. Customers are unique in their interactions with companies and they might have disparity of opinions on companies' behaviours. Because it is very difficult to motivate customers in the new era, customer acquisition plays a key role in advanced customer relationship management. In the competitive market, the majority of firms are vying to maximise profits and reputation by capturing a larger market share. To achieve this, most of their marketing budget is allocated to the acquisition of customers. Today, customer acquisition is frequently a matter of word of mouth which travels faster via cutting edge technology. The number of referrals by previous or current satisfied clients of the company may enhance the acquisition of future clients; whereas, negative word of mouth and fewer referrals may have a negative impact on future client numbers. Companies should always benchmark themselves and be prepared to change the rules according to the time and requirements of a specific area of work. The process of customer acquisition will ultimately benefit the customer retention process. In the majority of cases, customer acquisition is deemed as the most significant part of the customer relationship process, and if the company fails to create an appropriate process, it will be affected as a whole by increasing dissatisfaction, creating disloyalty and losing revenue, all of which will create a negative image for the company. Furthermore, the cost of acquiring a potential customer in a contemporary market is much higher than it is for retaining an existing customer. To grow and expand their businesses, firms need to find their customers in variety of ways using marketing techniques such as sending emails to their customers and asking them to distribute flyers to their friends and relatives, providing legitimate vouchers to customers, advertising via the electronic and print media, and on public billboards [1].

4.2.4 'Customer Satisfaction' as Defined for I-CRM

Customer satisfaction is, or should be, one of the major goals of organizations. Satisfied and happy customers are deemed as key success factors in any organization. To evaluate the overall level of customer satisfaction, the level of satisfaction for each individual needs to be ascertained. Customer satisfaction must be monitored on a regular basis, and companies should identify and respond to the complaints of unsatisfied customers. Evaluation of customer satisfaction will increase sufficiency of the clients, improve turnover of the customers, enhance technical operation of goods and services, and increase effectiveness. It also improves time management as well as the quality of goods and services. Customer satisfaction is described as the extent to which a product or a service can meet the expectation of a single customer to satisfy the requirements of the customer. Customer satisfaction is utilized to differentiate between particular products and services. A customer's feeling of happiness and satisfaction is the outcome of a comparison process between perceived accomplishment and performance and comparison standards such as interactivity and expectation. Customers will be satisfied if they feel that the product or service is useful and meets their needs and expectations at a particular time. The product or service must be equal to or exceed the customer's expectation. If the quality of the product or service exceeds the client's expectation, the customer will be positively satisfied. If the product or the service provided falls short of the client's expectation, the client will be dissatisfied and likely to shift to another service provider or product manufacturer. The level of satisfaction or otherwise is a customer's emotional response after assessing the perceived discrepancy between prior experience and expectations of services and products. In order to accurately gauge customer satisfaction, the system of measurement needs to be standard and flexible. Since satisfaction measurement is of great importance to a company, all inputs in the company must be related to satisfaction. To construct a measurement system for customer satisfaction, several factors need to be considered including product scope and availability, price, speed in providing product and service, commitments, accuracy, warranty, durability and meeting quality requirements. Using these factors, the company can establish a process to measure customer satisfaction which is the most important criterion in CRM.

To be able to acquire and satisfy more customers, the company needs to be interactive. To achieve this aim, the company needs to use sophisticated and interactive tools and strategies. Interactivity is one of the factors that may ultimately lead to customer acquisition and customer satisfaction, and it has the ability to affect loyalty. Based on the literature review, it appears that lack of interactivity is a major problem in most companies around the world as they cannot be responsive to the customers. Companies and research centres have failed to conduct comprehensive and applicable research on interactivity and its relationship with acquisition and satisfaction; hence, the notion of interactivity is little understood. In addition, unclear results have created difficulties and shortages in interaction and this has reduced the level of customer satisfaction and induced customers to change their preferences and tendencies. Using I-CRM, the focus of this study is on dynamic communication and interaction with customers and service providers.

4.2.5 'People, Technology and Process Integration' as Defined for I-CRM

People, technology and process are three inseparable factors that need to be integrated in order to meet the ultimate needs of the company. Without this integration, nothing can be achievable in a work setting. By acknowledging the importance of all three factors, the company can recognize the customers and acquire them appropriately. All three factors constitute customer relationship. As CRM provides a 360-degree view of customers, these factors must work together appropriately and accurately. People are the focal point since 70 % of CRM activities are being done by people. Processes are prior to functions and 20 % of CRM activities are being accomplished by processes. Finally, 10 % of the activities are being accomplished by the technology [2]. To strike a balance in procedures, create value for the customers and increase chance of using information to recognise the customer, these three factors need to be integrated [3].

People are defined as those individuals who participate and collaborate in a business process. They may be considered as customers or consumers, each of which has a different meaning. A consumer is the final user; however, the customer is the one who buys and may sell the product or service. People or individuals have a major impact on the trend of organizational improvement because they have the choice of simply changing their providers of products or services. With time, people are becoming more educated and they tend to use more sophisticated goods and services. To fulfil customer requirements, companies are expected to upgrade their products and services in an increasingly competitive market. Knowledge of customer needs and behaviours is vital, so companies need to become acquainted with people and to investigate consumer behaviour. Process is defined as a continuous and related procedure that aims to transform input into output. The output can provide appropriate or even inappropriate feedback which then will be used in iterations. Thus, the process is expected to be enhanced and rectified to address any emerging or previous issues so as to obtain the best result.

In customer relationship management, each process has the same performance as marketing and management processes do, and implies all sorts of activities of controlling, evaluating and planning the accomplishment and administration of a process. One of the key terms and concepts that every process needs is knowledge. Knowledge is about customers, whether potential or current. Knowledge can be considered as the customers' knowledge, as each customer has his/her own understanding and perception. Here, we discuss the knowledge about customers which companies need to acquire. Hence, prior to implementing CRM systems, companies should implement a customer knowledge management system (CKM) to create customer profiles, customer understanding, and customer preferences in order to rectify or enhance the company's point of contact with customers. Having a comprehensive knowledge about individual customers, the company can simply provide a good and secure customer experience for its clients.

CRM "is about customer acquaintance in an effective way and obtain, own a better experience to drive revenue growth and profitability". Knowledge consists of a company's collective policies, beliefs, opinions, insights and practices. CRM is a mixture of business, relationship marketing and information technology. In the process of CRM technology adoption by a company, it is essential to identify and administer the legitimate bonds, and acquire IT skills. The knowledge about customers improves with time and depends on the duration of the relationship. As the productivity and efficiency of CRM systems depend upon the characteristics and nature of the input, the effect of CRM implementation and its performance over a particular time must be evaluated. Knowledge is also defined as a body of information which consists of facts, ideas, theories, principles, and models.

Data are the raw facts that will turn into information and information is facts with context and perspective. Then information will be transmitted to knowledge which is finalized, or information with guidance. Knowledge alone needs to have specific characteristics: it must be reusable, easy to identify, easily stored, ready to be distributed, and measurable. Technology is defined as the knowledge of tools, strategies and the firm steps toward science and different approaches that an organization takes to resolve a particular problem. Using technology, companies will be able to help thousands of individuals, simultaneously. CRM is considered as one of the most significant technologies ever created. CRM technology has various characteristics and different components.

4.2.6 "Customer Retention" as Defined for I-CRM

Customer retention is defined as an activity undertaken by the service provider or a manufacturer in order to reduce deficiency. After we identify the customers, we must acquire them and understand their perceptions. The company must tactically use I-CRM to discover customer behaviours and to increase the scope of customer attraction and customer retention. If they fail to accomplish that, they will encounter customer dissatisfaction, and customers will be unwilling to purchase the company's goods and services. This may lead the firm to lose its reputation, image and profitability. The retention procedure requires companies to empathise with customers by providing different facets of information to them, inviting them to a number of important meetings and having open channels of communication for current and potential customers in order to understand customers' issues. Only in this way can companies identify their own customer-related problems. This requires effective programming in order to provide responses to complaints and involves processes such as receiving and responding, dynamic analysis of customer satisfaction information, and development of relationship with customers. Companies should meet the needs of the customers precisely. The company's goal should be to keep customers satisfied, which is the key to maintaining a long-lasting relationship with them. To do so, a company needs to have a strong commitment to the customers in terms of providing goods and services of the requisite quality, and also, the company alone should engender a proper trust. To retain customers, a company's activities and procedures must be transparent and re-engineered based on the needs of the customers. An efficient and productive customer satisfaction and customer relationship level enhances the quality of customer relationships. Customer satisfaction can also be integrated with the emotions of people, which makes it a critical facet of the CRM process. A company should adapt its strategies to ensure customer satisfaction based on customer emotions and their sensitivities. As different customers have different characteristics, each should be treated individually. CRM systems are responsible for manipulating these sorts of elements.

To improve customer retention, a company needs to react appropriately, take proper action, and provide timely feedback to customers. In this way, active customers will be retained and remain loyal. This is especially the case if we believe that retention will bring about profit and revenue and retention is better than acquisition. Today, a majority of organizations concentrate on customer acquisition and fail to have properly planned retention procedures. Companies should always evaluate the market, customers' preferences and choices, and introduce updated services and products to them, to remain competitive. Not only does customer satisfaction relate to customer retention; it also has a close bond with customer loyalty.

Meanwhile, government rules and regulations, offensive or defensive strategies and reactions from competitors affect the company and its customers, and sometimes shift the customers away from the company. As long as the company can retain the customers in an appropriate manner, the company may be regarded as a reliable source. Customer retention is an obscure concept that cannot be seen and is not easy to pursue. To estimate the customer retention rate and to evaluate it, we should have good customer retention management under our intelligent CRM supervision.

4.2.7 'Loyalty' as Defined for I-CRM

Loyalty is defined as a customer's faithfulness to a product or a service, evidenced in repeat purchases from the same company. Customer loyalty is an important issue in the field of marketing and CRM that is extensively applied and utilized by researchers. In addition, it is mostly about frequent buying behaviour of the customers and is dependent on the tastes of people regarding different services and products. What is more, loyalty is about purchasing the same brand in order to fulfil and satisfy the consumer's needs.

4.2.8 'Customer Life-time Value' as Defined for I-CRM

Customer life-time value is defined as the lasting relationship and value that a customer can achieve from owning or using a product or a service. Life-time value has become a highly useful method for directing marketing strategy. When a company takes steps towards establishing a customer-focused environment together with increasing its accessibility to customer data, it should also take an interest in understanding and evaluating customer life-time value (CLV). Customer life-time value indicates the current financial worth of the expected cash circulation and cash flows associated with a customer. Hence, the CLV of each customer should enable the decision-maker to enhance the process of individualised service and this alone will increase the level of customer retention, followed by good reputation and profits [4]. The next section presents a discussion of the features of the I-CRM which enable an evaluation of the quality of customer relationship management being achieved.

4.3 Overview of Conceptual Framework of I-CRM

Previously, customer relationship management attempt to solve the issues and halt their strategies. The objective of our framework is to turn customer complaints into combination of opportunities, unlike the traditional CRM in which the companies do not have a willingness to concentrate on customer complaints or they do not address the customer complaints in an effective way. Also, they do not take the components of the CRM. Likewise, they do not have proper perception of customer complaints. Additionally, based on traditional CRM, companies only look at complaints as consequences of the actions and as outcomes of the procedures, whereas, in this thesis we go further and not only look at customer complaints as opportunities but we attempt to follow them up and turn them to new opportunities in the work setting.

The following figure (Fig. 4.4) represents the abstract of Fig. 4.5.

An intelligent complaint analyser analyses customer complaints with the set of objectives, impact factors, strategies and actions. An intelligent customer relationship manager converts the customer complaints framework into an opportunity framework.

In the following approach, we go a step further and convert customer dissatisfaction to customer satisfaction. The ultimate expected outcome of this research is to turn the customer dissatisfaction to customer satisfaction. This is shown in level three of our conceptual framework which is depicted in Fig. 4.6. To the best of our knowledge, this is the first thesis that investigates customer complaints in I-CRM area in depth and addresses customers' issues using strategies and recommendations. Various approaches have been considered by different researches to identify and address complaints. Some companies even outsource and get third parties to manage their complaints and create customer satisfaction.

Figure 4.5 depicts the complete intelligent CRM framework. The top frame illustrates the Intelligent-Customer relationship management system. The above section (Fig. 4.6) of the conceptual framework which is the main part of this thesis focuses on (mainly negative) customer feedback. After defining, categorising and analysing customer feedback, strategies are developed to examine the problems,

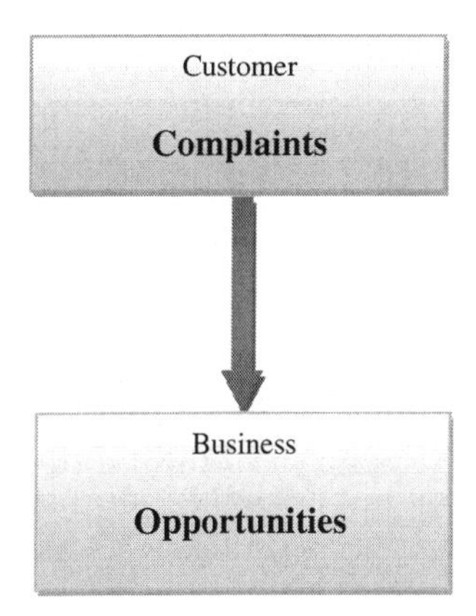

Fig. 4.4 Level one of proposed conceptual framework for I-CRM

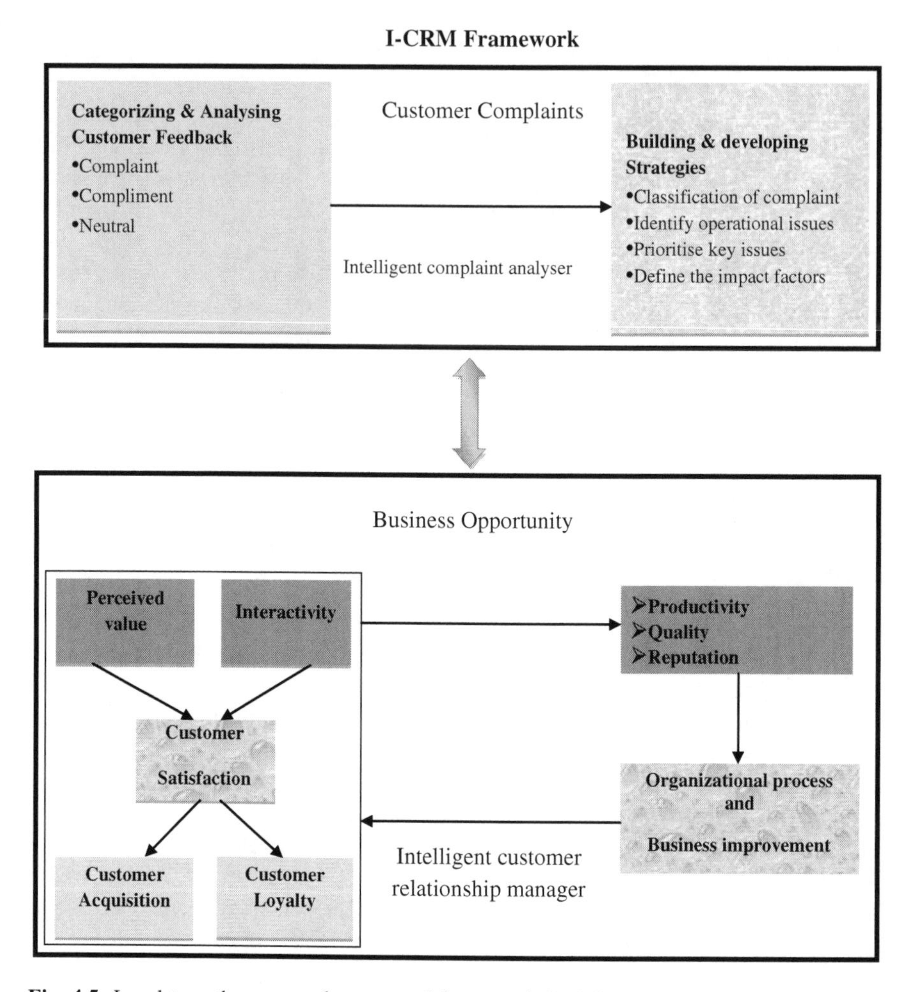

Fig. 4.5 Level two, the proposed conceptual framework for I-CRM

and prioritise them according to the CRM strategy and determine recommendations for resolving the issues. The bottom frame represents I-CRM evaluators that are intended to measure customer satisfaction and build upon the top framework to assess and decrease the level of complaints. The whole process is an iterative one intended to enhance loyalty, satisfaction, acquisition and productivity of the organization. The left side of both frames show the customer activities. The right side shows the different company activities. In the bottom frame, the ultimate objective is to increase productivity, quality and reputation of the company utilising customer service.

Meanwhile using perceived value, the values and experiences that customers obtain from the system will be collected and generated. This alone enables the system and company to acquire more customers and enhance the loyalty level.

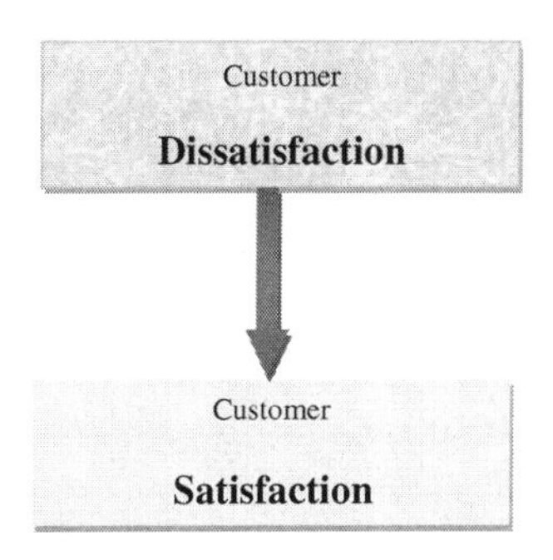

Fig. 4.6 Level three, the proposed model to generate customer satisfaction

Furthermore, the main reason for including perceived value is that there is the need to estimate the disparity between the amounts of value that company can obtain using the I-CRM methodology. Additionally, using interactivity, and attempt is made to improve the interaction and communication of the system with the company's customers and clients. Hence, the relationship between, and the impact of, interactivity on customer satisfaction needs to be determined. Based on the proposed conceptual framework, interactivity and perceived value will be both directly and positively associated with customer satisfaction. Both diagrams in the conceptual framework have a close mutual relationship and make a significant contribution to each other. The physical absence of each variable in both diagrams may create an issue within the proposed methodology. The whole conceptual model represents Intelligent-CRM. Cloud service architecture for this I-CRM can be adopted to create shared knowledge. Using the conceptual model and its components, a company can target its customers and overcome some of the challenges in a competitive market. However, the diversity of practices using this conceptual framework must be taken into consideration.

I-CRM initiates its activity with many customers into expand businesses. To employ this methodology, a company needs to acquire complete information about all customers and competitors in order to organise the information, analyse it, and find solutions to the issues. Implementation of this system will empower marketing activities, bring the customer into the central focus, and encourage the customers to be more committed to the company. Also, utilising I-CRM, the information from customers visiting a company's website can be captured. It also automates the processes thereby allowing customers to provide referrals to their friends and relatives and introduce the system to them. The main aim is to ensure customers' satisfaction and retain them.

This framework has many features. For example, it does customer complaints classification, identifying them and categorising them based on their importance. Then, it prioritises the issues followed by defining the impact factors. It also estimates perceived value and ensures the interactivity level of the company with its customers and tests the extent to which existing customers are satisfied with the services. Additionally, the I-CRM framework evaluates customer satisfaction using various approaches and creates customer loyalty and customer acquisition. Apart from ranking the customers, this model has the ability to find the areas of organizational practices which need to be improved. Predicated by this framework, customer complaints can be used in the customer satisfaction section to meld with perceived value, interactivity, customer acquisition and loyalty. Customer complaints can be evaluated with respect to these factors in order to generate rules.

4.4 The I-CRM

In this section, the upper level of the I-CRM Framework is presented together with a description and definition of each component of the I-CRM (Fig. 4.7).

The methodology must treat all complaints equally, ensuring that customers do not receive discriminatory or differential treatment. According to I-CRM, swift action is required to categorise the issues and build new strategies. Simultaneously, the company should assure the customers that their complaints are confidential and they will not be disadvantaged by submitting these complaints. Extreme care should be taken to avoid victimization. All of the complaints must be accurate irrespective of the sender's background and the company must keep track of all complaints and maintain records of them.

4.4.1 Analysing Customer Feedback

The categorizing and analysis of customer feedback are two separate processes which are accomplished using the qualitative methodology discussed further in Chap. 5. By analysing the grouped complaints, recommendations can be developed to enhance the performance of the system using various tools and strategies.

4.4.2 Complaint Feedback

Customers may have various reactions depending on the individual complaint. The reactions of a dissatisfied customer may range from taking private action by shifting from that company to another, to no longer cooperating with the company,

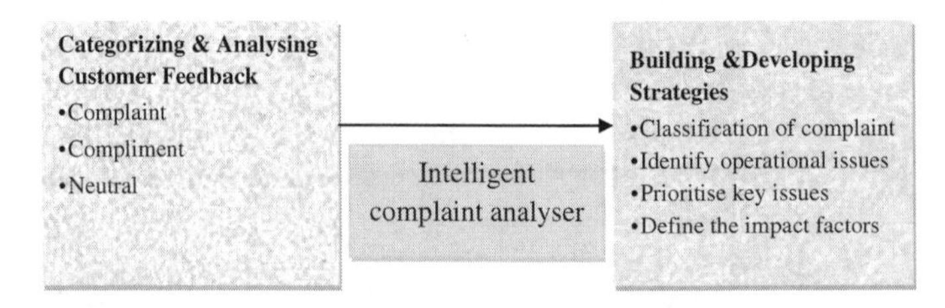

Fig. 4.7 I-CRM system

or creating negative word of mouth by discouraging friends and relatives. Some might not take any action on the complaint and this could create issues in the future. There are customers who choose public action such as a retaliative action against the company in the form of legal recourse or taking the complaint to a relevant institute or government body.

In this thesis, a complaint is defined as a negative feedback from customers about their experience regarding the product or the service they have received from a company. It is best to consider complaints as constructive and helpful information that can assist a company to rectify issues and improve customer service. Complaints can assist companies to recognise patterns of failure submit them to further investigation and address problems. Complaints alone can be a final solution to the problems and management can use them to investigate the problems and their source. Since, complaint handling is an important task; it needs to be undertaken by personnel in the customer service department who should address the filed complaint according to customer expectations.

4.4.3 Complimentary Feedback

A compliment is defined as positive feedback from customers and has an effective and important impact on the performance of the company. In this thesis, a compliment is considered to be the opposite of a complaint, and illustrates perceived value derived from the system and shows the level of customer satisfaction.

4.4.4 Neutral Feedback

Neutral is defined as neither positive nor negative feedback to the system and that the system cannot provide any positive or negative impact for the company. Neutral feedback might be given for different reasons, such as the customer being afraid of the people in charge of the system. Also, the complainant could face job loss or discrimination. Alternatively, the customer might not know how to provide feedback.

4.4.5 Building and Developing Strategies

Building and developing strategies are defined in I-CRM as the aim of track and trace of the issues and creating a solution based on customer relationship management approaches. Also, these approaches can provide strategic analysis and strategic goals to the system. They can also ensure that the developed strategies will contribute to later solutions. The establishment of new strategies may create more market share and increase customer satisfaction. Furthermore, they have the ability to enhance profitability.

4.4.6 *Classification of the Issues*

Classification of the issues refers to the approaches that researchers utilise to categorise the customer complaints into various groups to be able to clearly analyse them. Hence, we need to trace the issues and identify them from the source. Tracking the issues refers to the methods that the system or the company adopts to become aware of the issues and different sorts of feedback whether positive, negative or neutral. Therefore, a company must identify a strategic choice to trace the problems using various tools followed by assessing and choosing the most important options.

4.4.7 *Solution to the Issues*

A solution to the issues is defined as one that that provides the correct answer and solution to the problem or complaint of each individual who is dealing with system or the company. After tracing and identifying the issues, the company must listen attentively to the customers' feedback and they should always appreciate the comments that the customers provide. If necessary, they must apologise to the customers. The main goal of a company must be to find the best possible solution to satisfy customers and meet their expectations. Providing a solution will be impossible without any follow-up, as the company intends to make the customer happy and satisfied.

4.4.8 *The Impact Factors*

The impact factors are defined as the main intelligent customer complaint analysers or evaluators of intelligent customer relationship management. They are also components of CRM and include perceived value, interactivity, customer satisfaction, loyalty and customer acquisition.

4.4.9 *Intelligent Complaint Analyser*

The Intelligent complaint analyser refers to software, devices, mechanisms or methodologies used to evaluate and analyse the data in depth and turn it into usable and applicable information. By means of an analyser, new structures and patterns will be found that can clarify or explain the relationship between various elements of the information. Without the analyser, the trustworthiness of the data cannot be ascertained.

I-CRM will be performed at the forefront and it is expected to increase the feedback ranking and sorting. It can increase customer awareness about the system. As indicated by Fig. 4.5, the outcome of this research will be an intelligent customer relationship management system. As discussed earlier, after categorising and analysing customer feedback whether this be complaints, compliments or neutral comments, new strategies must be created for track and trace problems, to provide solutions for the issues, and lastly, to improve customer relationships. These steps will ultimately create an intelligent customer relationship management system. This is not going to be achieved unless all the procedures can be fully evaluated and validated. In the next section, there will be a discussion of the I-CRM evaluators which have a critical role in increasing I-CRM performance and making it grow faster.

To achieve the above objectives which have been mentioned in the I-CRM section of the conceptual framework, this research will be divided into four stages as discussed below.

(a) Categorise and analyse customer feedback.

Based on the data collected from qualitative interviews, customers' feedback will be categorised and analysed. Some feedback is complimentary, some is neutral, but the majority consists of complaints. In this stage, the following three steps will be carried out:

Step 1. Feedback data extraction from the interviews, meetings and company's database.

All the interviews, whether audio or video have been transcribed and turned into a comprehensive raw interview. In the second phase of this process, one of the content analysis rules was applied and abstractions of the interviews were made. The data was then categorised into sets, each with various sub-sets. Then, an analysis of the data will help to provide constructive and productive solutions for the company.

Step 2. Carry out qualitative analysis on customer feedback.

Qualitative analysis is conducted on the data extracted from interviews. In qualitative research, purposeful sampling is often used in a deliberate way with a specific intent or purpose. Qualitative analysis is concerned with the description and understanding of the situation behind the factors [5]. Specific software is used to sort and analyse the data. As the transcribed data is qualitative, software needs to be used for qualitative analysis. This transforms the outcomes into quantitive form; then, other software or certain statistical approaches will be used to obtain a numeric result. If at any stage an appropriate result cannot be obtained, the previous stage will be repeated, the issue will be rectified and then the following stage can proceed.

In this way, all the stages will be double-checked. At the same time, the results will be validated using industrial software to ensure the accuracy of the outcomes.

Step 3. Customize related service features and data collection mode.

An observation checklist will be utilised. The purpose of an observation checklist is to record observations as data, making them quantifiable. The types of data that will be collected will be determined by the research questions. Customer feedback is the procedure or particular example of preparing information for businesses about goods, services and customer services. Management, marketing and sales departments can all use customer feedback to simplify processes and improve profitability. Customer feedback produces invaluable information. Therefore, customer feedback is among the critical types of "social information" that potential customers can trust to decrease uncertainty and help in decision making. Currently, there is no systematic approach to dealing with customer feedback in CRM applications. One aim of this research is to address this shortcoming. Customer feedback will be clarified, categorised and analysed. The proposed model also has the ability to streamline businesses.

(b) Identify, address the issues with strategies.

This stage comprises two steps, namely:

Step 1. Develop a classification scheme for this feedback.
In this step, we review the classification schemes to demonstrate the variances, differences and potential incompatibility of customer feedback classifications and definitions. Such a scheme may use text mining techniques and business process analysis for each type of customer feedback. A search of similarities and commonalities between feedbacks, based on selected characteristics, will create an objective foundation for classification. Software will be used for text mining of interviews and expertly chosen software will be utilized in order to conduct a proper and valid comparison between factors and to validate the results.

Step 2. Address issues through solution proposal (Business Process improvement, productivity and quality).
It is intended to address the issues by proposing I-CRM strategies and the technologies underpinning the new I-CRM. Prior to that, the importance and severity of the issues will be ascertained using NVivo as the tool to validate the issues based on their weight and importance.

Step 3. In this step, the hypotheses will be tested and validated using different tools such as Minitab, SPSS and structural equation modelling (SEM).

Step 4. In this step, principal component analysis and data envelopment analysis will be used to find key customers. Then, the cost and time required to improve customer satisfaction is optimised. This will be a re-validation of the first section of the research hypotheses.

Step 5. In this section, using linear and non-linear models, the relationship between perceived value and interactivity with customer issues will be tested and validated to obtain the legitimacy of the relations of the major issues with customer acquisition and loyalty. By using linear modelling and formulating the relationships, the main hypothesis will be re-validated utilizing a fuzzy inference system, which is non-linear analysis, and rules will be generated to make new recommendations to rectify and ameliorate the system.

4.5 The I-CRM Evaluator

Figure 4.8, is a part of the main conceptual framework, and depicts the evaluators of intelligent customer relationship management. Utilizing these, the performance of each variable in the system can be estimated. As mentioned, this framework is an iterative process and in each repetition of the process, we will expect to build upon previous steps and enhance the level of customer satisfaction, customer acquisition, loyalty, and productivity. In the following section, each of the terms related to the model is defined. The major thing to be considered to enhance other factors is customer satisfaction which is highlighted in the conceptual framework. Perceived value in this conceptual framework is related to the price that customers pay and the performance that they obtain from using a specific service from a system or company. As previously mentioned, customer satisfaction reflects the number of customers whose expectations have been fulfilled and to what extent a company has satisfied their expectations. Furthermore, satisfied customers normally tend to be loyal customers. Not only do they purchase the same products and services from the same company; they are also likely to be price tolerant and to recommend the products and services to other individuals [6]. In this way, the company will be highly successful in customer acquisition. The majority of customers are quite intolerant of the new pricing strategies of their favoured companies. However, using a major customer satisfaction approach, it is possible to retrieve customers and encourage them to continue to purchase their products and services.

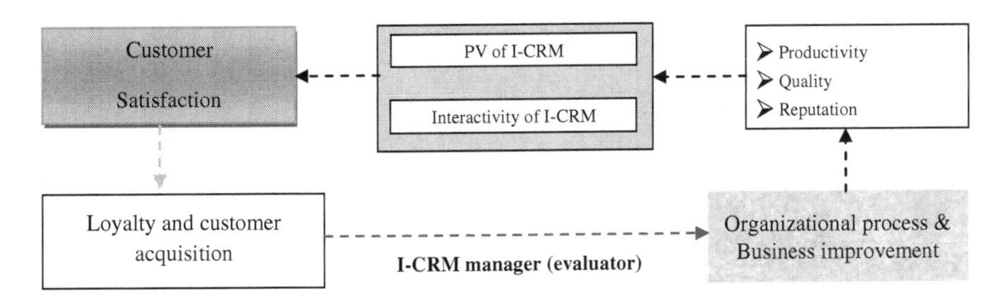

Fig. 4.8 The proposed model for I-CRM evaluators

4.5.1 I-CRM Evaluator (I-CRM Manager)

An I-CRM evaluator is a tool used to assess and estimate the efficiency and effectiveness of each factor involved in a process to maximize the performance of the system.

4.5.2 Perceived Value

Perceived value is defined as the value that a customer can obtain from owning or utilizing product. Perceived value determines the customers' overall evaluation of a product or a service. Using this, the customer can understand whether or not the cost of purchasing a product or a service is fair and legitimate.

Furthermore, perceived value is the rate that a customer wishes to pay based on his/her expectations and anticipates to be satisfied. Its structure creates chances to compare companies' performances. It must have a positive impact on customer satisfaction [6].

Perceived value is considered as a prerequisite for and antecedent of customer satisfaction and loyalty. Also, perceived value is a decisive and necessary strategy for various processes to find out the importance of different stages of a process and its outcome. While we give a valuable and state-of-the-art product or service to the customer, we are basically delivering a value to them. Besides, perceived value may occur in different parts of the business process before purchasing a product and it may be created after purchasing a product.

4.5.3 Interactivity

Interactivity is defined as an awareness and ability of an individual or a company to create and generate an active communication with the customers. Additionally, interactivity implies the dynamic and active essence of the involvement that occurs between a company and customers via the company's different methods of interactivity. In this process, two or more groups conjointly impress each other by utilizing information transmitted. Different researchers have illustrated the significance of interactivity to the loyalty of customers in the virtual or physical world. Lack of interactivity is a major issue for many companies and various websites. Interactivity is difficult to create and navigate, and there is insufficient information on how this can be promoted. Interactivity needs compatible and reciprocal communication functions, whether it is face-to-face communication or other communication settings. Interactivity includes three elements: ***speed,*** or the proportion at which input can easily convert to a mediated communication; ***scope***, which refers to the number of probabilities that need to be accomplished within a mediated environment; and ***mapping***, which is the capability of a system to map and scrutinize its monitors to alter in a mediated environment. Current descriptions of interactivity which are particular to the Internet have introduced several terms to better define interactivity: user control, time and active control, reciprocal communication, synchronicity and the inclination of communication [7].

Interactivity has various components. The first one is its monitoring aspect with which a user may collaborate and leverage in communication. The second facet is its creation of reciprocal communication that provides mutual flow of information; and last is the synchronicity of the interactivity which shows the velocity of an interaction [8].

4.5.4 Customer Service

Customer service refers to any services supplied to the clients in different steps of a business process to satisfy their requirements and keep them attached to the company. Customer service departments work at sympathetic activities to give motivation to the customers and prevent them from feeling neglected. From the company perspective, services provided to different customers should be customized according to their specific needs and wants since different customers require different types of customer service due to their many and various expectations. Customer service departments must consider various factors in order to better and more effectively serve customers. Companies should create comprehensive customer profiles in order to effectively monitor their customers' business transactions and buying behaviours. To do so, every step that a customer takes should not be overlooked by a company and it needs to take advantage of this knowledge. Company actions should parallel customers' expectations and to maximise their impact, companies should strive to exceed customer expectations. Advanced customer service does have the ability to keep the customers loyal and return them to the company to for repeat purchases. Positive feedback through word of mouth filters through to others and thence to society in general.

A good customer service department needs to cultivate certain characteristics and behaviours. For instance, they must always be good listeners and show empathy toward customers. They probably have listened to or read thousands of complaints, so they should be well-equipped to deal appropriately with a particular complaint; this can be supplemented by an attempt to please the customer. Not only should customer service personnel be knowledgeable and helpful, they should undertake regular training in order to better provide services to the customer. What is more, an effective customer service department should recognize customer needs and anticipate them. Of paramount importance is that they treat all customers fairly and equally. Finally, a company needs to give its customers regular feedback and should respond promptly to concerns or queries. As the function of a customer service department is completely based upon CRM rules, it should be customer-centric. Hence, the customer needs to come first and should be central to any issue.

4.5.5 Business Profitability

In essence, profitability implies the status of generating financial profit. This term encapsulates the ultimate desire of companies whose everyday work is focused on achieving maximum profit. Customer relationship management systems and strategies always assist companies and their managers to increase the level of profitability. CRM applications assist companies to assess the profitability, using a measurement of repetition of purchases, durability and lastingness as well as the amount spent on the purchase. Long-term profitability is a burdensome and complicated procedure which is being done fortuitously by companies across the globe. In order to increase profitability, companies need to make a wise investment from the first moment that they initiate the work, until the implementation of the systems.

4.5.6 Productivity

Productivity is defined as a measurement of a company's performance and effectiveness. It has always been one of the main concerns of managements. It can be regarded as a bond that a final product may create via its value. The primary objective of productivity is to provide better services, support and training to the company. Increasing productivity will give better insight to the company. It must be emphasised that productivity is not achieved by accident or good fortune. It is created in companies after years of putting up with hardships and discipline. Without productivity and effective planning, a company will go off track and will lose its control over the market. Using productivity as a measure, a company can compare its efficiency and outcomes with those of its competitors.

Two of the key elements that a company needs in order to increase its productivity are recruitment expertise and a qualified management team. The company should concentrate on ensuring continuous achievement and effectiveness rather than obtaining these only for the short-term. Another element that enhances productivity is the empowerment of labour using different approaches so that employees are motivated to be loyal to the company. As one of the business outcomes and value of the CRM systems is productivity, it is essential to invest in every company process. In this thesis, the aim is to increase satisfaction and productivity using an intelligent customer relationship management. Another focus will be the functionality of the Intelligent CRM. To increase business productivity, a management team must have a sound commitment and all the tasks need to be clarified to the clients as well as to the customers. In addition, increasing productivity will also decrease the level of cost. In this thesis, business productivity will provide a new solution to improve a company's performance and reputation.

4.5.7 *Quality*

Quality indicates that a product or a service is superior to others and has more or better attributes than those of the same product or service. Quality in the business sense has a practical meaning and is perceived differently by different people as everyone has a different attitude toward a service or product. The quality is mandatory and vital in every facet of customer relationship management and business processes. Quality may help to establish a productive relationship with customers as well as business partners. To provide a better service and product quality, a company must be armed with an interactive management system and understand clients' expectations.

4.5.8 *Reputation*

Reputation is defined as perceptions of customers about certain products and services as it resonates in people's memory. It is also defined as the idea of an individual about a matter. Also, it can be considered as an expected behaviour about a person, group or company. Reputation is being established when a company or a person is committed, responsible and reliable. Effectiveness and productivity usually lead to creation of a great amount of reputation or what we previously called fame. This criterion can exclusively differentiate a company or a group of people from others, and can give it a priority ranking in the eyes of potential and future customers.

Reputation can be considered as one of the outcomes of correct customer orientations. It is assumed to be accomplished by a company and builds up company strength by attracting more employees. It can also assist a company to maintain current customers. This characteristic will make the company stand out from its competitors.

A company with a good reputation can contribute to society because of its ability to recruit more people and contribute to resolving unemployment issues in society which in turn can reflect positively on the government of the day.

Hence, reputation depends on a company's strategy for promoting a high level of interactivity, its organizational process and its business development. In order to achieve a good reputation, a company should satisfy customers and ensure their loyalty using different tools and strategies. Companies must act vigorously to create product and service awareness for the customers and generate good experiences for them.

4.6 Analytics Solution for I-CRM

This section provides an explanation of the analytic recommendations and solutions for analyzing the stated variables, based on the conceptual framework. The solution lies in creating intelligence in the system. I-CRM is an initiative and in developing an analytic solution, both customers and their complaints should be prioritised.

4.6.1 The Distinguishing Features of I-CRM and Application in the Logistics Industry

In the logistics industry, all stakeholders such as managers and clients who are dealing with the company can be considered as customers and the business model for their interactions is identified whether it is business-to-business (B2B) or business-to-customer (B2C). In this thesis, using CRM and its variables, a platform will be created to address various issues predicated on what has been suggested in the logistics industry. It is initiated from customer touch points, brings about service management, and creates a life cycle for management. Also, it will streamline the booking system of the company and as CRM does manipulate all activities, it manages the time along with facilitating waiting time of the customers. A complete CRM system has an appropriate customer contact service and, therefore, it can listen to the customer's voice and provide instant feedback regarding the needs of the customers. The implementation of the CRM system creates an efficient, low-risk loyalty program [9]. To date, logistics companies have had issues with retention of drivers, but the CRM system and strategy, will assist in reducing driver turnover [10].

Also, there are business-to-business (B2B) and business-to-consumer (B2C) models. Since B2B conduct business by communicating with others physically or on the web, B2B solutions may provide effectiveness and productivity and can cover some B2C models [11]. The business model considered in this research relates to the logistics industries at Fremantle port which can be considered as being both B2B and B2C, as the port deals with other companies, competitors and customers separately. CRM has its own unique features and can be integrated with both B2B and B2C. Based on [11], business-to-business CRM can reduce administrative costs and improve responsiveness to the customers. As B2B and CRM have various advantages, if consolidated, they may complement each other and establish new, improved services for the customers.

Additionally, in this research, the truck drivers are the customers, so the types of services they require are in some ways quite unique. Normally, companies have customers from a range of socio-economic, ethnic and cultural backgrounds who differ in terms of gender, age, religion, occupation and nationality, and each may have different goals in terms of service expectations. However, in this particular study, customers are truck drivers who may be either male or female and all have more or less the same goals and objectives. Hence, the proposed I-CRM and the approach used to analyse customer complaints in this environment would be different. This postulation is supported by the comprehensive methodology presented in this thesis; no previous works have utilized the same integrated method in their researches.

4.6.2 Chocie of Case Study and Data Collection

The logistics and transport industries at the Fremantle port are used as the case study which is quite complicated. Due to the complexities of the environment, it was considered best to initiate the data collection process at the port itself and to concentrate on one area and its problems in order to establish a comprehensive methodology. After carefully examining various aspects of port operations in order to identify a research problem, it was discovered that truck drivers are highly disadvantaged in terms of services and facilities. Discussions led us to conclude that there were numerous dissatisfied customers, in this case the truck drivers, who had many and various complaints about the poor quality of port services. To track and trace the complaints and level of dissatisfaction, the company needs to spend a great deal of money to turn unpleasant and negative customer experiences into pleasant and positive ones. To provide a good customer experience, the company must have an accurate view of the customers and concern for them, and it must be accountable to the customers. Companies at the port must listen to the voices of the customers and create an environment in which they feel free to voice their concerns without fear of negative repercussions. To actualize an experience for the customers, particularly drivers at the port, the company must match its performance with customer expectations across various customer touch points. This will provide a reciprocal and long-term relationship between the company and its customers.

At the port, customers have various complaints which can all be categorised as dissatisfaction which will be further discussed in relation to the hypotheses in Sect. 4.8. In this thesis, we identify and provide a response to the customer dissatisfaction to create new recommendations and solutions.

4.6.3 Customers' Issues at Fremantle Port

As discussed in the literature, customers must be considered at the core of all activities by a business regardless of the transaction that is taking place. Additionally, companies need to concentrate on all customer-driven problems. Below, each of the issues raised at the port will be discussed.

1. Cost issues at the port in different segments.

Cost is considered as one of the most important issues at the port and in the following part, its significance and the need for companies to remediate the situation are discussed.

2. Booking system

A good booking system provides a flexible and efficient management process whereby a driver can simply make his/her reservation. A booking system identifies its customers and their profiles and customers must rely on the system. There is no fee-for-booking system at the port and people can simply register themselves. An efficient booking system will provide accurate pick-up and delivery times for the drivers. Also, they will have a timely and planned loading and unloading time slot. Customers will be able to operate more efficiently and without having to wait or being interrupted. Having a booking system may eliminate mistakes that occur through manual bookings. A booking system would allow people to refer to the system for information or advice, rather than making personal contact, provided that the system is secure. Currently, the main problem of the port, specifically regarding the companies who are working at the port namely BP World and Patrick Fremantle Stevedoring, is the lack of a vehicle booking system. A VBS or vehicle booking system creates an automatic means by which the truck drivers can be scheduled or re-scheduled for all sort of activities that are conducted at the port. Moreover, a VBS can help decrease traffic congestion on roads in and around the port.

3. Container issues

Truck drivers have issues regarding their containers; these include container sorting, container labelling and being able to locate containers without any delay. Also, some drivers encounter problems with containers that exceed the weight limit. While the port is developing in terms of environment, facilities and expertise inside, there should be some strategies in place to help drivers resolve their container issues. An analysis of the issues pertaining to containers indicates that this is a significant problem that must be considered and addressed.

4. Driver health and safety concerns

Truck drivers are often required to drive long distances, and when reaching their destinations, they have to deal with loading and unloading their cargo in different areas which may be urban, rural or remote locations. Appropriate health and safety rules and regulations at the port could contribute to ensuring drivers' peace of mind and mental health. The port and the companies operating within it need to have in place the means of promoting better health and safety for their customers. If drivers can be reassured about their health and safety at the port, they are less likely to suffer any trauma such as disease and stress. Sometimes, drivers have to operate in a polluted atmosphere, often being exposed to toxic chemicals that they inhale in the cabin. This issue should be dealt with to reduce risks to drivers' health.

Following is a discussion of some of the issues that emerged from statements made by drivers regarding their health and safety.

5. An inadequate infrastructure is one of the problems identified at the port.

Infrastructure is the means by which each industry provides its strategic requirements. In addition to other infrastructure problems, Fremantle port is facing the challenges of enormous issues regarding efficiency, complaints and dissatisfaction. Port authorities and stevedoring companies must propose strategic approaches to find remedies for infrastructure problems. Otherwise, infrastructural issues will impede Fremantle port's future development and progress. Issues in terms of infrastructure are various and must be addressed case-by-case at different ports.

6. Jurisdiction and policy issues

Along and parallel with managerial issues, policy and jurisdiction issues at the port present a major dilemma, since they deal with accreditation, among other things. Moreover, the involvement of politicians in the policies at the Fremantle port is also a cause of customer dissatisfaction.

7. Management issues

Management is an integral part of every organization and without it none of its sections will perform effectively. The main purpose of a management team in an organization is to direct and control procedures, and preside over operational, front office and back office activities.

An efficient management team at the port could prevent traffic congestion at the port and in its immediate vicinity. Furthermore, people would have a better environment in which to demonstrate their capabilities and behaviours. An effective management can optimise the performance of clients, ensure sustainability and make a social contribution. Likewise, having skilled management at the port is integral to all of the activities as it creates a better flow of trade and therefore the needs of businesses are more likely to be met. Moreover, it can establish the port as an invaluable facilitator and contributor in individual lives.

Conversely, poor or inefficient management may reveal unsavoury issues, or create scandals related to environmental contamination, for example, and other issues which have an impact on society.

8. Stevedore performance

A stevedore, otherwise known as a wharfie or dockworker, has various jobs including loading and unloading containers from trucks and ships. They work for big or small companies and may have various job descriptions based on the size of the port and its location in the world. Rules and regulations significantly affect the performance of a stevedore. A stevedore usually works at a ship's terminal and is responsible for the effective and safe loading and unloading of ships and trucks [12]. Stevedores must have good communication and team skills and demonstrate a high level of responsibility in the work setting. They are individuals with flexible working hours and must be able to perform under different conditions in terms of weather and location. To adequately meet working standards and customer expectations, stevedoring companies need to provide timely service to customers twenty-four hours a day. They must also provide flexible time slots for the customers.

9. Time management

Time management can be defined as the control of the allocation of time and spending time on various procedures and activities to enhance the productivity and efficiency of an organization. Many researchers concur that time management in an organization is vital and that an expert team must be in charge of this process in every organization. Based on [13], there is a significant lack of planning at the port in terms of the timing of loading and unloading activities. According to [14], time is an important factor in the administration of any company, and having a legitimate and just-in-time management system will improve workflow and decrease the cost of business procedures.

A time management system will enable the time needed for each activity to be calculated. Hence, an appropriate time can be allocated for each business process, thereby improving the workflow. Using time management, the duration of each process and the deadlines can be determined and the managers can trace the improvement or failure to improve the procedures or to rectify the issues [14]. Due to the time limitation of this thesis, it is not possible to address and solve all the problems and issues at the port. However, the most significant complaints will be targeted and an attempt will be made to find solutions for them. Also, in the analysis section, the key customers will be identified in order to address the major problems revealed by this thesis. Being the key customers, these individuals may provide more credible information regarding the issues.

4.7 Questionnaires and Data Collection in I-CRM

Firstly, a preliminary and introductory study of the problem domain was conducted through interviews and discussions with key stakeholders including stevedores, truck drivers and operators at the port. It was necessary to acquire a clear understanding of the matter and to select the approach that would best reveal and identify the issues through the interview process. In the next stage, another method of data collection was used, namely a questionnaire and group meeting, which are included in this research. In previous studies and theses, researchers were unable to find a precise method to address issues. Some of the issues have been identified such as excessive existence of complaints in the company, adopting CRM system and strategies. Having complaints in a system is the outcome of poor interaction with customers. When customers provide feedback, whether it is positive or negative, they expect a prompt response from the company. Today, due to the complexity of the issues and because the time allocated to each customer is limited, companies will not be able to interact with each customer properly and on time. There is a need for sophisticated systems and strategies to deal with these sorts of issues clearly, irrespective of the issue itself.

Using face-to-face interviews and by identifying the major stakeholders, a great deal of information was acquired. Prior to the interviews, different questions were devised and all questions were piloted by several experts at Curtin University. In the following section, some of the questions which were put to the interviewees are presented. The interview responses have enabled port issues to be identified and the key issue statement to be generated. The following table displays some of the main questions that the stakeholders were asked during interviews (Table 4.2).

Subsequently, the key statement is verified with key stakeholders via different important meetings. The meetings were scheduled with randomly selected individuals including stevedores, operators and managers. However, it was often difficult to conduct interviews in a timely manner since interviewees had their own responsibilities to meet in the workplace.

Table 4.2 Example questions put to workers in the Fremantle port

Questions	Sample questions
1	Are you happy with the management and improvement of Fremantle port?
2	Are you happy with the current operation of Fremantle port?
3	Do you have a problem with accommodation in Fremantle port?
4	Do you have an issue regarding container sorting at Fremantle port?
5	What comments do you have in terms of managerial issues at Fremantle port?
6	What can the Fremantle port do to make life easier at the port?
7	Does the Fremantle port need any improvement to the infrastructure?
8	Do you have time breaks during the day at Fremantle port?
9	What sort of problems do you have with operators at Fremantle port?
10	Do you have an issue regarding your working shifts at Fremantle port?
11	Do you have any problem with loading or unloading a container at Fremantle port?
12	What can Fremantle port provide in order to have a better flow of the processes at the port?
13	Do you have an issue regarding the booking system at Fremantle port?
14	What functions do you want to have for booking system at Fremantle port?
15	What are the requirements you have to increase for the efficiency of the work at Fremantle port?
16	What do you recommend for congested areas around the Fremantle port?
17	What functions and facilities do you want to have to help drivers at Fremantle port?
18	Do you have any interactivity issues with other drivers and employees at Fremantle port?
19	In terms of health and safety issues, do you have any concern at Fremantle port?
20	Have you got any issue regarding amenities at the Fremantle port?

4.8 The Interview and Questionnaire Approach in I-CRM

Regarding the interviews conducted, most of the issues were comprehensively identified and understood. In the next stage, the research questions were formulated, followed by writing detailed interviews and questions asked in the port. Different drivers were randomly chosen to be interviewed at the Fremantle port and a few of the drivers and operators were selectively chosen. Some interviewees brought up different issues, whereas other interviewees refuted others' arguments and challenged their opinions. The collected data needed to be categorised and analysed. The classification, abstraction and summary of key issues and key findings were conducted for all stakeholders.

In the next step, the key findings based on interviews with all stakeholders must be verified and validated. Hence, appointments were made with several of the key management personnel at the port to have question and answer meetings regarding the researcher's questions and to clarify research issues. Some of the experts in this field were invited to validate the data and to ensure that the materials are valid. In the next phase, issues were identified and prioritised based on their importance and their frequency of occurrence so that the more significant ones are dealt with first, and to avoid further failures. Once the main complaints were finalised, they could be prioritised in order of weight and importance. In the next stage, customer satisfaction questionnaires were devised based on the identified complaints; the second phase of the interviews is initiated to validate the interviews and obtain the reassurance that the existing complaints were valid enough to be analysed. Following this, tools and strategies were adopted to address each complaint as required by the customers. In the second phase of this study, the key customers, in this case the drivers were selected. Using the software, significant drivers in order of importance were obtained. Then, based on the feedback received from each of the stakeholders and the customer satisfaction analysis, a draft of the solution is proposed. At this stage, feedback was required from stakeholders to minimize the level of errors in the proposed strategies, solutions and further recommendations. After that, the model and solutions was presented to all stakeholders to rectify minor errors and improve the model. In the final stages, technical specifications were presented in an attempt to provide solutions for the issues in question.

4.9 Significance of the Proposed I-CRM Framework

I-CRM generates a consistent and integrated view of customer complaints and various customer profiles. Also, it facilitates interaction and gives customers a comprehensive view of the company irrespective of the way the customer interacts. Both parties will derive mutual value from the complete implementation of the methodologies. I-CRM enables customer feedback to be acquired and categorised using a qualitative approach. Then, it deploys the main variables and tests their relationships utilising a quantitative approach. Ultimately, all relations will be mathematically and quantitatively formulated in a methodical way.

This system and its methodologies will help companies to acquire more customers and to retain them. It constantly generates new strategies that will maintain or improve the company-customer relationship by creating the best possible experience and interactivity methods for the customers. I-CRM also consistently generates a customer profile based on customer feedback. I-CRM provides a methodology, a strategy and concept that uniquely and completely define it and highlight its significance.

4.9.1 *Unique Features of I-CRM*

Intelligent CRM has different core advantages, although the most critical are as follows. It has the capability of reducing customer service costs and increasing the revenue provided by satisfied customers. It promotes data sharing and improves customer service. With I-CRM, the relationship between buyers and sellers will be improved and it enables personalisation of products and services. I-CRM creates a one-to-one experience for the customers and support for product and service development. Long-term partnership and coordination of communication are other benefits of I-CRM. In addition, individual pricing and channel choice will be increased. Furthermore, cost per service and administrative duties will be decreased.

4.9.2 *Intelligent Customer Relationship Management as a System for Evaluation of Customer Relationships*

Intelligent CRM can create a competitive environment and enhance the competence and efficiency of organizations. So, it always needs to be updated in keeping with the requirements and expectations of customers in order to meet these appropriately. Hence, there should be a precise measurement system using legitimate information. The information can be sourced in part from customers and the rest would be derived by means of systems. To measure I-CRM: criteria must be defined, failure to recognize the possible dimensions (value must be considered based on the preferences of the customers) must be avoided, a proper weight for the dimensions must be found based on correct management, and the result should be compared with that of competitors [15].

4.10 Summary of the Proposed Conceptual Framework for I-CRM

In this section, the various stages of the research are presented. In order to analyse customer feedback and arrive at the ultimate result, the flowchart shown in Fig. 4.9 is proposed.

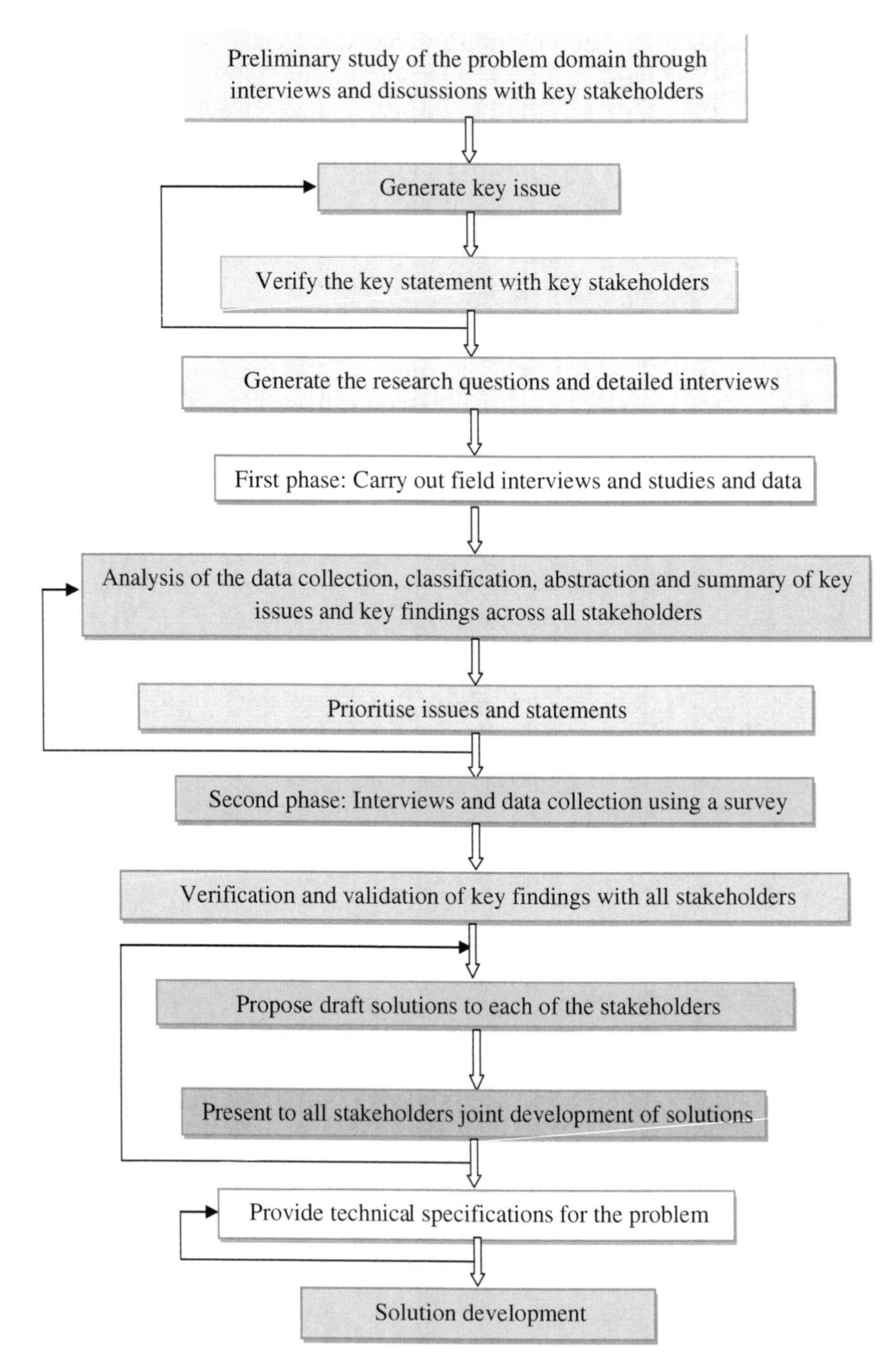

Fig. 4.9 Flow chart showing overview of solution for CRM and complaint solution

The conceptual framework is comprised of two different sections; however, they will be integrated together in order to have an iterative process. As illustrated in the previous chapter, the current research does not propose a comprehensive methodology for customer relationship management modelling. In this section, an overview is given of the complete solution for I-CRM modelling. In Sect. 4.4.8, an outline is provided of the individual solutions for all research issues discussed in Chap. 3. Figure 4.9 depicts the series of steps that will be taken to achieve the objectives based on the conceptual framework. However, each segment of the conceptual framework needs to be categorised and addressed based on the flow-chart, and will be discussed further in stage 1. The following flow chart faithfully represents the conceptual framework and shows the steps that will be taken to achieve the objectives of this research.

4.11 Conclusion

In this chapter, the initial step was to define intelligent customer relationship management and all its related terms. The software for coding and analysing definitions of CRM was used to create an integrated and unique definition for the purposes of this research. A solution was proposed for the problems that were addressed in Chap. 3. An outline was provided of various sub-categories of I-CRM, main CRM systems and their functions using their advantages. Next, the conceptual framework was proposed together with a flowchart to mind-map the process which will be undertaken to obtain the result. A brief definition was provided for each of the terms used in the conceptual framework. In the next chapter, a detailed solution will be provided to the research problems identified in this chapter.

References

1. Villanueva, J., Yoo, S., & Hanssens, D. M. (2008). The impact of marketing-induced versus word-of-mouth customer acquisition on customer equity growth. *Journal of Marketing Research, 45*, 48–59.
2. Limayem, M. (2006). *Customer relationship management: aims and objectives*. Tehran: Tarbiat Modares University of Tehran.
3. Payne, A., & Frow, P. (2005). A strategic framework for customer relationship management. *Journal of Marketing, 69*, 167–176.
4. Benoit, D. F., & Van den Poel, D. (2009). Benefits of quantile regression for the analysis of customer lifetime value in a contractual setting: an application in financial services. *Expert Systems with Applications, 36*, 10475–10484.
5. Chen, W. (2004). A paradigmatic and methodological examination of information systems research from 1991 to 2001. *Information Systems Journal, 14*, 197.
6. Bayraktar, E., Tatoglu, E., Turkyilmaz, A., Delen, D., Zaim, S. (2011). Measuring the efficiency of customer satisfaction and loyalty for mobile phone brands with DEA. *Expert Systems with Applications*.

7. Faed, A. (2010). A conceptual model for interactivity, complaint and expectation for CRM. Computer information systems and industrial management applications (CISIM), International Conference on IEEE, pp. 314–318.
8. LIU, Y. (2003). Developing a scale to measure the interactivity of websites. *Journal of Advertising Research.*
9. Ramakrishnan, R. (2010). The moderating roles of risk and efficiency on the relationship between logistics performance and customer loyalty in e-commerce. *Transportation Research Part E: Logistics and Transportation Review, 46*, 950–962.
10. Min, H., & Lambert, T. (2002). Truck driver shortage revisited. *Transportation Journal, 42*, 5–16.
11. Zeng, Y. E., Wen, H. J., & Yen, D. C. (2003). Customer relationship management (CRM) in business-to-business (B2B) e-commerce. *Information Management and Computer Security, 11*, 39–44.
12. P. o. B. P. Ltd. (2012). *STEVEDORE*. Available at http://www.portcareers.com.au/careers/stevedore/.
13. Li, J. A., Liu, K., Leung, S. C. H., & Lai, K. K. (2004). Empty container management in a port with long-run average criterion*. *Mathematical and Computer Modelling, 40*, 85–100.
14. Eder, J., Panagos, E., Pozewaunig, H., & Rabinovich, M. (1999). Time management in workflow systems. *BIS, 99*, 265–280.
15. Chalmeta, R. (2006). Methodology for customer relationship management. *Journal of Systems and Software, 79*, 1015–1024.

Chapter 5
Analytical Text Mining in I-CRM for Customer Complaint Analysis

Your most unhappy customers are your greatest source of learning.

Bill Gates
"Business @ the Speed of Thought"

5.1 Introduction

This chapter provides details of the interviews collected, and their subsequent categorisation and analysis using the text-mining tool. Several advanced concepts used in text mining analysis are defined, followed by qualitative data analysis. Issues are identified in detail and a scientific approach is presented to justify the issues identified as being those relevant to the port in terms of logistics and transport. We will conclude the chapter by providing significant analysis in terms of real-world problem definition. In the following sections, we define the relevant terms and concepts utilised in this chapter.

5.2 The I-CRM Concepts Data Analysis

5.2.1 Node

In a text analysis approach, "node" is defined as an idea or a topic derived from interview data and survey questionnaire. There are two node types: "Type 1" is an independent concept which has no relationship with other nodes and "Type 2" has a structural or association relationship with other nodes. As an example, a node with a tree structure is associated with a hierarchical relationship or inheritance relationship. If we are analysing interviews regarding customer complaints about the Fremantle port authority and logistics operators, we can create a node called "management issue".

5.2.2 Node Theme

"Node theme" is defined as the collection of nodes that represents the content.

A. R. Faed, *An Intelligent Customer Complaint Management System with Application to the Transport and Logistics Industry*, Springer Theses, DOI: 10.1007/978-3-319-00324-5_5,
© Springer International Publishing Switzerland 2013

5.2.3 Reference

"Reference" is defined as the percentage of repetition of a particular sentence that occurs overall in the transcribed interview document from a specific interviewee. In this thesis, it helps us to classify sources of information about the Fremantle port interviews. The source of reference could be in the form of a document, dataset, picture, audio, and video.

A collection of "references" represents the semantics of a particular node. For example, "Infrastructure issues" is considered as a node; hence, the collection of references could be as follows:

1. There is no place to have lunch.
2. The place is dusty and therefore harmful to the lungs.
3. The port has no clean and green area where one can rest and relax.
4. Truck queuing at the port sometimes takes up to 5 h and quite often there is a waiting time of 2 h.

5.2.4 Reference Coverage

"Reference coverage" is defined as the percentage of occurrence of an item within the whole interview text transcription. This value indicates the importance of a particular reference. Also, a thesaurus is used to ensure that the synonyms are identified. In the figure below, as an example, we model the relationship between node theme, node, reference and reference coverage.

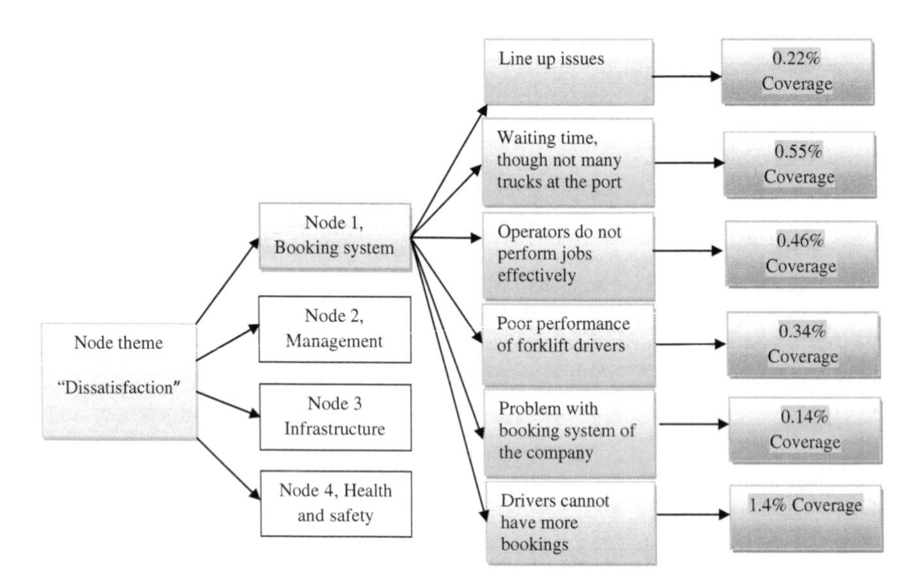

Fig. 5.1 Relationship between node theme, nodes and references

Using Fig. 5.1 which is the advanced concepts, the major issues that emerge from the interview transcripts can be categorised. Customer complaints are prioritized prior to identifying patterns through text analysis.

5.3 Real-World Case Study

According to [1], a case study is an inclusive analysis and problem-solving approach which explores similar circumstances in other companies that have the same essential issues as those being addressed in the existing situation. A case study requires that hypotheses be established and that issues be investigated. The case study chosen for this research is both authentic and unique, and therefore challenging.

5.3.1 The Fremantle Port Case Study

Figure 5.2 depicts the Fremantle port is one of Australia's most important docking facilities and is located close to Perth in Western Australia. According to [2], the Fremantle port performs its work utilising two separate harbours, an inner harbour and an outer one. The inner harbour controls and administers all types of activities associated with the facilitation of imports and exports. On the other hand, the outer harbour which is located in Kwinana is one of Australia's major bulk cargo ports that sorts bulk commodities including petroleum, liquid petroleum gas, alumina, and mineral sands, to name a few. Fremantle port also controls shipping pathways. Moreover, it works with Commonwealth agencies and is responsible for customs, quarantine and maritime safety.

There are also several jetties at Fremantle port, three of which in the outer harbour are operated by private companies: Alcoa, BP and CBH. Fremantle Ports runs the Kwinana Bulk Jetty and the Kwinana Bulk Terminal [2]. However, the key issue in Fremantle Ports is that productivity is very low, which means that container movement operations are not up to standard and are therefore not

Fig. 5.2 Fremantle port at a glance

keeping pace with local and national economic development. Furthermore, this issue has been an ongoing one for the past 10 years.

In this study, fifty interviews were conducted in order to collect raw data from drivers, operators and stevedores. The interview process faced several challenges. Firstly, the port's operational environment meant that the majority of people approached were not interested in being interviewed since truck drivers have limited or little time available due to shipping operations. Furthermore, some were not willing to be interviewed because they have little faith in research; others were not interested because they believed that, from their past experience, no one can improve the conditions in the Fremantle port area. Then there were those drivers who believed that their knowledge of the port and its operations far exceeded that of any researcher. Since they apparently knew what to do and what can be done, they did not bother to talk to the researcher. Yet another type of truck driver/company did not want to participate in interviews for fear of disclosing data. Despite these challenges, over a period of 6 months, fifty detailed interviews were conducted.

All interviews, whether audio or video were transcribed and recorded in order to conduct a comprehensive study of the text. The interviewees, mostly comprising drivers and operators, discussed the issues and challenges they confront in their daily work. During the transcription process, care was taken to ensure that the main issues and important points raised were faithfully recorded. For the purposes of this thesis, the population includes the clients and stakeholders at the Fremantle port, as the reasons for the dissatisfaction need to be investigated. After becoming familiar with the issues at the Fremantle port, we selected the drivers at the Fremantle port as the main sample. Truck drivers at the Fremantle port are the core of container logistics and transportation operations. Most of them are employed by small to medium enterprises or are owner/operators themselves. In this WA port alone, there are over 240 logistics operators and they move about 220,000 containers each year. According to international port logistics operational standards, they should be able to move more than 800,000 containers each year.

5.3.2 *Qualitative Research*

To the best of our knowledge, this thesis is the first attempt to categorize and analyse customer complaints in the logistics and transport industry at the port, using CRM initiatives in qualitative research. Data is collected using two different approaches, namely face-to-face interviews and questionnaires. To analyse the interview texts, a qualitative scientific approach is taken which enables us to explore problems, and identify, understand and respond to some of the queries. Furthermore, using qualitative research, we can intuitively become aware of people's concerns and wishes. Also, we can leverage the policy-making and decision-making processes of the companies. Using qualitative analysis, we will be able to identify the focal point and have a better interaction and communication with the customers.

The aim of this study is to establish the focal points of the research and narrow down the port problems in order to identify the main issues.

5.3.3 Qualitative Analysis of the Interview Data

A detailed study needs to be carried out and a structure imposed on the unstructured interview data. The information obtained needs to be classified and a reference has to be created for each statement made by interviewees. This analysis gives better meaning to the interview data and facilitates accurate data interpretation.

The qualitative research and the interviews are conducted twice. The purpose of the initial interviews is to collect the necessary information and become familiar with the issues. The interviewees are well positioned to clarify the situation for the researcher and provide an insight into issues. The first part of the analysis involves an examination of the interview texts in order to identify complaints, compliments or neutral feedback from the interviewees. In the second part of the interview process, the previously identified issues are validated and a text analysis is conducted in order to pinpoint the areas of dissatisfaction and identify the key issues.

5.3.4 Text Analysis of Interview Data

Using qualitative research, the transcribed interviews are imported into the software, and a text search query is conducted. For the purpose of analysis, matches between similar words are considered; for this reason, stemmed words are selected. Using an analytical and qualitative approach, nodes (codes or themes) are created based on the frequency results. However, nodes must be placed into a hierarchy and be based on the research. Then, related statements and words are transferred under each node. Also, various subsets are created for some of the nodes. To the best of our knowledge, no such work exists in the literature and this is the first work of its kind to investigate customer complaints via interviews and an analysis of the transcribed interviews. The ultimate result of the analysis would be a prioritization of complaints based on their importance and weighted percentage.

The following section presents a discussion of the issues that were gathered using qualitative research and text analysis. Logistics operators face a variety of issues in their working environment at the Fremantle port. The issues are classified and ranked according to their significance.

5.4 The Analytic Findings on Fremantle Port Issues

In this section, we discuss the major issues that create dissatisfaction at the Fremantle port according to text-mining analysis. Each issue is based on the facts and results derived from analysis.

5.4.1 Cost Issues at the Port in Different Segments

Based on Fig. 5.3, the following results are derived from our analysis of the interview text, and the word length, word count and word frequency that the software generated:

Based on Fig. 5.3, reference 1, shows that roads are 20 % more expensive than rail and this is supported by the fact that we derived from our analysis which is 0.19 % of out text coverage. Based on reference 2 and 0.17 % of our interview coverage, operators always charge their customers too much, which is not fair according to the drivers. According to reference 3, which is 0.17 % of the text coverage, the port and the companies working within it must provide some techniques and strategies to eliminate or reduce the cost in order to benefit both clients and customers.

Furthermore, based on reference 4 which comprises 0.79 % of the interviews, if the port increases the rents for the cafe, restaurant and petrol station, the owners have to increase their prices as well. They are not willing to do so because they want to retain their customers. Hence, if proprietors increase prices, the customers (drivers) will not wish to use the services. Based on reference 23 and 0.44 % coverage of the text mining analysis, Fremantle port does not have a proper cafe and restaurants which is another cause of complaint since drivers, passengers and others who need to have decent food, a cold drink or coffee, need to have a good cafe and restaurants on the road.

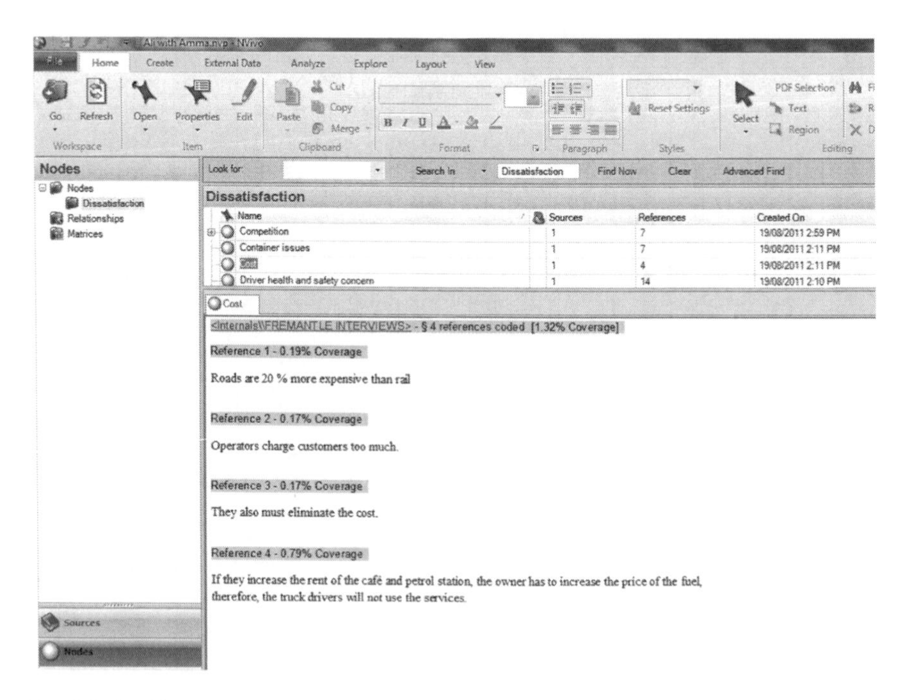

Fig. 5.3 Cost issues

5.4.2 Booking System

Based on the analysis conducted using the text analysis method on Fig. 5.4, references 1, 2, 3, 4, 9, 10, 11, 12 confirm that the customers at the port have major concerns with the booking system. Each reference generated by the software is a sole statement. The drivers cannot arrange their shifts properly. An efficient booking system will expedite procedures, eliminate the need for paper documents, support electronic commerce and provide a means whereby customers can source information and be made aware of different situations.

According to reference number 2 in the interviews which covers 0.22 % of the text, currently, drivers need to line up for their bookings. And based on reference 3–0.55 % coverage, the driver who works on Tuesday's needs to wait in line for more than an hour although there are not many trucks on that day and there is less congestion.

Based on reference 5–0.34 % coverage, the lack of an appropriate booking system is confirmed by fork lift drivers. Sometimes they do nothing; at other times, they unload the containers. Reference 6–0.19 % coverage asks why it takes so long to do things at the port. A comprehensive booking system will improve operations, and increase driver efficiency and performance.

Reference 7–0.79 % coverage comments that the companies operating at the Fremantle port have been in trouble for the past 14 years regarding booking, and this situation has never improved despite the fact that every day they are challenged to provide a remedy. In the past, researchers and experts have attempted to initiate and implement various systems and strategies at the port, but without success. With DP World, waiting time is very long. However, Patrick is more organized due to their better booking system. Drivers still have to wait in line, but the truck will be gone after twenty minutes.

Another problem that they have regarding the booking system at the port is that, with DP World, when a driver wants to make a booking s/he has to be called. However, with Patrick, the driver can make the booking, take it in and go straight through. Based on reference 12–0.73 %, into efficiently get the containers off the wharf, the drivers need to have a booking system. Also, based on reference 13–0.17 % coverage of the text, drivers have many time slots and an accurate booking system would solve the current problem.

Reference 14–0.25 % coverage mentions that a booking system provides a fixed program for a port customer. According to reference 16–0.78 % coverage, drivers are often behind in loading due to the lack of a booking system. For example, a driver arrives at the port from Bentley (an area near Perth, WA) with two empty containers and leaves the port with still-empty containers at 6:00 pm because s/he cannot get a booking on that day.

Based on reference 17–0.50 %, the driver mentions that operators should increase their peak time, for example, in the morning from 7:30 am to 9:00 am and in the afternoon from 3:30 pm to 6:00 pm. This is not possible without having a booking system at the port.

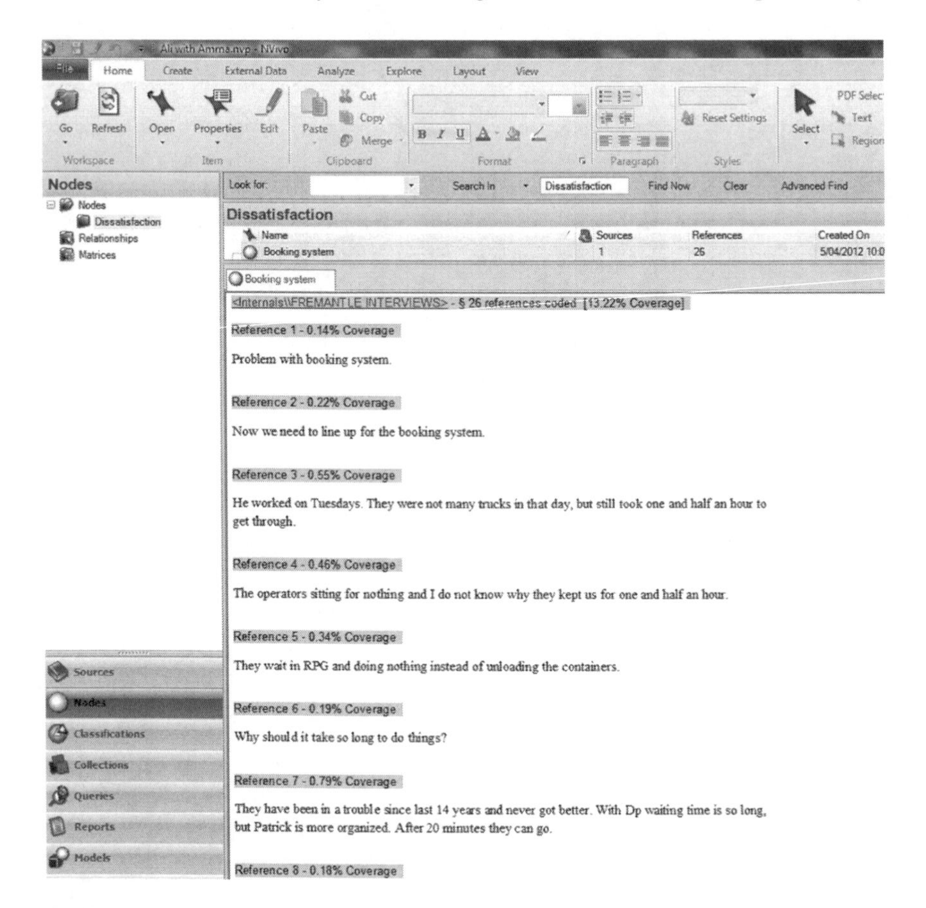

Fig. 5.4 Booking system issues

Based on references 19, 20, 21 and 22 and the coverage of 0.62, 0.45, 0.34, 0.91 % respectively, the longer the driver has to wait, the higher is the chance that s/he may lose money and not be paid. However, with a booking system, it is unlikely that a driver will waste as much time. Moreover, a booking system could make the customers more punctual.

Based on reference 23—coverage of 1.31 % of the interviews, the driver is often in a queue for more than an hour and when finally getting to the front, s/he may be asked to go and take care of documentation (paper work). And, when the driver gets back in the queue, another long wait begins.

Reference 24–1.33 % of coverage mentions that Fremantle port is supposed to operate 24/7; however, if someone wants to collect a container at the weekend, the person may encounter closed doors and no available service. With a proper booking system, people can set their schedule ahead of a time.

5.4.3 Container Pick-up and Drop-off Issues

Figure 5.5 illustrates the following outcomes:

Reference 1-0.24 % coverage mentions that as there is just one place to load or unload the containers, there is always a bit of confusion in finding the containers.

Reference 2–0.29 % coverage states that there should be a line up for loading containers. Reference 3–0.28 % coverage claims that empty containers are the main issue since some drivers come to the port every day from 6:00 am to 6:00 pm with empty containers. Reference 4–0.24 % states that the capacity of containers presents a problem. Companies could easily increase the capacity of the trucks, thereby decreasing company costs and increasing their productivity.

Reference 5–0.57 % is concerned that trucks are not designed to drive over speed humps which may cause the load to become misaligned. At the same time,

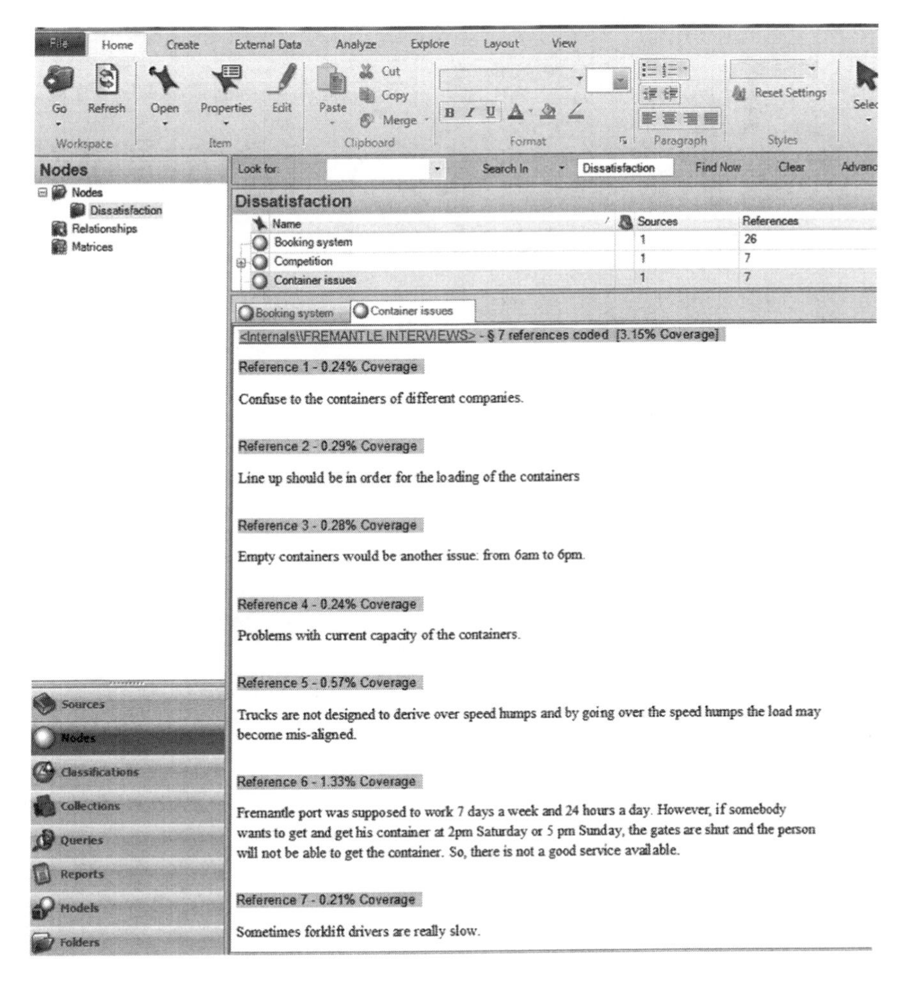

Fig. 5.5 Container issues

this could cause damage to trucks with the driver incurring the cost of repairs. Therefore, there should be some strategies and regulations in place concerning the containers in order to prevent danger and risk of an accident. This may also be a managerial issue and will be discussed in the relevant section.

5.4.4 Health and Safety Issues at the Fremantle Port

According to Fig. 5.6, reference 1 and 0.90 % coverage of the interviews, on some weekdays particularly during lunch time when it is dusty, the working environment becomes quite unbearable and detrimental to the lungs. When trucks come

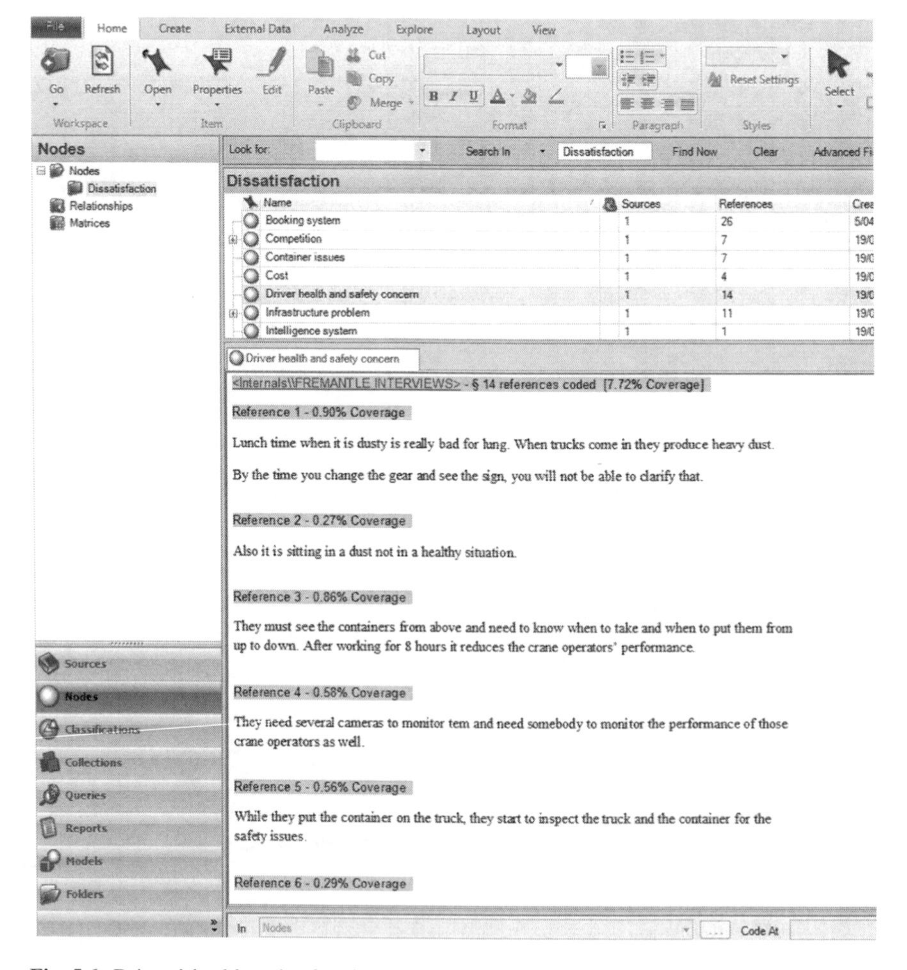

Fig. 5.6 Drivers' health and safety issues

into the port, they create heavy dust conditions. Based on reference 6 which is 0.29 % of the coverage, dust poses a major health risk. Moreover, it is impossible for drivers to open the windows of their cabins to better direct their vehicle. According to the same reference, in the port there should be some highly visible signs and arrows pointing to the direction that drivers should take. This will be a significant safety measure since many drivers are exhausted on arrival at the port, and would appreciate signage, particularly if they have poor eyesight or reduced visibility due to the dusty conditions.

Based on reference 5–0.56 % coverage, when operators load containers on trucks, they inspect the truck and the container to ensure safety; this is a good sign but the procedure should follow a systematic and consistent set of rules and regulations.

Based on reference 7–0.53 %, the fumes generated by the asphalt plant burn the lungs, thereby posing a threat to the health of the clients and drivers. This particular driver would rather smoke than breathe in these fumes.

Reference 8–0.71 % of the coverage believes that dehydration is another big issue for drivers. DP World can predict when hold-ups will occur and should provide water for the drives.

Another increasing and disturbing trend is the number of accidents on the roads leading to the port; in recent times, this has increased dramatically due to poor supervision of the roads. Based on reference 9–0.51 % coverage of the text, the majority of accidents occur when other cars on the road cut in front of trucks, causing accidents and risking being run over by the heavy vehicles. As a remedial action, there should be an extra road for the use of trucks alone.

Reference 10–0.60 % coverage, states that when the temperature exceeds 40 degrees, drivers who work on machines and are employed by one company can simply go home; where as, other drivers have to stay and finish their work. This may create a health issue for the driver for which no-one is accountable. Also, based on reference 32 and 0.50 % of the text, bike riders and runners are using the same road which is extremely dangerous.

Reference 11–0.38 % claims that, in order to be healthy, ready to work, and happy, drivers need to have access to good services and a roadhouse. In this way, they are unlikely to feel exhausted and run-down as they can have a good meal while waiting at the port and a proper rest. Reference 12–0.38 % coverage, mentioned that when the drivers get out to check or to secure the load, they are told to stay in their trucks. This has the potential to be very dangerous and must be considered as an issue of great importance. Not only is an unsecure load highly dangerous, but according to reference 13–0.76 % coverage, the drivers themselves can become dehydrated in the hot, confined space of the cabin, particularly when temperatures are high. For this reason, drivers should have access to a drink or to the rest room, but at present they are not permitted to go to the toilet or get a drink. Finally, according to reference 14 and 0.26 % coverage of the text, rain creates muddy conditions at the port. Hence, the port should build pavements and establish new roads in the port area in order to reduce the hazards created by muddy surfaces.

5.4.5 Infrastructure Problem

Based on the text mining analysis of infrastructure issues in Fig. 5.7, according to reference 1 which comprises 0.33 % of the text, the Fremantle port and the companies that operate there often face 3–4 m of train derailment which stops everything. To address this issue, companies must cooperate in order to ensure ongoing rail maintenance. The necessity for this was demonstrated in 2009 when the derailment of a freight train occurred near a residential area, which was potentially highly dangerous for the residents [3].

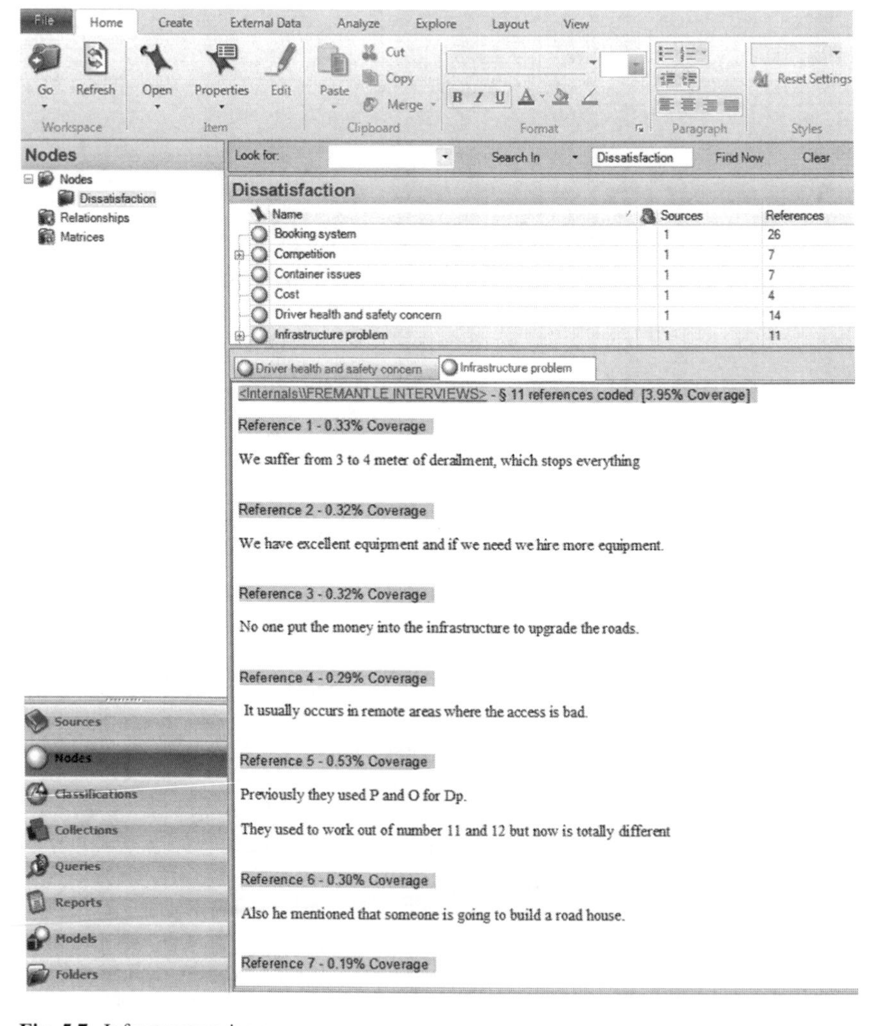

Fig. 5.7 Infrastructure issues

Figure 5.8 displays how derailment may occur in reality. The derailment has been occurred at the Fremantle. According to reference 2 and 0.32 % coverage of the interviews, the interviewee made the positive comment that Fremantle port has good equipment and if necessary, is able to hire more.

Fig. 5.8 Fremantle derailment 2009

The interviewee in reference 3, 0.32 % coverage mentioned that no-one has put the money into the infrastructure to upgrade the roads. It seems that other companies need to help the Fremantle port to move it forward. Reference 4, another interviewee, mentioned that infrastructural issues and limitations usually occur in remote areas where there is poor access to the equipment and facilities.

Reference 6 which is 0.30 % of the text coverage based on text mining, emphasized that the Fremantle port must build roadhouses for those drivers who do not have accommodation in Perth and need to stay overnight to rest in order to minimise the risk of an accident when they resume driving.

Reference 7 which is 0.19 % of the text claims that customers, mainly comprising drivers, experience significant problems with forklift drivers and their performance. It appears that forklift drivers work during set hours and then take set breaks throughout the day. Regardless of whether they have finished, just begun, or are in the middle of a task, they leave the job and take their break. Drivers regard this as unfair and feel that they are discriminated against in this environment. A recommendation and possible solution is that stevedoring companies provide more forklifts and roster forklift drivers in various shifts. This will ease the tension between them and the drivers, in addition to improving efficiency.

Currently, there are not enough roads into and around Fremantle port; consequently, the number of accidents on roads leading to the port is increasing. Based on reference 8 which captures about 0.15 % coverage of the interviews, the drivers have asked for a new road for the trucks so as to minimise the risk of accidents. The various companies and the port authorities need to take action to improve this situation.

The lack of ready access to technical support is another important issue that restricts port operations and activities. According to the interviewees' responses in reference number 9 which has 0.50 % of the text, when a machine breaks down at the port, it takes lot of time for that machine to be fixed. As a recommendation, the port authority or the stevedore companies must provide the road operators and truck drivers with adequate technical support.

Reference 37 and 0.50 % of the text mentioned that drivers need to have access to good services, including another petrol station and a roadhouse.

Reference 10 which is 0.40 % of the coverage, comments that stevedoring companies need to build another road to separate heavy and light vehicles. Hence, trucks need another road exclusively for them in order to alleviate present dangers posed by the current congestion. Also, according to reference 15 and 0.34 % coverage, when the freight train arrives and wants to cross the road, the heavy traffic and congestion is worsened.

Another issue which exacerbates the current traffic problem at the Fremantle port is, according to reference 11 and 0.62 % coverage, the increase in traffic congestion caused by cars and trucks going to the ferry from the city and from the port to the city (on a two-way road); moreover, this makes it extremely difficult and dangerous for pedestrians. Based on reference 10 and 0.36 % coverage of the text, another reason for the necessity of having a new road is the fact that sometimes truck queues reach back as far as the highway and residential area, inconveniencing local residents and posing some danger to them. This must be taken into account in future planning.

There are various other concerns and limitations at Fremantle port related to the subset of infrastructural problems. These are discussed in detail below. Based on the text mining analysis, reference 1 and 0.65 % of the text discusses the difficulties of finding directions at the Fremantle port. The interviewee mentions that there should be arrows and signs to direct people, especially the drivers, to appropriate areas in order to avoid confusion regarding the containers of different companies.

Reference 2 on the limitations at Fremantle port which is 0.17 % of text analysis, mentioned that there is no sufficient pavement and walking space for the pedestrians. The companies must build pathways for pedestrians for reasons of convenience and safety, especially in high-traffic areas.

Based on the 3rd reference and 0.25 % coverage, the drivers have problems accessing a cold beverage such as water in hot weather. In addition, food and beverages can be bought in only one restaurant located at the Fremantle port, where prices are higher than usual. There should be a solution for this issue. At present, this restaurant monopolises the business. There should be other cafes and restaurants at the port to provide competitive pricing. Also based on reference 8 and 0.53 % of the coverage, if trucks stand in the queue, the drivers cannot go to the toilet and there is no toilet around the marshalling area. The port authority must review its regulations and strategies to address this matter. Reference 39 and 0.60 % of the analysis further clarifies this issue and confirms the need for a new food and beverage outlet.

Although some of the logistic and transport companies such as Steel Haul provide their drivers with a mobile phone and radio so as to be in touch with each other and their families, the mobile phone credit is limited to a certain amount—often not enough to give drivers the peace of mind they desire. Based on reference

4 and 0.69 % of the text coverage, there are interaction and communication issues, and companies must propose a solution for them to effectively facilitate the procedures.

If a container is imported, as a quarantine requirement for the ships and boats, clients who are in charge are required to fumigate the container for health and safety reasons. This may prolong the truck's stay at the Fremantle port and limits its efficiency because drivers are kept waiting for long periods. Sometimes, the drivers miss their booking and are then required to queue up again. Based on reference 5 and 0.29 % of the text, the companies which import the containers must have them fumigated, although this takes time and is unavoidable.

As discussed previously, there are two stevedoring companies at the Fremantle port which load the containers and unload the ships using RTGs and cranes. Mostly, it takes one to 2 h to unload and load one container which is very hard to tolerate. Patrick's system is better according to reference 6 and 1.44 % of the text. At DP World, the driver is required to sit in the cabin of the truck and wait to be loaded or unloaded for approximately 2 h, sometimes longer. However, the problem is that the driver is not permitted to get out of his/her truck for safety reasons.

The recommendation and solution is to provide an environment like Melbourne port and let the drivers leave the cabins of their trucks. This is a better option in terms of the drivers' health and safety.

Based on reference 9 and 0.83 % of the text coverage, it takes a long time for a container to be unloaded from one side of the road, and loaded on the other side. A strategy is needed to reduce the time that it takes to do all this. Also, there are several cameras at the Fremantle port to monitor the drivers and the flow of operations. However, drivers have expressed their great concern that they cannot clearly see the different sides of their trucks and the road, and they have asked that cameras be installed in different parts of the road. Furthermore, based on reference 16 and 1.03 % coverage, there is very little room for a truck to turn at the Fremantle port, thereby causing delays and increasing the risk of accidents.

According to reference 14 and 1.09 % of the interview coverage, there are not enough traffic lights in Fremantle port itself or in the neighbouring streets. The installation of more traffic lights in and around the port area is an obvious solution. Furthermore, the same interviewee observed that the roads need to be widened. At present, intersections are so small and narrow that cars can only move slowly.

Reference 24 and 0.38 % of the text states that a bigger parking area is needed in the Fremantle port. Drivers are not allowed to park their trucks by the roadside; if they do, they risk a fine and their truck may be towed away.

Another significant difficulty for drivers is that there are no stores or shops within the port area, and those shops in the vicinity of the port close at 5:00 pm. Items that need to be bought are very expensive and drivers prefer to go without. Based on references 25, 26, 27 and 29, 0.31 %, 0.27 %, 0.57 % and 0.19 % coverage of the text analysis respectively, drivers have requirements while they are at the port. However, they do not have ready access to items they need, and besides, these are more expensive in the local shops than they are in the city. What is more, shops are closed on the weekends.

The solution is to build a new store at the Fremantle port with prices that are comparable to those in major chain stores. Hours of trade could also be extended.

Based on reference 28 and 0.67 % of the text analysis, consideration and priority should be given to those drivers who transport frozen foods and meat to the Fremantle port.

5.4.6 Jurisdiction and Policy Issues

Based on the facts derived from the analysis in Fig. 5.9, the following jurisdiction and policy issues emerged. Due to road regulations, drivers are very restricted when attempting to carry out their tasks properly, based on reference 1–0.23 % coverage. No-one has put the money into the infrastructure to upgrade the roads and this could be because of poor regulations at the Fremantle port. Based on reference 2–0.90 % coverage, the working environment is always dirty with rubbish in areas where drivers need to park their vehicle. This is because each company has its own rules and regulations. They also have several unwritten laws at Patrick and DP World and although there are people who oversee these 'laws', there is no formal authority with overall control of operations.

According to reference 4–0.17 % coverage, accreditation is a huge issue. The accreditation of drivers should be uniform throughout Australia, so that it is recognised nationally throughout all states. The Heavy Vehicle Accreditation (HVA) is difficult to obtain, with numerous forms to be filled out and submitted. However, even though WA drivers obtain accreditation, as soon as they cross the state border, the certificate is not acceptable. On one side of the truck is a sticker confirming vehicle accreditation. The eastern states vehicles have a green sticker (NVA: national) which is difficult for WA trucks to acquire. There is no reciprocal recognition of accreditation. WA's HVA is not accepted in other states, but the latter are accepted everywhere. So, other states' accreditations are recognized nationally, but WA is recognized only in WA. Furthermore, the whole port system is set up by politicians and bureaucrats who know nothing about transport.

5.4.7 Management Issues

According to the analysis displayed in Fig. 5.10, reference 1–0.17 % coverage of the text, even though the operators may have several years of experience, promotions are few even though some managerial roles need replacements. Companies could implement a system of job rotation so that managers in one section can transfer their experience to another department. According to this interviewee, it would be a good idea to recruit new managers who are willing to make changes to

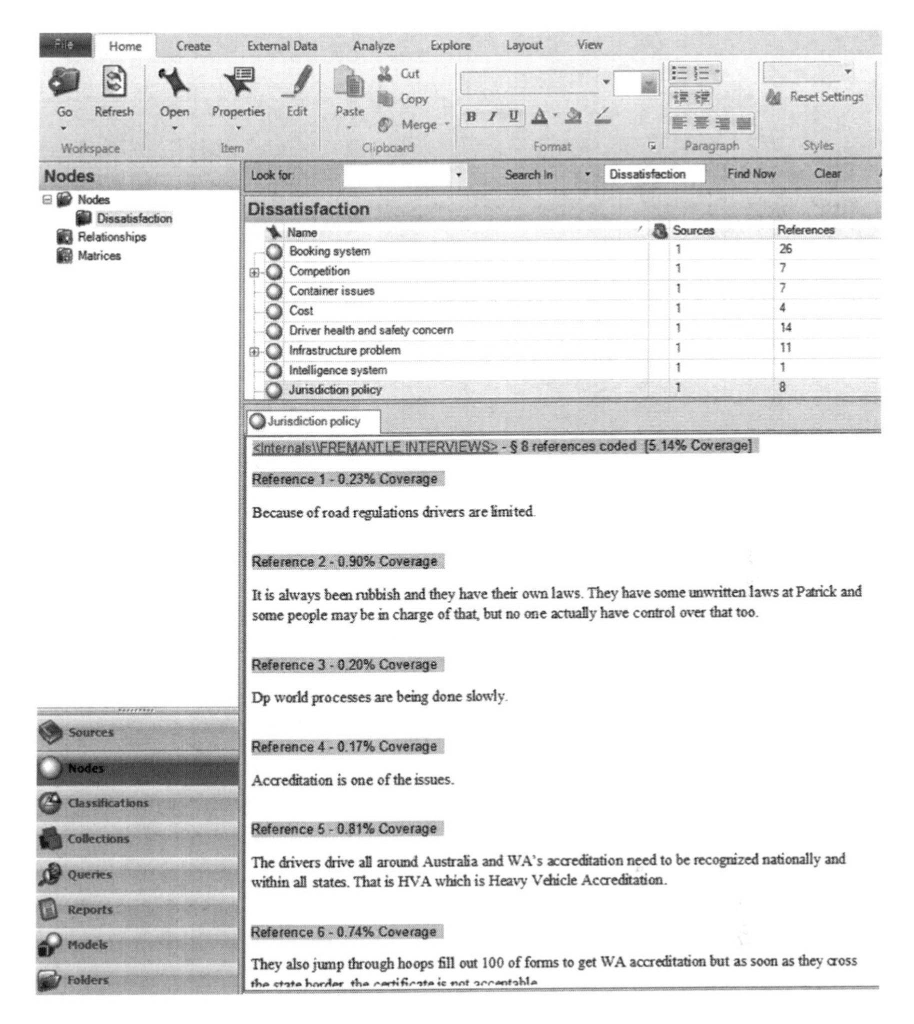

Fig. 5.9 Jurisdiction issues

improve the flow of operations. In Patrick, when they go over to load a ship they may get charged. When a ship docks, Patrick may get charged, so they need to unload the containers quickly. To load and unload a ship, Patrick is charged thousands of dollars—much more than for loading and unloading trucks—hence, they make truck drivers wait. Even if drivers are waiting for a booking, they are charged; and even if they make it on time, they are still made to wait. All these are managerial issues that must be addressed.

If a management team cannot sort out the current problems, it is inevitable that drivers will choose a more efficient port in other states. Already, many drivers have gone north or to the eastern states, preferring not to work in Western Australia. An efficient management team will be able to streamline company operations and

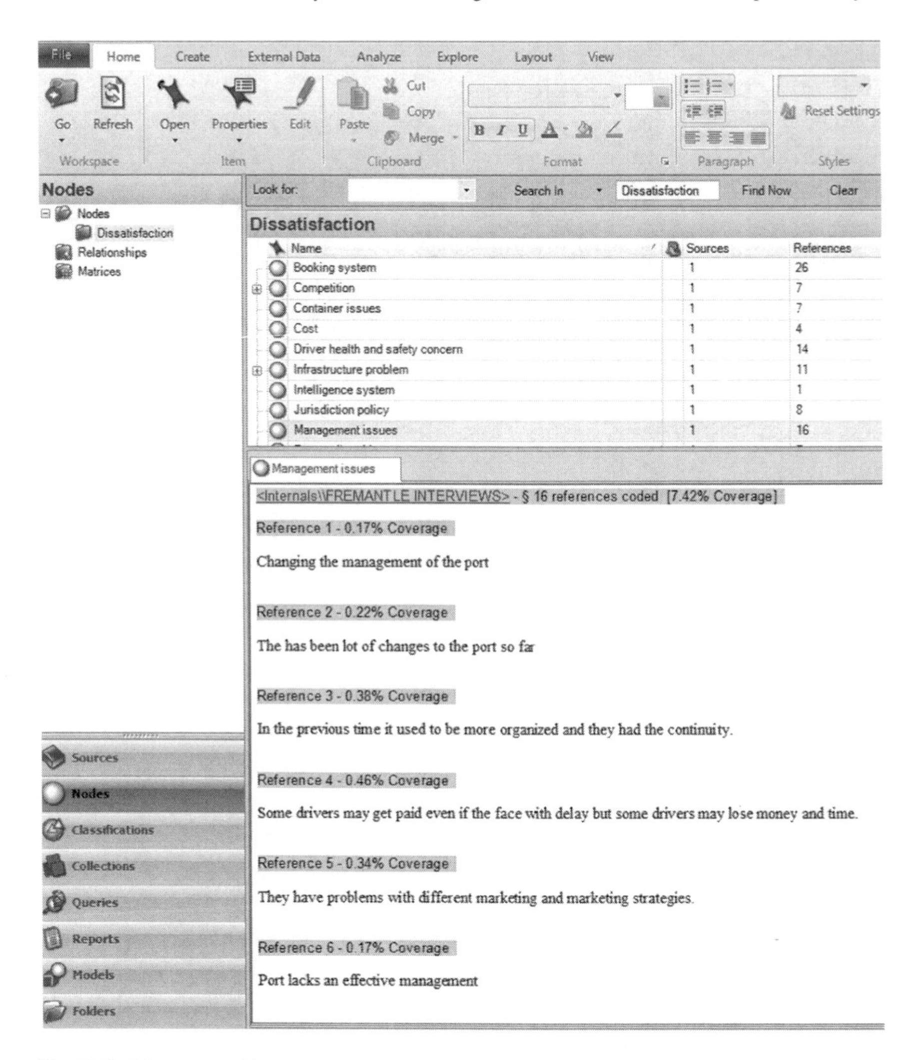

Fig. 5.10 Management issues

minimize delays and procrastinations. By addressing the various issues, it can reduce the number of complaints and save the customers both time and angst. Since customers are central to all the company's activities, they should be given certain rights and priorities. Only in this way can a company can survive. Some of the drivers' complaints concerned the shortage of personnel with a sound knowledge of the transportation industry and related fields. This extended to the managers who seemed to lack both knowledge and experience. According to one interviewee, 'it has always been rubbish and they have their own laws' such as previously mentioned in the case of Patrick. Several drivers expressed their opinion that the various companies have their own rules and regulations and there are no firm, consistent rules or guidelines by which they must all abide. In other words, there is no formal authority to which all companies are accountable. Apart from all of the managerial issues, some of the drivers claim that the port is administered by bureaucrats and politicians rather than a legitimate management team.

Based on reference 3–0.38 % coverage, a proper managerial team could improve the organisation of the port, ensure consistency and continuity. Good management can facilitate the work and decrease the waiting time to the minimum. Moreover, it could substantially increase company profits. While the company has a really educated management team with good characteristics, they will be able to manipulate all environments around the port and optimize the level of performance.

Based on reference 5–0.34 % coverage of the text, Fremantle port lacks appropriate marketing and management strategies which can be rectified and improved with a professional management team.

This is confirmed by reference 6–0.17 % coverage, who maintains that currently the companies in the port have problems because they each have different marketing and marketing strategies. An effective overall management team is needed that can ensure consistency and establish policies to address shortcomings.

Furthermore, according to reference 9–0.32 % coverage, a good management team is needed to generate competition at the Fremantle port which will improve conditions for both employees and drivers.

Based on reference 10–0.23 % coverage, due to inadequate management, the port authority and police department do not collaborate effectively, and this alone may increase the risk of traffic and other accidents at the Fremantle port and on roads leading to it.

Based on references 12–0.34 % and 13–0.38 % respectively, another issue that drivers are challenging and which definitely goes back to the lack of good management is that drivers are not allowed to leave the cabins of their trucks for any reason, even if it is to check on the load. When in the queue, drivers are required to remain in their vehicles.

According to reference 14–0.42 % coverage, if a company cannot handle tasks efficiently because of poor organisation, another stevedoring company should be engaged to do the work. However, such a move needs an effective and empowered management.

Based on reference 15–0.26 % coverage, good management must provide a solution to remedy the muddy conditions at the port on wet days which create additional hazards for drivers and employees.

Based on reference 16–0.79 % coverage, if the port authority increases the rent of the café and petrol station, the owner has to pass on this increase to its customers; consequently, the truck drivers will not use the services. This is another managerial issue which needs to be resolved.

Other issues:

Also, here in the Fremantle ports, they have not had any change at all, and a new management team could bring about that overdue change to provide better quality services to the customers.

A new and younger management team could introduce the job to young people who are currently not attracted to driver jobs because of the complexities and hardships. A young team of drivers who are healthy and have physical stamina will not tire as easily as older drivers. Additionally, young drivers have fewer expectations in comparison with experienced drivers. Another problem is that if vehicles are in a queue the drivers cannot go to the toilet and there is no toilet around the marshalling area. So, companies need to provide these amenities. Many drivers are heading up north for better money and this is because of the poor management system. The drivers do not have fixed time slots for arrival and departure and a good management team could provide this. Companies need to increase the capacity of containers as the ones in use are too small. At DP World, all processes are being carried out slowly. A new, effective management team is needed to revitalise the operations and improve efficiency.

5.4.8 Stevedore Performance

Figure 5.11 shows the Fremantle port activity which is performed by stevedores in Western Australia.

Fig. 5.11 Fremantle Harbour, Stevedores operation, Perth, WA

Based on Fig. 5.12, the following results have been obtained:

Reference 1–0.05 % coverage, stevedores must be accountable for the safety of the Fremantle port as well as customers. Based on reference 2 and 0.05 % of the interviews, stevedores, like other workers at Fremantle port, have issues with congestion at the Fremantle port. Based on reference 4 and 0.04 % of the text, stevedores may cause difficulties for Western Australian road transporters. Hence, the port authority needs to deal with stevedores with more flexibility.

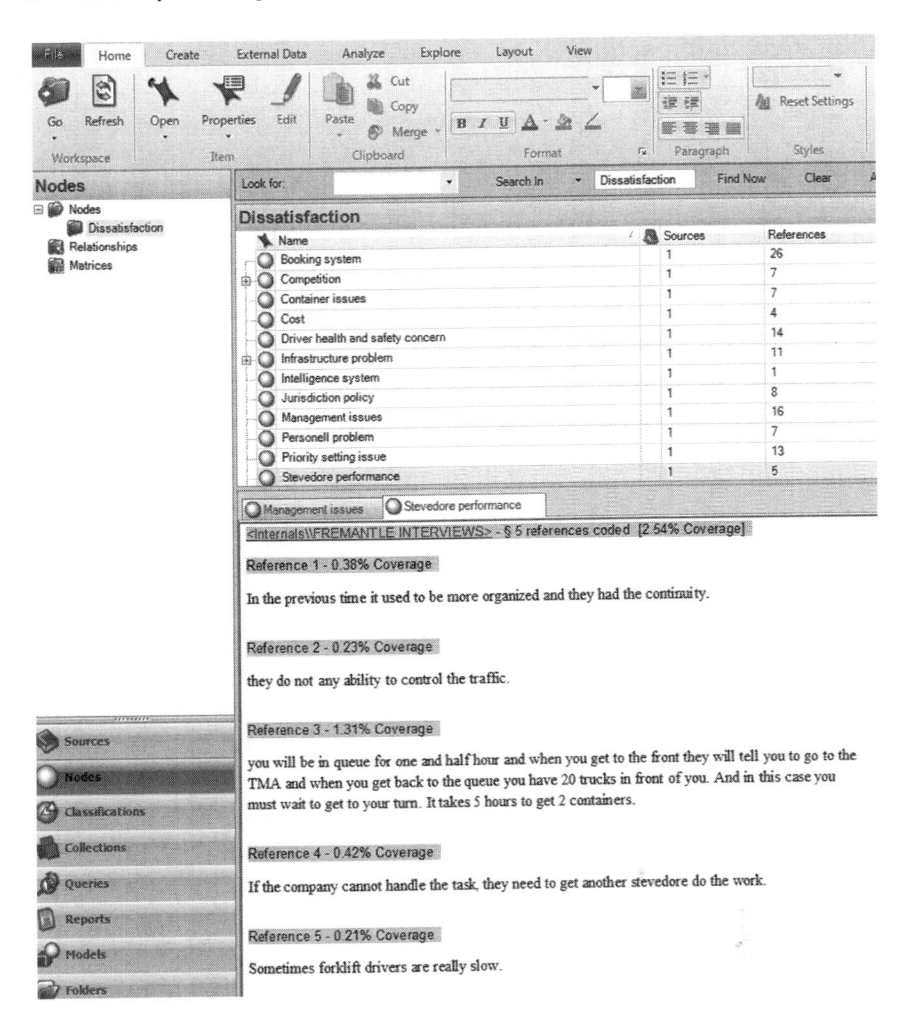

Fig. 5.12 Stevedore performance

Authorities at Fremantle port must liaise with stevedores and make precise arrangements for tasks to be carried out. Also, they know the reason for delays and timing and should let the transport and logistics companies know about appropriate procedures. Previously, stevedore performance used to be more organised and it used to have continuity; however, currently, stevedores are not performing efficiently. At the same time, customers are dissatisfied with the stevedoring companies and their relationship has deteriorated.

5.4.9 Time Management

According to the analysis derived from Fig. 5.13, reference 1 which has 0.95 % coverage of the text clarifies that, it takes one to 2 h to unload one container. There are two major stevedoring companies working at Fremantle port. The performance of DP World is poorer than that of Patrick. For example, a driver must wait at DP World for 2 h and the truck driver is not permitted to leave the cabin of his vehicle. Reference 2–0.37 % coverage states that stevedores at Patrick work from 7 am to 4 pm, and that this amount of time needs to be increased. Reference 3–0.18 % coverage, maintains that Fremantle port should establish various time slots for the drivers. Reference 4–0.16 % coverage of the text mentions that the Fremantle port

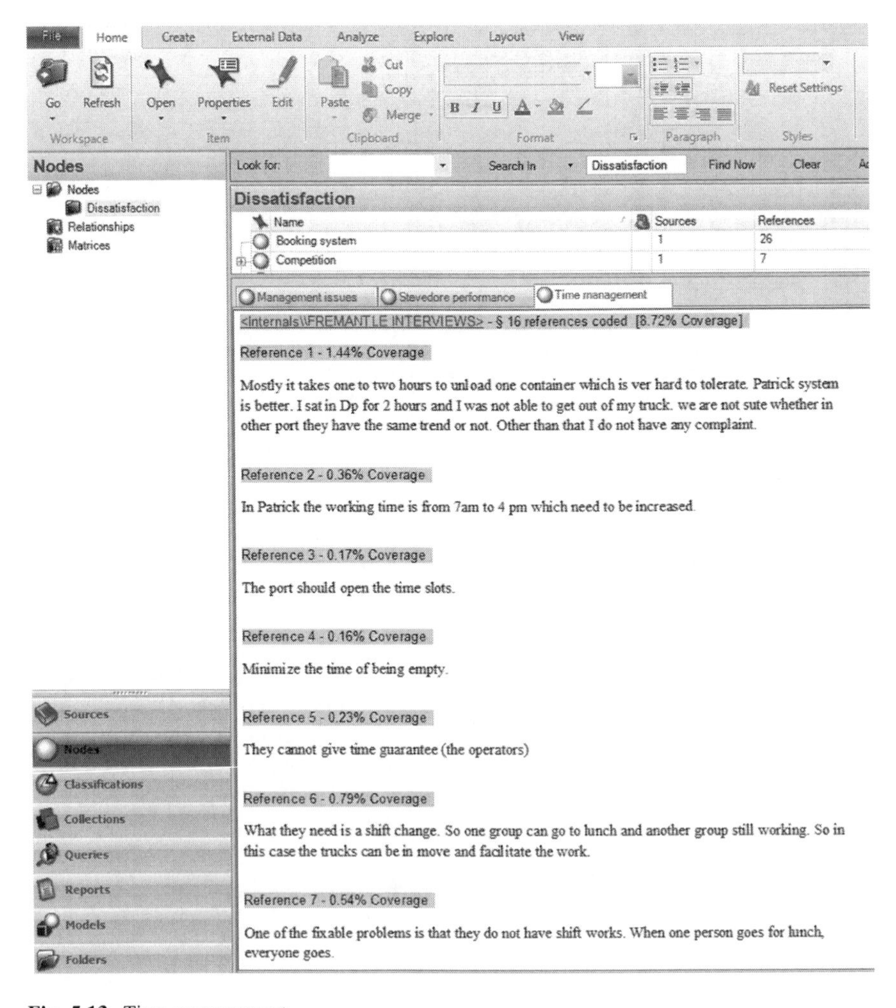

Fig. 5.13 Time management

authority must minimize the time of being empty. Reference 5–0.23 % coverage mentions that operators cannot guarantee the time that it will take for a task to be completed. Reference 6–0.79 % coverage, ascertains that drivers need a shift change, so that one group can go to lunch and another group can work and in this way, drivers might have better motivation to work. Also, this will keep the trucks moving and the work will be carried out more quickly. Reference 7–0.55 % coverage, states that one of the biggest problems that greatly affect drivers is not having flexible shifts. The question remains: why is it that when a client at Fremantle port goes for lunch, everyone else goes as well?

Reference 9–0.25 % coverage claims that even in the middle of their work, employees, especially forklift drivers stop working, regardless of whether or not the job is finished.

Reference 10–0.48 % coverage wonders why clients and operators need so many breaks. Reference 11–0.92 % coverage believes that trucks are always held up for a minimum of an hour each day, thereby wasting a great deal of time.

5.5 Frequency of Keywords, Weight and Importance Used to Help Identify the Issues

Text analysis was used to find word frequencies in the interview texts, the length and significant count of words related to the research area, and the weighted percentage and importance of each word based on the frequency.

The 'length' refers to the number of letters or characters in each word. Length is always generated by the software, and this is one of the options that the software allows the user to select in the text analysis [4].

The 'count' is the frequency of a particular word within the whole text. When running a word frequency query, the output is the occurrence of each word throughout the text.

'Weighted percentage' is defined as "the occurrence of the word relative to the total words counted". The weighted percentage allocates a number of the word's frequency to each category in order not to exceed 100 % [4].

The 'importance' of each word is defined as the relative weighted percentage of each word with respect to the total number of words. This must be estimated for each of the nodes in order to establish the priorities.

As seen in the following table, drivers, truck, container, operators, booking, management, Patrick, road and waiting are the most important words generated by software analysis. The data input to the software were the transcribed interviews that were collected in the first phase of the data collection process. The very first four words—drivers, truck, container and operators—indicate items of importance in the infrastructure. Booking relates to the importance of a booking system at the

Fremantle port. Furthermore, Management refers to the significance of the management at the Fremantle port. The word Patrick was among the outstanding words based on the analysis; however, as Patrick does have better performance in terms of productivity and adeptness, the software selected this to outline the most significant achievement. Also, the right hand column in the table below shows words similar to those on the left, and may be a different part of speech or number (i.e., singular or plural).

Importance % = Weighted percentage/Total weighted percentage

Word frequency is utilized to generate a list of the most frequently used words in the texts transcribed from the interviews. This extracted Table 5.1 is vital for further analysis, as it counts the words and creates a weighted percentage for each separate word. Apart from that, it generates similar words. Utilizing the above table, it will be possible to create a tag cloud, cluster analysis, and tree map. Also, Table 5.1 helps to reinforce the issues identified in the early sections. The number of words has been screened in order to achieve a certain and stringent result in our text analysis. In obtaining the word frequency, the stems of words were considered in order to find similar words. Also, to be able to derive the most frequent words, 200 were selected. This provided better screening, and unnecessary words were excluded. Furthermore, in order to generate words with a minimum length of 5 letters, number 5 was selected. This function generates words of 5 or more letters such as CARGO (5) and COMPLAINT (9).

5.5.1 Tag Cloud on the Fremantle Port Issues

The tag cloud in Fig. 5.14, shows up to 100 words in alphabetical order in different font sizes, where frequently occurring words are in larger fonts. A tag cloud assists us to visually identify and find the word in terms of their importance. A tag cloud or word cloud is utilized to present the significance of the words and weighted list of the words in a meaningful way to gain the attention of the audience. Also, based on QSR, a tag cloud is the ultimate result of word frequency. It develops a means of visualization in order to provide a snapshot of an issue [5]. A tag cloud enables one to visually identify the importance of a word in terms of an issue. The tag cloud above produced by the text analysis enables us to simply identify the word order and its importance based on the size of the word. This tag cloud shows that the word '**drivers**' has priority in terms of strength, indicating that the main issues at Fremantle port concern them. The word, '**truck**' is a dependent criterion and sub-category of time management, and management issues. The word '**booking**' illustrates the importance of a booking system and the strength of its relevance among other issues at the Fremantle port.

Table 5.1 General information regarding weight and importance of words

Word	Length	Count	Weighted (%)	Importance (%)	Similar words
Drivers	7	101	3.55	0.16	Driver, drivers, drivers'
Truck	5	62	2.18	0.10	Truck, trucks
Container	9	61	2.14	0.10	Container
Operators	10	35	1.23	0.06	Operation, operator, operators, operators'
Booking	7	33	1.16	0.05	Booking, bookings
Management	10	27	0.95	0.04	Manage, management, manager, managers, managing
Patrick	7	22	0.77	0.04	Patrick
Roads	5	17	0.60	0.03	Roads
Waiting	7	16	0.56	0.03	Waiting
Traffic	7	13	0.46	0.02	Traffic
Services	8	12	0.42	0.02	Service, services
Change	6	11	0.39	0.02	Change, changes, changing
Crane	5	10	0.35	0.02	Crane, cranes
Customers	9	9	0.32	0.01	Customer, customers
Stevedores	10	9	0.32	0.01	Stevedore, stevedores
Congested	9	8	0.28	0.01	Congested, congestion
Forklift	8	8	0.28	0.01	Forklift, forklifts
Competition	11	7	0.25	0.01	Competition, competitive
Effective	9	7	0.25	0.01	Effective, effectiveness
Infrastructure	14	7	0.25	0.01	Infrastructure
Queue	5	7	0.25	0.01	Queue
Shift	5	7	0.25	0.01	Shift, shifts
Control	7	6	0.21	0.01	Control, controlling
Intelligence	12	6	0.21	0.01	Intelligence, intelligent
Managerial	10	6	0.21	0.01	Managerial
Shops	5	6	0.21	0.01	Shops
Accident	8	5	0.18	0.01	Accident, accidents
Cameras	7	5	0.18	0.01	Cameras
Drink	5	5	0.18	0.01	Drink, drinking
Government	10	5	0.18	0.01	Government
Marketing	9	5	0.18	0.01	Market, marketing
Parking	7	5	0.18	0.01	Parking
Toilet	6	5	0.18	0.01	Toilet
Arrows	6	4	0.14	0.01	Arrows
Cargo	5	4	0.14	0.01	Cargo
Complaints	10	4	0.14	0.01	Complaint, complaints
Lines	5	4	0.14	0.01	Lines, lining
Process	7	4	0.14	0.01	Process, processes
Safety	6	4	0.14	0.01	Safety
Signs	5	4	0.14	0.01	Signs

(continued)

Table 5.1 (continued)

Word	Length	Count	Weighted (%)	Importance (%)	Similar words
Times	5	4	0.14	0.01	Times
Dehydration	11	3	0.11	0.01	Dehydration
Derailment	10	3	0.11	0.01	Derailment
Dusty	5	3	0.11	0.01	Dusty
Health	6	3	0.11	0.01	Health
Interaction	11	3	0.11	0.01	Interaction
Accommodation	13	2	0.07	0.00	Accommodation
Jurisdiction	12	2	0.07	0.00	Jurisdiction
Logistic	8	2	0.07	0.00	Logistic
Petrol	6	2	0.07	0.00	Petrol
Profitability	13	2	0.07	0.00	Profitability, profitable
Quality	7	2	0.07	0.00	Quality
Competitors	11	1	0.04	0.00	Competitors
Complaining	11	1	0.04	0.00	Complaining
Facility	8	1	0.04	0.00	Facility
Relationship	12	1	0.04	0.00	Relationship
Tolerate	8	1	0.04	0.00	Tolerate
Trend	5	1	0.04	0.00	Trend
Total:	465	613	21.64	1.00	

Fig. 5.14 Tag cloud for Fremantle port issues

Table 5.2 Emerging issues at Fremantle port

Emerging issues	Number of references
Management issues	16
Health and safety issues	14
Time management	16
Competition	7
Booking system	26
Jurisdiction issues	8
Container issues	7
Priority setting issues	13
Stevedore performance	5
Intelligent system	1
Cost	4
Infrastructure issues	11
Limitation at the port	43
Total	171

The tag cloud achieves the following:

1. It provides effective visualization of the issues and enables the audience to obtain a clearer perspective of the problem(s) [5]. It also defines the main words in Table 5.2.
2. The length of the words varies from 5 to 14 letters.
3. Also, the count or the frequency of the words varies from one word such as 'facility' to 101 for the word 'driver'.
4. The weight percentage which is the occurrence of the words has a maximum of 3.55 % for 'drivers' and minimum of 0.04 % for 'trend'.
5. The two most important words based on our formulation have been estimated as 0.16 for 'driver' followed by 0.10 for 'truck'. Also, twelve words have been considered as being the least important: accommodation, jurisdiction, logistics, petrol, profitability, quality, competitors, complaining, facility, relationship, tolerate and trend.

In the following section, each of the categories and dissatisfactions will be analysed separately. Using a graph, each of them will be interpreted.

5.5.2 The Issues Identification Through Frequency of Keywords Analysis

Based on Fig. 5.15 and using a qualitative approach, the text-mining method was applied to the collected and screened interviews. The stringent results reveal that customers have many issues and complaints about the Fremantle port. Using the text analysis approach, the complaints were categorized and analysed. Text

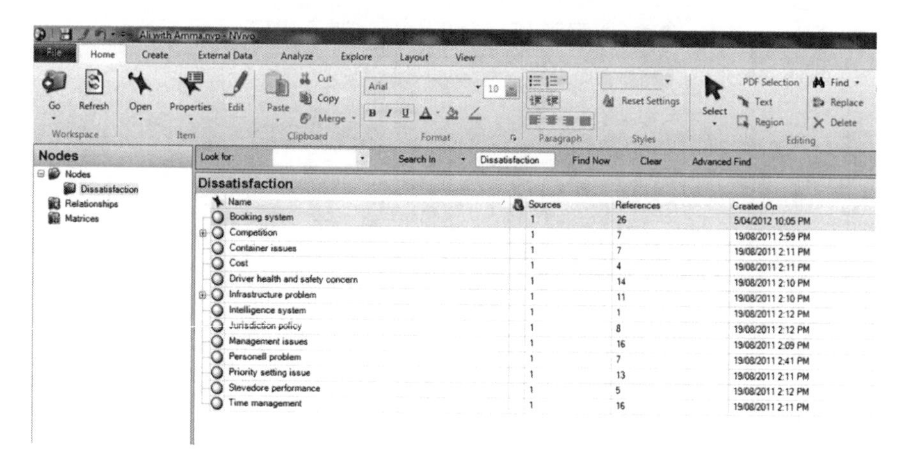

Fig. 5.15 Dissatisfaction node theme

analysis also generates results regarding reference numbers and the amount of repetition of each issue within the whole text. In total, there were 171 references covering thirteen issues (complaints). For example, regarding the management issue, there were sixteen references, meaning that sixteen interviewees or customers discussed this node which was considered as an issue, and this highlights the significance of the finding. Once transcribed, the data was exported to the text-mining tool which formatted and shaped the references. For each reference, the software generated the name, number and percentage of the repetition of the source that was coded as a node. When each separate node is opened, it contains references coded as that node.

Based on the interviews that were conducted in late December 2010 and January 2011 and in accordance with evidence and established facts, the following charts and statements were created. Based on the conceptual framework of this research that was presented in Chap. 4, the complaints were categorized and imported to text analysis software to obtain a comprehensive text analysis of the interview data. The numbers in front of each category indicates the number of references that are related to each of them. The above chart clearly shows that port limitation, which alone is one of the sub-categories of infrastructural issues, is the main cause of driver dissatisfaction. Therefore, it is the most significant complaint. The lack of a proper booking system is the second most important complaint that the drivers have.

Based on Fig. 5.15 and Table 5.2, the following results have been generated respectively:

As discussed, each reference relates to an interviewee, and is in no particular order, there are 16 references related to the management issues, 14 references to health and safety issues, 16 references to time management, 7 to the competition at the port, 26 references discuss the pros and cons of a booking system of the Fremantle port, 8 references present the points for jurisdiction issues and failures in policies at the Fremantle port, 7 references discuss container issues that concern drivers and stevedoring companies, 13 references discuss the priority issues regarding waiting time, lineup, containers' priorities, 5 references mentioned the importance of stevedore performance and the issues, 1 reference talks about the importance of having an intelligent system and as it is different from a booking system.

Drivers who are hardworking clients of the Fremantle port deserve to have essential services such as toilets, and take-away food shops, technical services for vehicle breakdowns and failures, accommodation, and an adequate 24/7 h store at the Fremantle port. In addition, 4 references mentioned the costs and trend of increasing costs at the Fremantle port. Additionally, 11 drivers discuss the main shortcomings of an infrastructure that has been neglected for some time, although some concede that some of the infrastructure is acceptable. Forty-three references discuss the limitation issues at the port which need to be integrated with infrastructural issues.

The various issues concerning drivers are presented in Fig. 5.16. This figure demonstrates that infrastructure issues which are linked with limitation at the Fremantle port are by far the most pressing issues of all and they are raised by 58 references in total. It can be noted that the booking system issue is the second most problematic issue with 25 references. Management and time management remain constant and have an equal number of references of 15 and together are counted as 30 references. Priority setting issues followed by container sorting can be considered under one category and account for 20 references. Jurisdiction is another important issue which comes after priority setting issues that is mentioned by 8 references; however, we considered jurisdiction after container sorting based on the previous statement. Next, and arranged in order, are the issues of competition, stevedore performance, cost and intelligent system.

Figure 5.17, is based on Fig. 5.16.

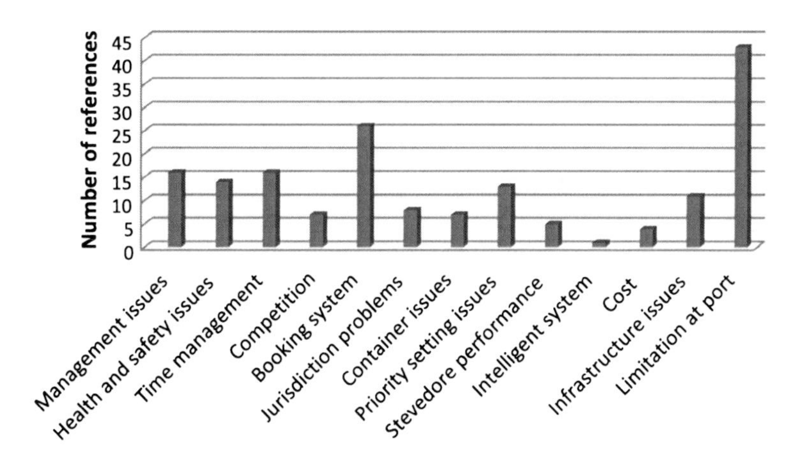

Fig. 5.16 Emerging issues at the Fremantle port based on the importance

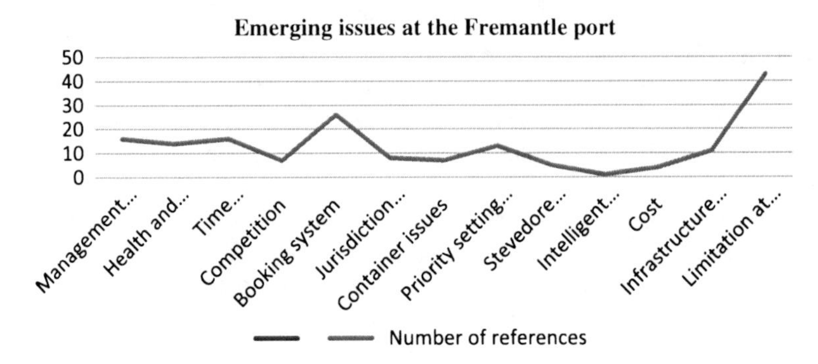

Fig. 5.17 Emerging issues at the Fremantle port

Figure 5.17 is based on the results of the text analysis. Complaints regarding the major problems at the Fremantle port, which translate into dissatisfaction, show the shortcomings of each of these areas of operation which must be addressed.

5.5.2.1 Conceptual Representation of the Fremantle Port Issues Based on Text Analysis

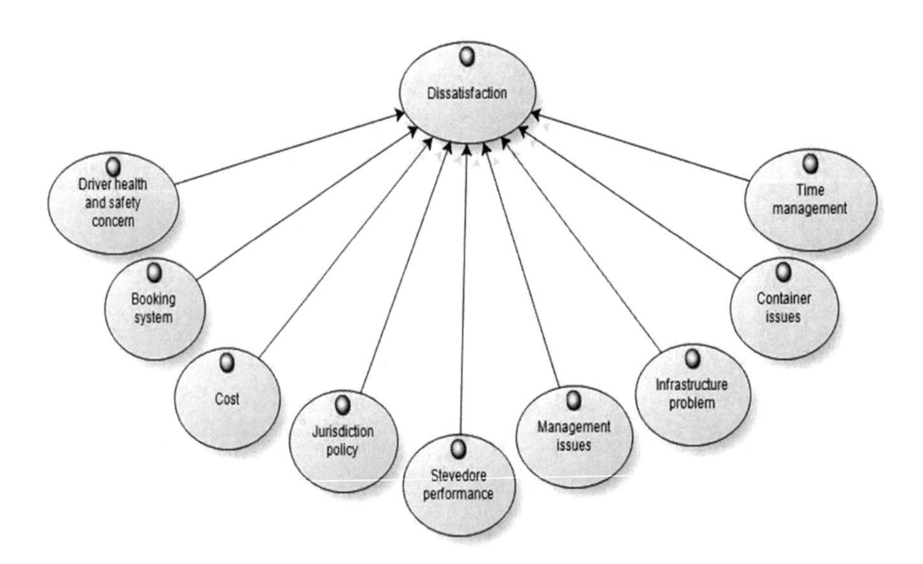

Fig. 5.18 Model of key issues at the port

The above model [6] illustrates the issues of concern at the Fremantle port. The model depicted in Fig. 5.18 was obtained from software. As discussed in Chap. 3 (Problem definitions), the project was carried out at the Fremantle port to evaluate customer complaints and customer satisfaction using customer relationship management strategies. The diagram summarizes the various customer issues which emerged after their feedback was categorized, grouped into complaints, compliments or neutral comments. Based on the problem defined in Chap. 3, only customer complaints were considered and evaluated. Recommendations will be made with the aim of reducing the number of customer complaints and increasing customer satisfaction. The complaints which our model identifies as the main causes of dissatisfaction with Fremantle port are: drivers' health and safety, booking system, cost, jurisdiction, stevedore performance, management, infrastructure, container and time management.

The above model is derived from the text analysis. Applying a lab software text-mining analysis tool to the transcribed interviews, a certain number of words (sub-nodes) was established which related to the main node. These words are associated with dissatisfaction and each has a one-way relationship with the dissatisfaction node. Put another way, these nodes have a direct correlation and association with the dissatisfaction node. The above model displays the results of a word frequency query in a tag cloud and reinforce the tag cloud [5].

5.6 Confirmation of Issue Interpretation Through Weight and Importance

In this section, we discuss in detail the most significant customer complaints derived from the analysis. We utilize a table and figure for each major complaint to illustrate the significance of including relevant words and to make them easy to interpret.

5.6.1 Management Issues

Management is an integral part of each organization, without which no section can perform effectively. The main objective of having a management process in an organization is to implement and control procedures. It can also control operational, front office and back office activities.

An efficient management could prevent traffic and operational congestion at the port. Furthermore, people will have a better environment in which to develop and display better interpersonal skills. An effective management can optimize the performance of clients and promote sustainability and social contribution. Furthermore, having knowledgeable management at the port is integral to all of the activities as it creates better flow in trade and businesses whose needs will be more efficiently met. It can establish the port as an invaluable facilitator and contributor in individual lives. Management issues may reveal new aspects of an organisation such as scandals and environmental contamination, and these alone may impact on society.

To achieve the significance of each element involved in management issues, the following formula is used:

Importance of each word = weighted percentages of each word/total sum

$$X_{\text{importance}} = \%_{\text{Weight}} / \text{Sum}_{0-\infty}$$

Table 5.3 Management issues

Management Issues	Weight	Importance
Accident	0.18	0.02
Process	0.14	0.02
Marketing	0.18	0.02
Drivers	3.60	0.42
Control	0.21	0.02
Container	2.17	0.25
Competition	0.25	0.03
Queue	0.25	0.03
Complaint	0.18	0.02
Change	0.39	0.05
Management	0.96	0.11
Total	8.51	1.00

Table 5.3 illustrates the pertained words to the management issues followed by their weights and importance. As can be seen in Fig. 5.19, the graphs have various fluctuations which appeared mostly in drivers, container, competition and management. Based on the text analysis, these four words have the highest importance in relation to the rest of the words. 'Driver' has the highest frequency, indicating

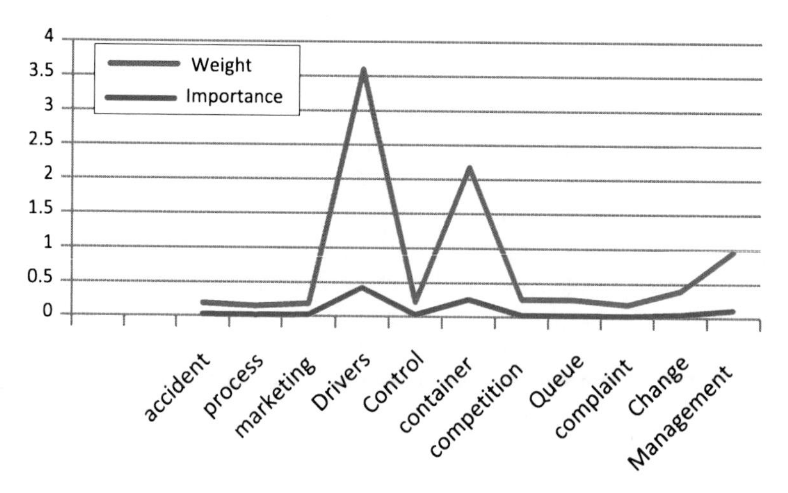

Fig. 5.19 Weight and importance of management issues at Fremantle port

that logistics companies must pay more attention to them and care for their welfare. Also, the majority of the sample customers were drivers who had problems with their containers, hence its importance as the second highest priority.

5.6.2 Health and Safety Issues

Truck drivers generally drive considerable distances and on reaching their destination, they are required to deal with loading and unloading cargo in different areas which may be urban rural or remote. In order to assist drivers to have peace of mind and mental well-being, the port must establish and implement responsible and adequate health and safety measures. The port authority, and the companies which operate within the port, need to take action to create better health and safety conditions for their clients. If the drivers are able to work in a safe and healthy environment, they have a better chance of resisting stress and stress-related diseases. Sometimes, the drivers' environment is polluted by toxic chemicals which they inhale. There should be preventative measures put in place to minimise the amount of toxins that can infiltrate the cabins.

In order to ascertain the importance of each factor pertaining to health and safety issues, the following formula is applied:

Importance of each word = weighted percentages of each word/total sum

$$X_{\text{importance}} = \%_{\text{Weight}} / \text{Sum}_{0-\infty}$$

Table 5.4 Health and safety issues

Health and Safety Issues	Weight	Importance
Drink	0.18	0.25
Dehydration	0.11	0.15277778
Toilet	0.18	0.25
Health	0.11	0.15277778
Safety	0.14	0.19444444
	0.72	1

According to the results derived from our analysis and based on Table 5.4 and Fig. 5.20, shortage of drink and lack of enough toilets at Fremantle port are of highest significance. Safety is the second important issue in this part. This is followed by concern at the lack of availability of appropriate beverages, which could lead to dehydration especially in hot weather. What is more, drivers do have several health issues at the port which need to be investigated and addressed.

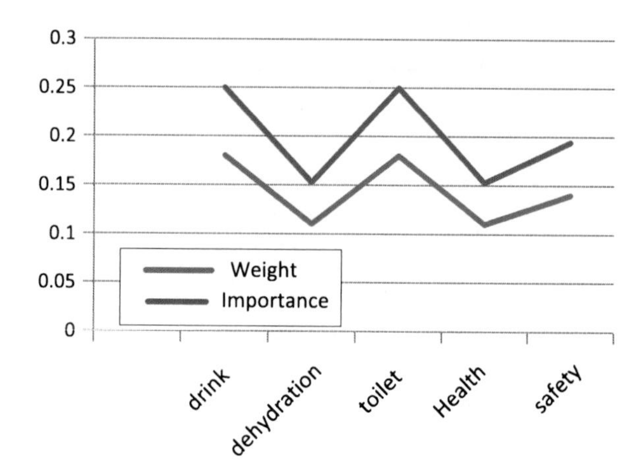

Fig. 5.20 Weight and importance of health and safety issues in Fremantle port

5.6.3 *Time Management*

Time management can be defined as a control over the allocation of time and spending time on various procedures and activities to enhance productivity and efficiency of an organization.

There are collaborative evidences that state that an expert time management team must be in charge of this process in every organization. Based on [7], decision makers are falling short in their planning in terms of loading and unloading. According to [8], time plays a crucial role in administering business procedures and having a legitimate and just-in-time management system will help a company to improve its workflow and decrease the cost of business procedures. The time that needs to be taken for each procedure can be calculated and an appropriate workflow scheme can be set up for each business process.

Using time management, the duration of each process and deadlines can be determined and managers can trace any improvements or failures to improve the procedures or to rectify the issues [8]. Due to the time constraints for this thesis, it is not feasible to propose solutions to all the issues which emerged during the course of this research. However, the most significant complaints are targeted and solutions are proposed. In the analysis section, we try to find the key customers to address the major problems of individuals. Key customers may provide more credible information regarding the issues.

To determine the importance of each related factor in time management, the following formula is used:

Importance of each word = weighted percentages of each word/total sum

$$X_{\text{importance}} = \%_{\text{Weight}} / \text{Sum}_{0-\infty}$$

The table and figure above clearly depict the issues involved in time management. Table 5.5 displays the related words produced by the analysis software. Figure 5.21 demonstrates that the most time consuming issues at Fremantle port

Table 5.5 Time management

Time management	Weight	Importance
Tolerance	0.04	0.00501253
Change	0.39	0.04887218
Shift	0.25	0.03132832
Truck	2.18	0.27318296
Facility	0.04	0.00501253
Drink	0.18	0.02255639
Booking	1.16	0.14536341
Queue	0.25	0.03132832
Forklift	0.28	0.03508772
Interaction	0.11	0.01378446
Container	2.14	0.26817043
Time management	0.14	0.01754386
Parking	0.18	0.02255639
Camera	0.18	0.02255639
Traffic	0.46	0.05764411
Total	7.98	1

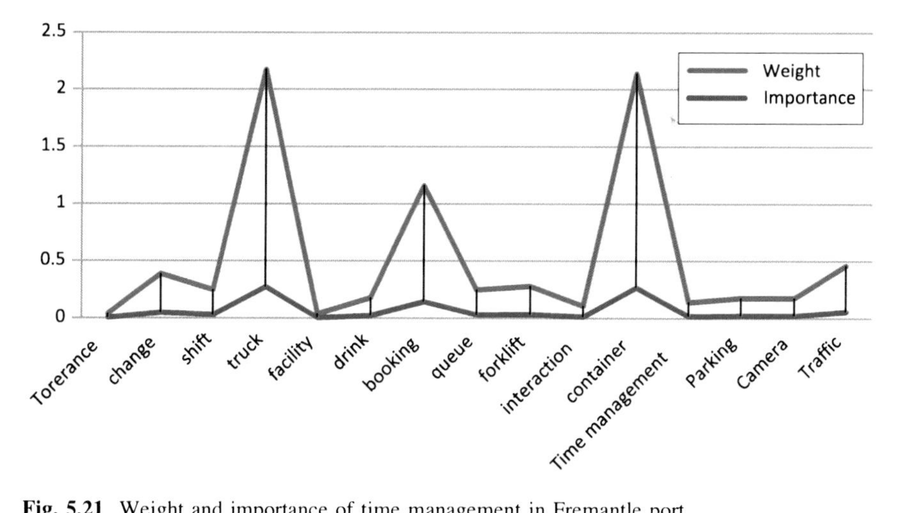

Fig. 5.21 Weight and importance of time management in Fremantle port

are to do with trucks and containers. Booking is the second most important thing that must be planned in advance.

5.6.4 Infrastructure Issues

Infrastructure is the base for each industry to provide strategic requirements of that industry. Like other infrastructure problems, Fremantle port is being challenged by enormous issues, complaints and dissatisfactions. Port authorities and stevedoring companies must propose strategic approaches to find remedies for the various undesirable situations. Infrastructural issues can prevent the ports and wharfs at Fremantle port from moving forward into the future. Infrastructure issues vary with different ports. In order to obtain some insight into infrastructure issues, the frequency of related words and related limitations was obtained. Then, the following formula was applied to obtain the importance of each factor.

Importance of each word = weighted percentages of each word/total sum

$$X_{\text{importance}} = \%_{\text{Weight}} / \text{Sum}_{0-\infty}$$

Table 5.6 Infrastructure issues

Infrastructure issues	Weight	Importance
Truck	2.18	0.24088398
Container	2.14	0.23646409
Road	0.60	0.06629834
Traffic	0.46	0.05082873
Services	0.42	0.04640884
Crane	0.35	0.03867403
Forklift	0.28	0.03093923
Shop	0.21	0.02320442
Camera	0.18	0.0198895
Drink	0.18	0.0198895
Parking	0.18	0.0198895
Toilet	0.18	0.0198895
Arrow	0.14	0.01546961
Cargo	0.14	0.01546961
Line	0.14	0.01546961
Sign	0.14	0.01546961
Derailment	0.11	0.0121547
Accommodation	0.07	0.00773481
Petrol	0.07	0.00773481
Facility	0.04	0.00441989
Congested	0.28	0.03093923
Waiting	0.56	0.06187845
Priority no. 1	9.05	1

Table 5.6 and Fig. 5.22 list the most related words derived from text mining. ‘Truck’, ‘container’, ‘road traffic’, ‘waiting time’ and the congested area around Fremantle port are among the most significant issues that need to be rectified. The remaining issues (words) are integrated into the major issues in the following:

Basically, truck drivers are facing a plethora of challenges and difficulties. For instance, based on Fig. 5.22, drivers are paid by the cargo load, not by the hour. They do not have competent and appropriate services at Fremantle port such as rest rooms, so the port authorities must provide amenities for the drivers who plan their trip based on assigned stops.

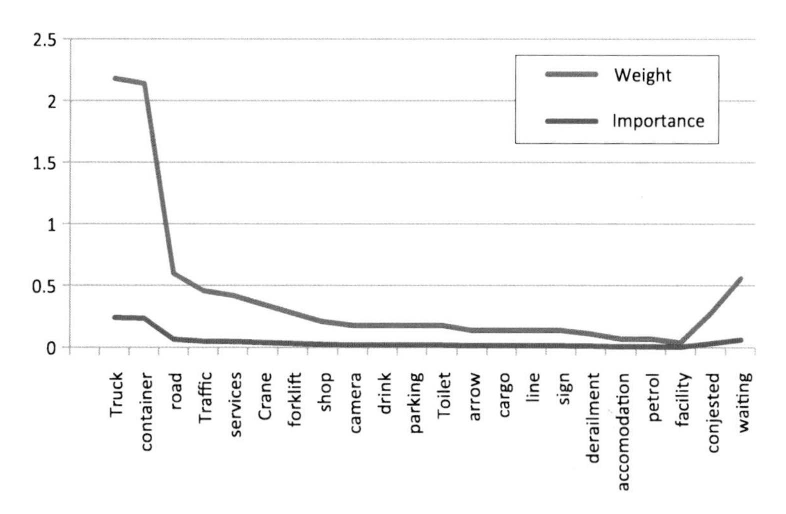

Fig. 5.22 Weight and importance of infrastructure issues in Fremantle port

Fig. 5.23 Container loading

Drivers also have various container issues. One of the significant priorities that the authorities must attend to is trucking efficiency and productivity. What is more, those who are in charge of loading and unloading containers must be aware of the container weight and standard capacity of the trucks, as some of the trucks are overloaded. However, in Fremantle port, we can notice trucks with weigh-down beds, as most of the trucks are flatbed semi-trailers; there should be someone to check the extra weight of the trailers. Also, there are tandem tractor trailer that their weights need to be standard for the sake of road safety and avoid danger. A forecasting team is needed to address this issue.

Figure 5.23 shows a clear sample of loading a container using a forklift at the Fremantle port.

Another problem for all drivers is that there are not enough parking lots. There is the need for a new parking lot to be built inside the port. Also, authorities could extend working hours and a better line-up system for truck activities. Another issue is congestion and traffic on the roads leading to Fremantle port. Traffic always needs an expert management system to ensure smooth traffic flow.

A separate parking area for employees and truck drivers would instantly alleviate road congestion as well as costs. There would no longer be the need to park in residential and industrial areas, and the problem of drivers incurring parking fines at the port would be eliminated. Authorities need to address the traffic issue by providing parking areas within the port precinct or building separate roads for heavy vehicles; traffic congestion on roads would be decreased and the surrounding residential areas would be more livable.

There should be ample clear signs and arrows to guide drivers to various locations around and inside the Fremantle port.

Last but not least, the waiting time that drivers have to put up with is a highly significant issue and a major cause of dissatisfaction. Not only do drivers have to wait in a queue to get their containers; they are also required to stay in their dock when loading or unloading their trucks.

5.6.5 Booking System

A good booking system creates flexibility and proper management and a driver can simply make his/her reservation. A booking system identifies customers and their profiles and customers must rely on the system. There would be no fee for a booking system at the port and people could simply register themselves.

A booking system will provide accurate pick-up and delivery times for the drivers. Also, they will have a planned loading and unloading time which will improve the efficiency and continuous performance of both port employees and drivers.

A booking system may eliminate the mistakes that previously occurred with manual recording. It will allow people to simply refer to the system for information, rather than having to contact relevant personnel, provided that they secure the system. The current lack of a vehicle booking system is one of the major problems at the port, particularly for companies such as DP World and Patrick.

A VBS or vehicle booking system provides an automated process by which the truck drivers can be re-scheduled for all sorts of activities that are carried out at the port. Furthermore, a vehicle booking system can help decrease congestion and the traffic in the roads leading to and from the port. The following formula is used to obtain the significance of each factor in the booking system:

Importance of each word = weighted percentages of each word/total sum

$$X_{\text{importance}} = \%_{\text{Weight}} / \text{Sum}_{0-\infty}$$

According to Table 5.7, ‘drivers’ followed by ‘container’ and ‘booking system’ are among the most important related words. As in Fig. 5.24, these words have the

Table 5.7 Booking system

Booking system	Weight	Importance
Lines	0.14	0.01601831
Waiting	0.56	0.06407323
Booking	1.16	0.13272311
Forklift	0.28	0.03203661
Patrick	0.77	0.08810069
Driver	3.55	0.40617849
Container	2.14	0.24485126
Complaint	0.14	0.01601831
Total	8.74	1

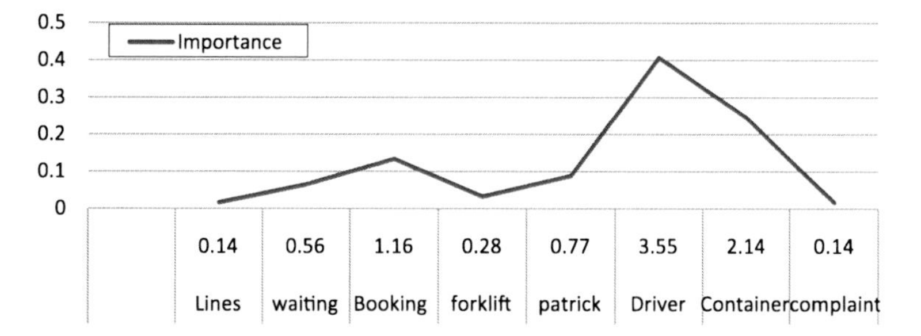

Fig. 5.24 Importance of booking system in Fremantle port

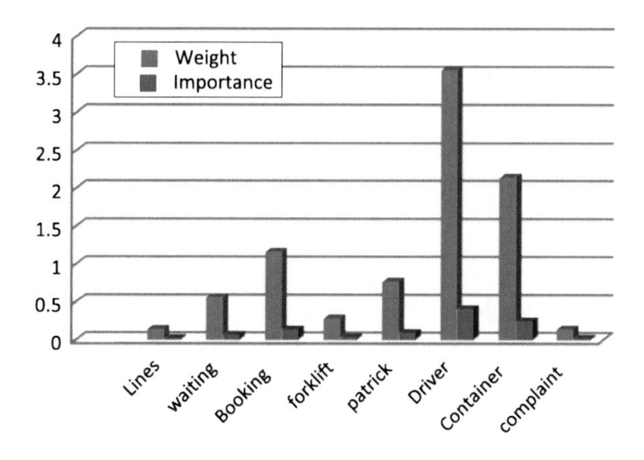

Fig. 5.25 Weight and importance of booking system in Fremantle port

highest importance based on the result produced by the software. Fremantle port and logistics companies must provide strategies to improve the workplace.

Figure 5.25 depicts the weight and the importance of each factor (word) in the booking system issue.

5.6.6 Cost

Cost has always been considered as one of the most important issues at Fremantle port and in the following, we will discuss the reasons why it is significant and why companies should remedy the various situations.

To ascertain the importance of each element included in cost issues, the following formula is applied:

Importance of each word = weighted percentages of each word/total sum

$$X_{\text{importance}} = \%_{\text{Weight}} / \text{Sum}_{0-\infty}$$

Table 5.8 Cost

Cost	Weight	Importance
Shop	0.21	0.28
Drink	0.18	0.24
Dehydration	0.11	0.14666667
Interaction	0.11	0.14666667
Accommodation	0.07	0.09333333
Petrol	0.07	0.09333333
Total	0.75	1

Table 5.8 and Fig. 5.26 indicate that words such as ‘shop’ and ‘drink’ are of higher superiority. Drink is a common word related to other issues too. In fact, lack of a proper drink especially in hot weather, is a big problem that must be recognised by those in charge. Lack of a proper drink may have negative impact and side effect on the customers like dehydration which have been discussed previously. Therefore, those in charge should provide cold and hot water, and other beverages. One solution would be to install vending machines at different parts of the port. Furthermore, interaction and communication is a common problem and often explains why companies cannot respond appropriately to customer needs. A concerted effort must be made to establish various connections and good relationships with customers, regardless of the initial costs.

Accommodation is another issue for the drivers, as those who do not have overnight accommodation need to find a decent place to stay and have a safe rest. However, it would be very costly for Fremantle port authorities to establish accommodation for the drivers. Finally, the petrol station at Fremantle port also presents a cost issue, and again, the building of a new petrol station would be very expensive.

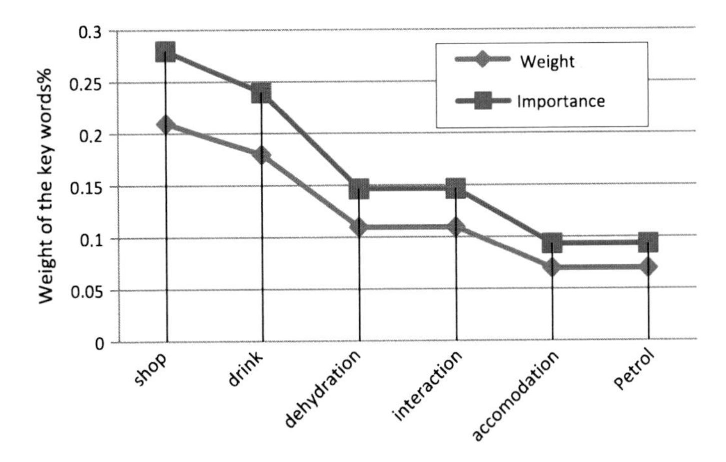

Fig. 5.26 Weight and importance of cost in Fremantle port

5.6.7 Container Issues

Truck drivers have issues regarding their containers including container sorting, container labelling and finding the container without any delay. While the port is developing in terms of environment, facilities and expertise inside, there should be some strategies to help drivers with their container issues. Also, some of the drivers may encounter problems with overweight containers.

Based on our analysis, there are some significant problems that need to be considered in order to improve container issues.

In order to calculate each factor in container issues, the following formula is used:

Importance of each word = weighted percentages of each word/total sum

$$X_{\text{importance}} = \%_{\text{Weight}} / \text{Sum}_{0-\infty}$$

Table 5.9, shows the most important and related words based on weight and importance.

Table 5.9 Container issues

Container issues	Weight	Importance
Container	2.14	0.59279778
Waiting	0.56	0.15512465
Crane	0.35	0.09695291
Forklift	0.28	0.07756233
Cargo	0.14	0.03878116
Safety	0.14	0.03878116
Total	3.61	1

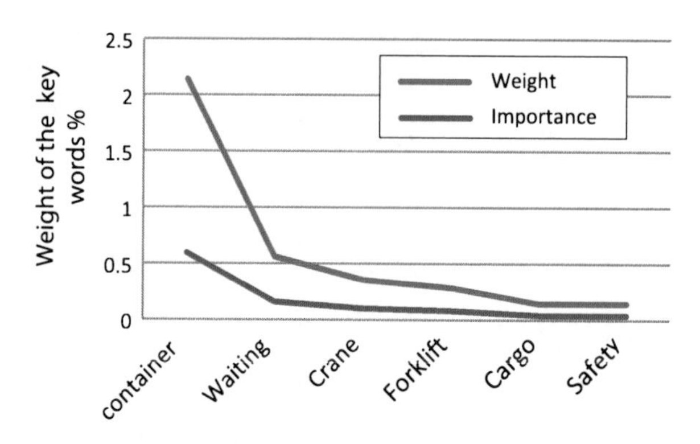

Fig. 5.27 Weight and importance of container issues at Fremantle port

Fig. 5.28 Siwertell unloaders, Kwinana Bulk Jetty

Regarding Fig. 5.27, in the container issue section, words such as 'container', 'waiting', 'crane' and 'forklift' have high importance and issues to do with waiting and resting must be addressed using updated strategies and methods. 'Container' is a factor with the highest trend of all. The problem with the cranes and similar infrastructure equipment is that they are mostly owned by the government, whereas if a private company owned the cranes, this would provide better competition at Fremantle port and might decrease their cost. Figure 5.28 shows Siwertell unloaders at Kwinana bulk jelly. They are mainly designed for handling cement. They are being utilised as a mechanical screw type pneumatic and mobile installation. It completely meets the requirements of the Ports at Western Australia.

5.6.8 Stevedore Performance

A stevedore is a wharfie or dockworker who has various jobs such as loading and unloading containers from trucks to ships. Stevedoring companies can be big or small and may be responsible for a carrying out a variety of jobs according to the

Table 5.10 Stevedore performance

Stevedore performance	Weight	Importance
Stevedore	0.32	0.22535211
Forklift	0.28	0.1971831
Effective	0.25	0.17605634
Queue	0.25	0.17605634
Control	0.21	0.14788732
Quality	0.07	0.04929577
Relationship	0.04	0.02816901
Total	1.42	1

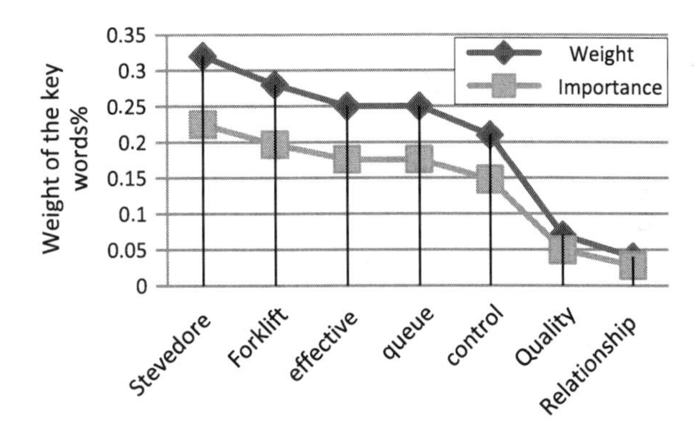

Fig. 5.29 Weight and importance of stevedore performance in Fremantle port

size of the port and its location in the world. Port and company rules and regulations significantly affect the performance of stevedores.

A stevedore normally works at a ship terminal and is responsible for the effective and safe loading and unloading of ships and trucks [9].

Stevedores must have good communication and interpersonal skills and demonstrate high standards of accountability in the work setting. They are individuals with flexible working hours and must be able to perform in different conditions in terms of weather and location.

To enhance the working standards, stevedoring companies need to provide timely service to the customers 24 h a day. Also, they must provide time slots for the customers and offer flexibility. The total importance of stevedore performance is arrived at using the following formula and the total importance based on percentage must equal 100 %.

Importance of each word = weighted percentages of each word/total sum

$$X_{\text{importance}} = \%_{\text{Weight}} / \text{Sum}_{0-\infty}$$

Based on Table 5.10 and Fig. 5.29, 'stevedore' (0.32) and 'forklift' (0.25) have the highest weight in comparison with the other words. The word 'effective' is another important word that illustrate that there is lack of efficiency and effectiveness in Fremantle port. Simultaneously, 'queue' is a word which is common in various issues and is a problem that seems to be inevitable and intensifying with each day and is different in different seasons. 'Control' is another word related to this issue. For better management of the Fremantle port, there must be a stronger security system to control different areas of the port and strengthen security. Even the staff in the security department must be stringent in order to further improve the quality of the services.

In the interviews, some drivers complained about the quality of the services provided at Fremantle port. Ultimately, the relationship between employees and customers needs to be improved. This can be done using customer relationship management strategies which have been outlined earlier in this chapter. Various methods must be applied to improve customer satisfaction. Organizations need to identify the basis of each complaint and address the issue, thereby improving their relationship with customers.

5.6.9 Jurisdiction Issues

Proper, consistent jurisdiction and regulation measures are vital at the Fremantle port. Currently, there are a variety of policies at the port but these are inconsistent depending on the companies concerned, with some drivers being disadvantaged. As an example, drivers with Western Australia's accreditation are not recognized interstate. Also, different states have various rules and regulations and the truck drivers want consistency.

Table 5.11 and Fig. 5.30 show the words related to jurisdiction at the Fremantle port. As can be noted, 'truck' (2.18) and 'Patrick', (0.77) are the highest weights among the components of jurisdiction followed by 'road', 'process', 'competitors' and 'facility'. These are the most significant words in jurisdiction issues. In order to provide improvement in terms of jurisdiction, the organizations need to effectively work with government to improve efficiency. This cannot be achieved without having intelligent processes along with knowledge and expertise at the Fremantle port. There is a need for new strategies, rules and regulations to ensure the satisfaction of the truck drivers. One way to achieve this is to establish new, or

Table 5.11 Jurisdiction issues

Jurisdiction issues	Weight	Importance
Truck	2.18	0.567708
Road	0.60	0.15625
Jurisdiction issues	0.07	0.018229
Patrick	0.77	0.200521
Competitors	0.04	0.010417
Process	0.14	0.036458
Facility	0.04	0.010417
Total	**3.84**	**1**

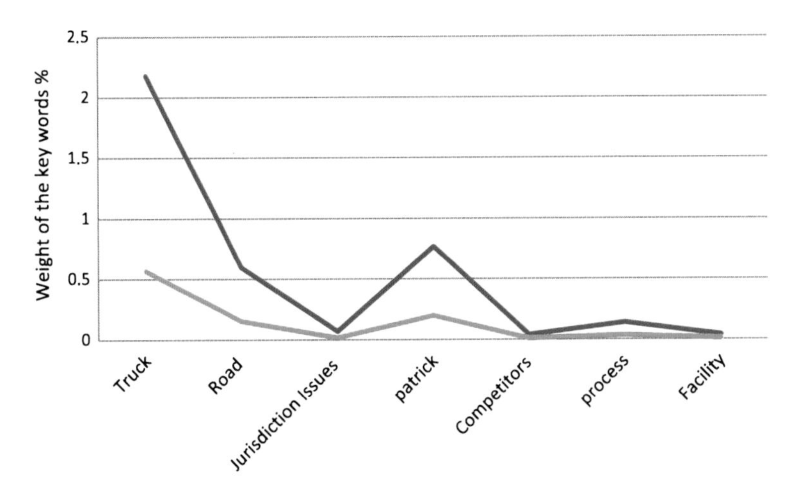

Fig. 5.30 Weight and importance of jurisdiction issues

improve existing, facilities for drivers at the port. Moreover, WA's HV accreditation needs to be recognized nationally and by all states.

5.7 Significance of Text-Mining Analysis

Text analysis is a text mining methodology. A qualitative model was developed and the impact of each customer complaint was ascertained. Therefore, the significance of the contribution can be summarized as:

1. Extensive interviews were conducted with customers of the Fremantle port to help identify customer dissatisfaction.
2. Issues were computed in natural language using text analysis techniques.
3. An advanced concept related to the text analysis was applied in analysis of interview texts.
4. The impact of each issue generated was qualified according to the number of references for each of the issues.
5. A lab software tool was utilized to assist with reference and the coverage definitions.
6. Each of the issues was analysed and its impact factor represented.
7. Word frequency query was used to confirm the length, count, weighted percentage, importance and similar words related to each issue.
8. A tag cloud was generated to visually present words according to their level of importance.
9. Issues were automatically prioritized so that they could be used in satisfaction modelling (Chap. 7); they were selected in this chapter in order to prevent bias.

5.8 Conclusion

In this chapter, we used the interview approach to obtain information from our interviewees who are mainly clients of the Fremantle port. We completed the data collection, then text analysis. We categorized the interview data, then identified and ranked the most important issues at Fremantle port. We determined the weight and importance of the key words, under each category. We examined the relationship between the key issues. We also performed a word frequency query in order to generate a tag cloud, and visually mapped the nodes. Utilizing tables and charts, we separately illustrated each of the complaints with their associated words and analysed them in detail.

The next chapter provides a further analysis of the issues, and the hypotheses discussed in Chap. 4 are tested.

References

1. Sekaran, U. (2006). *Research methods for business: A skill building approach.* India: Wiley.
2. Port, F. (2012). *How the port works?* Available: http://www.fremantleports.com.au/About/WorkingPort/Pages/default.aspx.
3. WABusinessnews. (2009). *Derailment.* Available: http://www.wabusinessnews.com.au/user?destination=node/73915.
4. Q. International. (2011). *Run a word frequency query.* Available: http://help-nv9-en.qsrinternational.com/nv9_help.htm#concepts/about_cluster_analysis.htm.
5. Q. International. (2011). *NVIVO 9.* Available: http://download.qsrinternational.com/Document/NVivo9/NVivo9-Getting-Started-Guide.pdf.
6. Faed, A., Hussain, K., & Chang, E. (2012). *Linear and fuzzy approaches for customer satisfaction analysis, service oriented computing and applications.* New York: Springer, Under review.
7. Li, J. A., Liu, K., Leung, S. C. H., & Lai, K. K. (2004). Empty container management in a port with long-run average criterion*. *Mathematical and Computer Modelling, 40*, 85–100.
8. Eder, J., Panagos, E., Pozewaunig, H., & Rabinovich, M. (1999). Time management in workflow systems. *BIS, 99*, 265–280.
9. P. o. B. P. Ltd. (2012). *STEVEDORE.* Available: http://www.portcareers.com.au/careers/stevedore/.

Chapter 6
I-CRM Validation of Customer Issues and Definition of Their Impact Factors Through Hypotheses Formulation

6.1 Introduction

In the previous chapter through text mining analysis of interview data, the main issues that produce customer dissatisfaction with the Fremantle port operations were identified. Furthermore, the major issues were ranked according to their importance and weights. In this chapter, the identified issues and their impact factors are validated through hypotheses formulation and proof of concept. Analysis will be carried out using variance and structural equation modelling. In the previous chapter, nine issues were ranked which are considered as variables and these are used as inputs. Statistical testing is performed to ascertain the impact of certain factors on CRM performance through these variables.

6.2 Questionnaire to Help Hypotheses Formulation

In order to formulate the hypotheses, firstly, data was collected from Fremantle port by means of a questionnaire survey. Two sets of hypotheses were tested. One set related to the issues at the Fremantle port and the other set related to the hypotheses and I-CRM variables.

The proposed hypotheses are related to the nine variables or issues that were identified in the last chapter. Each hypothesis is considered to be a one-way relationship between dissatisfaction and issues at the Fremantle port, namely the issues regarding: management, booking system, infrastructure, health and safety jurisdiction, stevedore performance, container sorting, costs and time management.

In order to formulate the hypotheses, a questionnaire survey was conducted to acquire more reliable information related to these nine variables. Likewise, more detailed information was needed about I-CRM variables such as perceived value, interactivity, customer satisfaction, loyalty and customer acquisition. Hence, questions were formulated pertaining to each variable.

A. R. Faed, *An Intelligent Customer Complaint Management System with Application to the Transport and Logistics Industry*, Springer Theses, DOI: 10.1007/978-3-319-00324-5_6,

© Springer International Publishing Switzerland 2013

The questionnaire has six sections with a total of sixty-seven questions related to the Fremantle port issues. For each section of the questionnaire, we achieved different Cronbach's alpha to ensure the reliability of the questionnaire and scales. Also, in order to ensure the validity of the questionnaire, the opinions of experts, professors and supervisors were sought, and their input was used to validate the content. Below, we discuss each section of the questionnaire in depth.

Section 1—Demographic: In this section, the respondents are required to choose one of several alternatives. This section has eight questions pertaining to age, gender, educational background, position, employment, vicinity to the port, familiarity with the culture, and personal characteristics. Two examples are provided below. The complete questionnaire is attached in Appendix 4.

- What is your position?

1. A novice
2. Driver and operator
3. Experienced driver
4. Driver and owner.

- How familiar are you with the working culture in WA?

1. Poor
2. Fair
3. Good
4. Very good
5. Excellent.

Section 2—Customer satisfaction: This section is the focal point and focus of our study. It has thirty-seven questions and each question is related to an issue. In this section, respondents need to choose only one alternative based on a 7-point Likert scale ("1" as Strongly dissatisfied, "2" as Mostly dissatisfied, "3" as Somewhat dissatisfied, "4" as Neutral, "5" as Moderately satisfied, "6" as Very satisfied and "7" as Extremely satisfied). 7-point likert scale has been chosen to measure the variables involved in conceptual framework. However, there is only one question which is related to the willingness of the truck drivers to remain in the queue and must be answered based on four different answers (the question will be provided in the section below). In this section, questions without any particular order are related to management, time management, booking system, jurisdiction, infrastructure issues, costs, health and safety issues, stevedore performance, and container sorting which have been derived from the text analysis presented in Chap. 5 and which will be validated in Chap. 6. In this thesis, due to the time limit, our focus is mainly on customer satisfaction. The reliability achieved for this section based on Cronbach's alpha was 0.809 which is reliable. Below, we provide some sample questions from this section. The full questionnaire is available in Appendix 4.

- How long are you willing to stay in a queue while you are waiting?

1. 15–20 min
2. 30 min
3. More than 30 min
4. More than 1 h.

Numbers	Level of satisfaction based on 7-point Likert scale
1	Strongly dissatisfied
2	Mostly dissatisfied
3	Somewhat dissatisfied
4	Neutral
5	Moderately satisfied
6	Very satisfied
7	Extremely satisfied

- The existing health and safety support at Fremantle port.
- The road regulation and the sufficiency and flexibility of operational rules and regulations at Fremantle port.

Section 3—Customer loyalty: The questions here are completely related to customer loyalty. The questions are intended to improve the loyalty of the truck drivers at the Fremantle port. The five questions in this section are intended to elicit responses regarding customer referral, switching to a new company, current employment status, the probability that the customer will continue working with the same company and word of mouth. The respondents need to answer the questions based on a 7-point Likert scale ("1" as Extremely unlikely, "2" as Very unlikely, "3" as Somewhat unlikely, "4" as Neutral, "5" as Somewhat likely, "6" as Very Likely and "7" as Extremely likely) that are provided in the section below. Also, the reliability for this part is 0.91 which is consistent and reliable. Below, we provide a sample question from our questionnaire. The complete questionnaire is available in Appendix 4.

Number	Customer loyalty
1	Extremely unlikely
2	Very unlikely
3	Somewhat unlikely
4	Neutral
5	Somewhat likely
6	Very likely
7	Extremely likely

- Would you recommend the stevedores to your friends who are seeking a career?
- How likely are you to switch from your present job?

- Are you employed? How long have you been with your current employer/3rd transportation provider?

1. less than 1 year
2. 1–4 years
3. 5–10 years
4. more than 10 years.

Section 4—Perceived value: In this section, the questions pertain to the knowledge of employees at the port, provision of information, cleanness of the port, salary of the truck drivers, provision of services, quality provided, and improvement at the port. We have 7 questions in this part. In this section, we also have two questions which are related to infrastructure and management at the port. We used a 7-point Likert scale to measure each question ("1" as Totally unacceptable, "2" as Unacceptable, "3" as Slightly unacceptable, "4" as Neutral, "5" as Slightly acceptable, "6" as Acceptable and "7" as Perfectly acceptable). The reliability measure for this section of questionnaire was done separately and the result was 0.76 which is reliable.

Number	Perceived value
1	Totally unacceptable
2	Unacceptable
3	Slightly unacceptable
4	Neutral
5	Slightly acceptable
6	Acceptable
7	Perfectly acceptable

- The personnel at the port have enough knowledge about their jobs.
- The personnel provide necessary information to the customers and make them content.
- How clean and spacious is your work space?

Section 5—Interactivity: The seven questions in this part are related to the relationship between customers and employees at the port, or collegiality at the port. Also, one question was intended to measure the adequacy of interaction between the staff and customers at the port. We were also curious to know whether customers received from the staff prompt responses to queries or complaints. Respondents answered questions using a 7-point Likert scale ("1" as Strongly disagree, "2" as Mostly disagree, "3" as Somewhat disagree, "4" as Neither agree or disagree, "5" as Somewhat agree, "6" as agree and "7" as Strongly agree). The reliability of this section is 0.88. Below are two sample questions regarding the interactivity section. The complete questionnaire is provided in Appendix 4.

Number	Interactivity
1	Strongly disagree
2	Disagree
3	Somewhat disagree
4	Neither agree or disagree
5	Somewhat agree
6	Agree
7	Strongly agree

- The relationship between you and stevedore personnel/other employees in your organization is good.
- Collegiality/work atmosphere at Fremantle port.

Section 6—Customer acquisition: The last part of the questionnaire concerned customer acquisition and comprises limited and independent questions to address the issues at the port. This part consists of four questions, three of which are answered using a 7-point Likert scale ("1" as Strongly disagree, "2" as Mostly disagree, "3" as Somewhat disagree, "4" as Neither agree or disagree, "5" as Somewhat agree, "6" as agree and "7" as Strongly agree). The remaining question will be discussed later. The Cronbach's alpha for this section is 0.7 which indicates reliability.

Number	Customer acquisition
1	Strongly disagree
2	Disagree
3	Somewhat disagree
4	Neither agree or disagree
5	Somewhat agree
6	Agree
7	Strongly agree

- Your organization endeavours to attract more drivers.
- My organization works or endeavours to take my concerns into accounts.

6.3 Hypotheses Formulation Related to the Fremantle Port Issues

The results of questionnaire analysis of each of the nine satisfaction categories are compared using the analysis of variance F-test to determine if there is a significant difference in the level of customer satisfaction between categories. The experiment should be designed such that any variability arising from extraneous sources can be systematically controlled. Here, each customer is the source of variability in the experiment that can be systematically controlled through blocking [1]; therefore, a blocked design of analysis of variance may be applied. The experiment designed in this study is a randomized complete block design (RCBD). The blocks provide sufficient replications in the experiment. In this case, the interactions between treatments and blocks are treated as the random error component.

Following the discussion presented in Chap. 2 and early in Chap. 3, it is apparent that the organization chosen as the case study has a great number of dissatisfied customers. Due to the causes of numerous complaints, many of the company's stakeholders have become disloyal and discouraged. According to the findings obtained from the text mining analysis, there are nine significant categories of complaints which are central to a variety of issues. Hence, nine hypotheses are presented in order to determine the relationship between dissatisfaction and customers' priorities.

The hypotheses are:

$$H_0 : \mu_1 = \mu_2 = \mu_3 = \mu_4 = \mu_5 = \mu_6 = \mu_7 = \mu_8 = \mu_9$$

$$H_1 : \mu_i \neq \mu_j \quad i,\ j = 1, 2, \ldots, 9 \text{ and } i \neq j$$

μ_1 = High dissatisfaction is due to the poor management performance at Fremantle port
μ_2 = High dissatisfaction is due to the lack of an effective booking system at Fremantle port
μ_3 = High dissatisfaction is due to the infrastructural issues at Fremantle port
μ_4 = High dissatisfaction is due to the poor time management at Fremantle port
μ_5 = High dissatisfaction is due to the health and safety problems at Fremantle port
μ_6 = High dissatisfaction is due to the poor stevedore performance at Fremantle port
μ_7 = High dissatisfaction is due to the high costs at Fremantle port
μ_8 = High dissatisfaction is due to the container sorting issues at Fremantle port
μ_9 = High dissatisfaction is due to the jurisdiction issues at Fremantle port.

where μ_1, μ_2, ... μ_9 are the average degrees of satisfaction obtained from all related questions for all customers in each satisfaction category.

In the previous chapter, a lab software text mining analysis tool was utilized to discover the major issues in terms of their significance and ranking in order of importance. In Sect. 6.3, previous findings will be validated using Analysis of variance.

Analysis of variance is a process to test the null hypothesis that the means of all population are equal. Since all the stated hypotheses in this study are one-way, and dissatisfaction has a one-way relationship with the issues, then one-way analysis of variance is selected based on the following assumptions:

1. One-way analysis of variance tests allows us to determine whether one given element, such as "infrastructure issues", has a significant impact on dissatisfaction.
2. The population from which the samples are drawn is assumed to be normally distributed.
3. The population has the same variance.

Later in this chapter, the second set of hypotheses is tested using the structural equation modelling approach. These were derived from the second part of the conceptual framework and are discussed in Sects. 6.5 and 6.6. A sample of the actual questionnaire and data collection method will be provided in Appendix 4.

In order to carry out analysis, validation of dissatisfactions and their impact factors through hypotheses formulation, the following steps are taken:

Hypotheses are proposed in order to test the relationship between the major issues. In order to measure the variables and validate the relationships, analysis of variance and structural equation modelling are applied. If the null hypothesis is accepted, this means that there is no significant difference in the level of satisfaction between categories. Otherwise, if the null hypothesis is rejected, it means that there is significant difference in the level of satisfaction between categories. From this, the next step is to determine the category in which customers are more satisfied, and which category is the source of customer dissatisfaction.

A detailed description of the procedure of analysis of variance for RCBD, mainly adopted from Montgomery [1], will be given here. The procedure is usually summarized in an analysis of variance table (Table 6.1).

Table 6.1 Analysis of variance for a randomized complete block design

Source of variation	Sum of squares	Degrees of freedom	Mean squares	F-value	P-value
Treatments	$SS_{Treatments}$	$a-1$	$\frac{SS_{Treatments}}{a-1}$	$\frac{MS_{Treatments}}{MS_{Error}}$	$P(F-value > F_{\alpha,(a-1),(a-1)(b-1)})$
Blocks	SS_{Blocks}	$b-1$	$\frac{SS_{Blocks}}{b-1}$		
Error	SS_{Error}	$(a-1)(b-1)$	$\frac{SS_{Error}}{(a-1)(b-1)}$		
Total	SS_{Total}	$ab-1$			

In RCBD of Table 6.1, there are a treatments, b Blocks and ab observations in total because there are no replications in the design. $F_{\alpha,(a-1),(a-1)(b-1)}$ is a value from *F-distribution* with $(a-1)$ degrees of freedom in nominator and $(a-1)(b-1)$ degrees of freedom in denominator and the probability that F variable is greater than $F_{\alpha,(a-1),(a-1)(b-1)}$ is α.

The formulas for computing the sum of squares (SS) in Table 6.1 are presented in Eqs. 6.1–6.4.

$$SS_{Total} = \sum_{i=1}^{a}\sum_{j=1}^{b} y_{ij}^2 - \frac{y_{..}^2}{ab} \quad (6.1)$$

$$SS_{Treatments} = \frac{1}{b}\sum_{i=1}^{a} y_{i.}^2 - \frac{y_{..}^2}{ab} \quad (6.2)$$

$$SS_{Blocks} = \frac{1}{a}\sum_{j=1}^{b} y_{.j}^2 - \frac{y_{..}^2}{ab} \quad (6.3)$$

$$SS_{Error} = SS_{Total} - SS_{Treatments} - SS_{Blocks} \quad (6.4)$$

where y_{ij} is the observation in *i*th treatment and *j*th Block. Dot (.) means sum over the index as $y_{..}$ is the sum of all y_{ij}s, $y_{i.}$ is the sum of y_{ij}s in ith treatment, and $y_{.j}$ is the sum of y_{ij}s in jth Block.

6.4 Category Indicators and Abbreviations

In Table 6.2, an acronym is used for each input or variable. However, in Table 6.3, indicators are presented in a horizontal manner and in Table 6.5, they are vertical. In Table 6.2, abbreviations are:

Table 6.2 Category indicator and abbreviation

Category Indicator	Abbreviation
Management	MNG
Booking System	BS
Infrastructure	INF
Health and Safety	HS
Container Sorting	CONS
Cost	CO
Jurisdiction	JUR
Stevedore Performance	STP
Time Management	TMNG

6.5 Input Factors Resulting from Variance Analysis

Table 6.3 presents the degrees of satisfaction obtained from all related questions in each satisfaction category.

The data in Table 6.3 are used to perform analysis of variance in MINITAB. The results of ANOVA are presented in Table 6.4.

As the results of Table 6.4 shows, the null hypothesis is rejected because the *P* value is too low (less than 5 %). We can conclude that there is a significant

Table 6.3 Satisfaction degree obtained from all related questions in each satisfaction category

DMU	Indicator actual values obtained from PCA								
	MNG	BS	INF	HS	CONS	CO	JUR	STP	TMNG
Customer 1	3.8	4.0	4.1	3.8	5.0	4.0	3.0	3.3	3.6
Customer 2	3.8	3.0	3.3	4.5	5.0	2.6	4.7	3.7	2.6
Customer 3	3.6	3.3	3.3	4.0	4.0	2.6	4.0	4.3	3.4
Customer 4	3.5	3.7	3.1	4.3	4.0	2.8	4.0	4.0	3.7
Customer 5	4.0	3.3	3.6	5.3	4.0	2.4	4.0	4.7	3.1
Customer 6	3.0	2.9	2.6	5.3	2.0	2.6	3.0	4.0	3.7
Customer 7	3.5	4.3	2.9	4.3	6.5	3.0	5.3	4.7	3.0
Customer 8	4.1	3.6	4.0	4.5	5.5	3.6	4.0	4.7	3.9
Customer 9	3.4	4.0	3.9	4.0	6.0	3.8	3.0	4.0	3.4
Customer 10	4.1	3.7	4.1	3.5	4.5	4.2	4.3	5.0	4.1
Customer 11	2.4	2.2	2.6	4.5	3.0	3.0	2.3	3.0	2.0
Customer 12	2.8	2.7	2.6	2.8	4.5	1.6	1.3	3.3	2.6
Customer 13	3.3	2.7	3.4	3.8	3.5	4.2	3.3	4.0	3.6
Customer 14	4.4	4.2	4.1	4.8	4.0	3.6	4.3	4.3	3.7
Customer 15	5.4	5.6	4.2	4.0	6.5	3.2	6.7	6.3	5.0
Customer 16	3.2	2.5	3.0	4.8	3.0	2.2	3.7	3.3	2.7
Customer 17	3.1	3.0	2.9	4.5	3.5	2.8	3.0	3.0	3.4
Customer 18	3.2	3.3	3.3	5.3	4.0	2.4	2.7	2.7	2.9
Customer 19	3.9	3.9	3.6	3.0	4.5	2.8	4.3	4.3	3.3
Customer 20	3.5	3.0	3.4	4.8	5.0	3.0	3.7	4.7	3.7
Customer 21	3.1	2.7	2.3	5.5	4.0	1.2	2.7	3.3	3.0
Customer 22	3.0	2.4	2.4	5.0	4.0	1.6	3.7	3.3	2.7
Customer 23	2.2	2.1	2.2	3.8	2.0	2.4	2.0	2.0	2.6
Customer 24	4.5	4.1	3.1	5.5	6.0	1.8	5.7	6.0	2.7
Customer 25	2.6	2.2	3.3	3.3	2.5	3.0	1.0	2.3	3.4
Customer 26	3.5	3.1	3.4	4.3	3.0	3.6	3.7	3.3	3.4
Customer 27	4.1	3.6	4.4	5.5	4.0	3.8	2.7	4.0	3.7
Customer 28	3.7	2.6	3.8	2.8	3.0	3.4	4.0	4.3	3.7
Customer 29	3.0	3.0	3.2	3.8	4.0	3.2	3.7	2.7	2.9
Customer 30	3.6	3.9	2.9	4.8	4.0	2.4	3.0	4.0	4.3
Customer 31	3.9	3.6	4.1	3.5	4.5	4.2	3.7	5.0	4.0
Customer 32	3.2	3.7	3.0	5.0	5.0	2.6	3.0	3.7	3.1
Customer 33	3.6	3.3	3.4	5.0	5.0	3.2	3.3	3.7	3.0
Customer 34	3.7	3.5	3.0	5.0	5.0	2.6	4.7	3.7	3.4
Customer 35	4.1	3.6	3.1	5.5	4.0	2.8	4.0	4.3	4.1
Customer 36	4.0	3.7	3.3	5.5	3.5	3.0	3.7	3.3	3.9
Customer 37	3.6	3.6	3.3	4.8	4.5	3.8	5.3	4.0	3.9
Customer 38	3.2	3.5	3.1	4.8	3.5	3.0	3.0	2.7	3.3
Customer 39	3.1	3.0	3.4	5.3	4.0	3.4	3.3	3.7	3.1
Customer 40	3.6	3.1	3.4	4.0	5.0	2.0	2.7	2.7	2.7
Customer 41	3.4	3.1	3.3	3.5	6.0	2.0	3.3	2.3	2.6

(continued)

Table 6.3 (continued)

DMU	Indicator actual values obtained from PCA								
	MNG	BS	INF	HS	CONS	CO	JUR	STP	TMNG
Customer 42	3.9	3.9	3.8	5.3	5.0	3.2	4.0	3.7	3.0
Customer 43	3.9	4.1	3.9	4.3	5.5	3.0	3.7	4.0	3.1
Customer 44	3.8	4.1	3.6	4.8	3.5	4.0	4.0	4.3	4.6
Customer 45	3.0	2.6	2.9	3.3	3.5	3.4	3.0	2.7	2.9
Customer 46	3.3	2.9	3.1	4.8	2.5	3.4	3.7	4.3	3.3
Customer 47	3.9	4.0	3.3	4.5	5.0	3.2	4.7	4.3	4.0
Customer 48	4.2	3.8	3.6	4.3	4.0	3.6	4.3	5.0	4.3
Customer 49	4.2	3.4	3.7	4.0	2.5	2.8	3.3	4.0	3.7
Customer 50	3.7	4.4	3.4	4.3	5.5	2.8	4.0	3.3	4.0
Customer 51	3.4	3.1	3.5	3.0	3.5	3.0	4.3	3.0	3.0
Customer 52	2.8	2.4	2.9	4.3	3.0	2.4	2.7	3.0	2.3
Customer 53	3.8	3.2	3.9	3.8	4.0	3.0	3.7	4.3	2.9
Customer 54	3.7	3.5	3.8	4.0	4.0	3.4	3.3	4.7	3.3
Customer 55	3.6	3.5	3.7	5.5	5.0	2.8	3.0	3.3	3.0
Customer 56	4.0	3.9	3.4	3.8	3.0	3.2	5.0	3.0	3.7
Customer 57	3.4	3.5	3.1	3.8	5.5	3.0	4.7	3.3	2.9
Customer 58	3.4	3.1	3.7	4.5	4.0	4.4	3.0	2.0	2.6
Customer 59	3.8	2.9	3.8	4.8	3.0	3.2	4.3	3.3	2.7
Customer 60	3.8	3.1	3.1	3.3	3.5	2.4	3.3	3.3	3.0

Table 6.4 Analysis of variance for categories of customer satisfaction

Source of variation	Sum of squares	Degrees of freedom	Mean squares	F-value	P-value
Satisfaction category	90	8	11.25	29.3	0.000
Customer	120.9	59	2.05		
Error	181.1	472	0.38		
Total	392.1	539			

difference between levels of customer satisfaction within different groups of satisfaction i.e. MNG, BS, INF, HS, CONS, CO, JUR, STP, and TMNG. If we consider the average value of PCA satisfaction degrees as a common measure of satisfaction in each category, then we will be able to rank the categories in an ascending order from the 'most satisfied' category to the 'least satisfied' category. Table 6.5 gives the average degree of satisfaction for each category. The detailed questionnaire is provided in Appendix 4.

As the results in Table 6.5 show, customers are more satisfied with the HS and CONS systems. They are least satisfied with INF, CO, BS, MNG and TMNG systems. These results are consistent with the results obtained from text mining since in text mining analysis the majority of complaints were in INF, BS, MNG and TMNG. However, cost (CO) was eliminated from this analysis because, from the interviews, insufficient data was obtained for cost-related issues. The correlation

Table 6.5 The average satisfaction degree of each category

Satisfaction category	Number of customer	Average satisfaction degree	Rank average	Rank text mining
MNG	60	3.5717	5	6
BS	60	3.3617	6	7
INF	60	3.3600	7	8
HS	60	4.3733	1	4
CONS	60	4.1667	2	2
CO	60	2.9933	–	–
JUR	60	3.6467	4	3
STP	60	3.7400	3	1
TMNG	60	3.3200	8	5

between the ranks from text mining and analysis of variance is 70 %, illustrating the consistency of text mining and analysis of variance results. This fact can be served as verification and validation of the questionnaire results and is a proof for the hypotheses.

In this section, we tested and validated that the six categories or inputs have a positive relationship with dissatisfaction at the Fremantle port, and that these variables are positively associated with perceived value, interactivity, customer satisfaction, loyalty and customer acquisition. In the next section, we try to prove that the conceptual framework described in Chap. 4 is logical and legitimate.

6.6 Impact Factors Mapping to I-CRM Evaluator

As discussed in Chap. 4, the I-CRM evaluator has major components including perceived value, interactivity, customer satisfaction, loyalty and customer acquisition. An attempt will be made to map this conceptual framework and the relationship between the variables utilised in this chapter. Utilising a methodology, we examine the mapping between the variables and I-CRM framework. We need to test whether the variables have a significant link with each I-CRM component, using hypotheses.

6.6.1 Structural Equation Modelling for Validation of Mapping

We use the model in our research with two examples. (1) We predicted over customer values and proposed hypotheses. Utilising structural equation modelling, the relationships between customer value anticipation, satisfaction and loyalty are discovered. It was proven that customer satisfaction will lead to long-term commitment from the customers. (2) After the variables have been introduced, the hypotheses are proposed. SEM is enhanced for customer satisfaction in order to evaluate quality of digital content, because, structural equation modelling is formulated by two kinds of equations: measurement equation and construct equation. Measurement equations can be applied to find the relationship between observed variables and latent variables. Construct equations can be utilized to evaluate the relationship between hypotheses [2].

6.7 Hypotheses Formulation Related to I-CRM

In this section, the second set of hypotheses is formulated in order to further clarify and refine the issues and ascertain the relationship between hypotheses in terms of the port operators.The second set of hypotheses to be tested is as follows:

H1 Perceived value has a direct relationship with customer satisfaction
H2 Interactivity has a positive relationship with customer satisfaction
H3 Customer satisfaction has a positive impact on customer acquisition
H4 Customer satisfaction has a positive relationship with customer loyalty.

6.7.1 Hypotheses Testing Related to the Fremantle Port Issues and I-CRM

As discussed previously, in this stage the process begins by testing the correlation between every two variables using SPSS software; then, the Spearman's Rank Correlation Coefficient must be used to disclose the power of the bond between two sets of data. If a significant relationship is found, then structural equation modelling can be used. Using this software, the causal and caused relationship between variables is being tested. In the following section, we will provide a correlation coefficient between variables which will be utilized to find the relationships as stated in the hypotheses. The results of the correlation coefficients generated by the software are provided in Appendix 2.

Based on the proposed conceptual framework, four more hypotheses need to be tested. As the system is vulnerable due to the environmental reactions, we must constrict and prevent the decreasing propinquity or elimination of each of the elements. To better exploit the system, we try to identify the relationship between the factors of intelligent customer relationship management evaluators.

The aim is to increase the level of customer satisfaction by decreasing the level of customer complaint. As a preliminary step, the hypotheses have been tested and proven to be valid.

It is envisaged that researchers will benefit from noting each of the factors used in our conceptual framework. However, we introduced a new model which is comprised of concepts applied in previous researches as discussed in the literature review.

6.7.2 Conceptual Diagram Derived from Structural Equation Modelling

From this section onward, causal relationships will be tested and evaluated utilizing structural equation modelling. The diagram (Fig. 6.1), which is derived from our analysis, illustrates the relationship between the main latent variables and observed variables.

In order to find the relationships and complete the analysis, we need to import the data gathered from the respondents to the structural equation modelling. As discussed at the beginning of this section, each latent variable is dependent upon various observed variables or questions. Here, we defined each question which is related to the main variables as an observed variable. In the following, we provide in brief the instances or representatives of each question applied in the questionnaire:

1. "Perceived variable" is related to knowledge, necessary information, cleanness, service, quality, and improvement.
2. "Interactivity": is dependent on relations, work atmosphere, insufficient interaction, quick feedback and interaction.
3. "Customer satisfaction" is related to multiple observed variable such as management, running the port, organized, management strategy, transport, traffic, waiting time, staying in cabin, queuing time, cost, health, road rules, competition, vehicle booking system, intelligent system at the port, stevedore performance, container sorting issue, speed and access to the container, signs and arrows, performance of forklift drivers, separate road, danger on the roads, new store, facility, parking, pavement, enough cameras at the port, to be monitored by security cameras, working procedure, petrol station, accommodation, responsiveness, technical support, respect, time slots and shifts.
4. "Loyalty" is related to words such as recommendation, switch, having the same employer, tendency to work with the same company, and word of mouth.
5. "Customer acquisition" is linked to attractiveness, incentives, responsible, and number of drivers.

By means of arrows, the conceptual diagram in Fig. 6.1 clearly depicts the main variables and their relationships. Each individual variable is related to multiple observed variables. The conceptual diagram for the structural part of the model shows that provision is made for the acceptance and recognition of the latent variables to be correlated with the observed variables [3].

Figure 6.1 presents an overview of the relationship between the major I-CRM factors (latent variables) and their observed variables which characterize the questions related to each I-CRM variable. In this diagram, the boxes ▭ outline latent variables and the ellipses ⬭ depict observed variables. The arrow → shows the one-way relationship between one variable and another.

On the left side of the diagram, we have independent latent variables with their observed variables namely, interactivity and perceived value. To the right of the

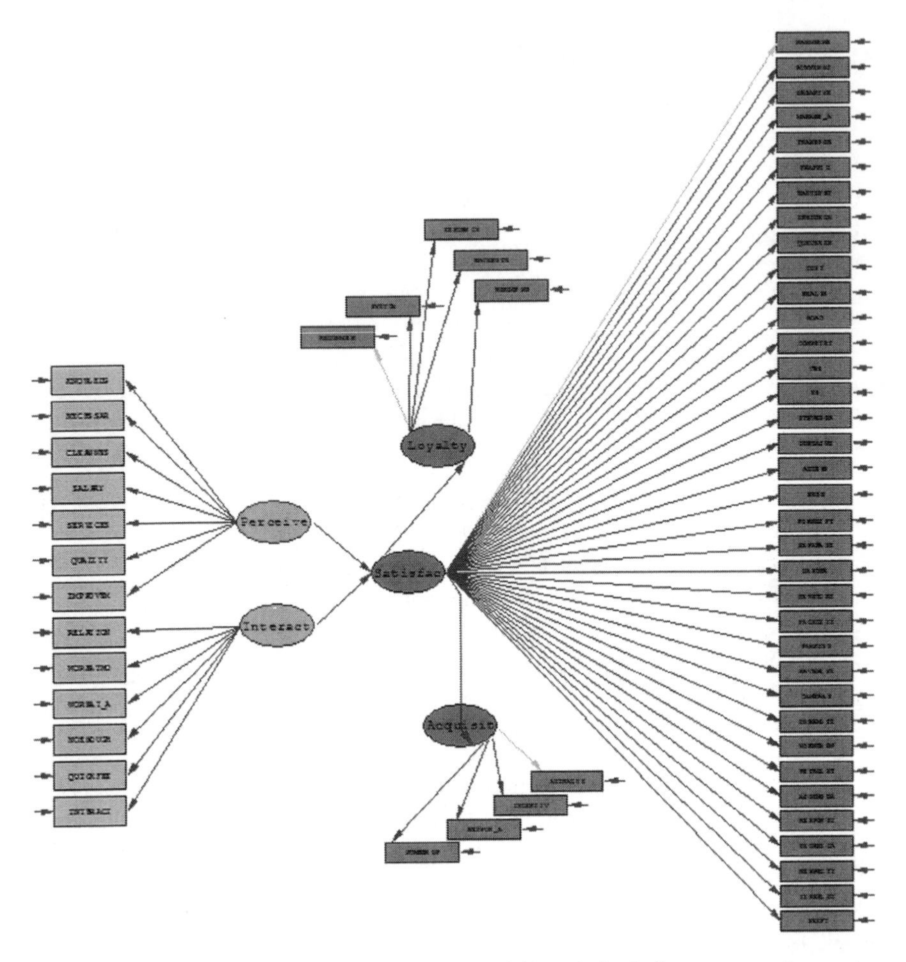

Fig. 6.1 The causal relationship between the variables derived from structural equation modelling in conceptual diagram state. *Note* The full view of the details of this figure is shown in Appendix 6.

diagram, we have customer satisfaction, loyalty and customer acquisition which are outlined as dependent variables or latent variables. Each latent variable has a relationship with its related observed variables.

6.7.3 Normalising the Coefficient to Show Relationships Between Variables

We need to standardize coefficients. Standardised solution or coefficients are the outputs from an analysis of variables that have been standardized in order to have a variance of one.

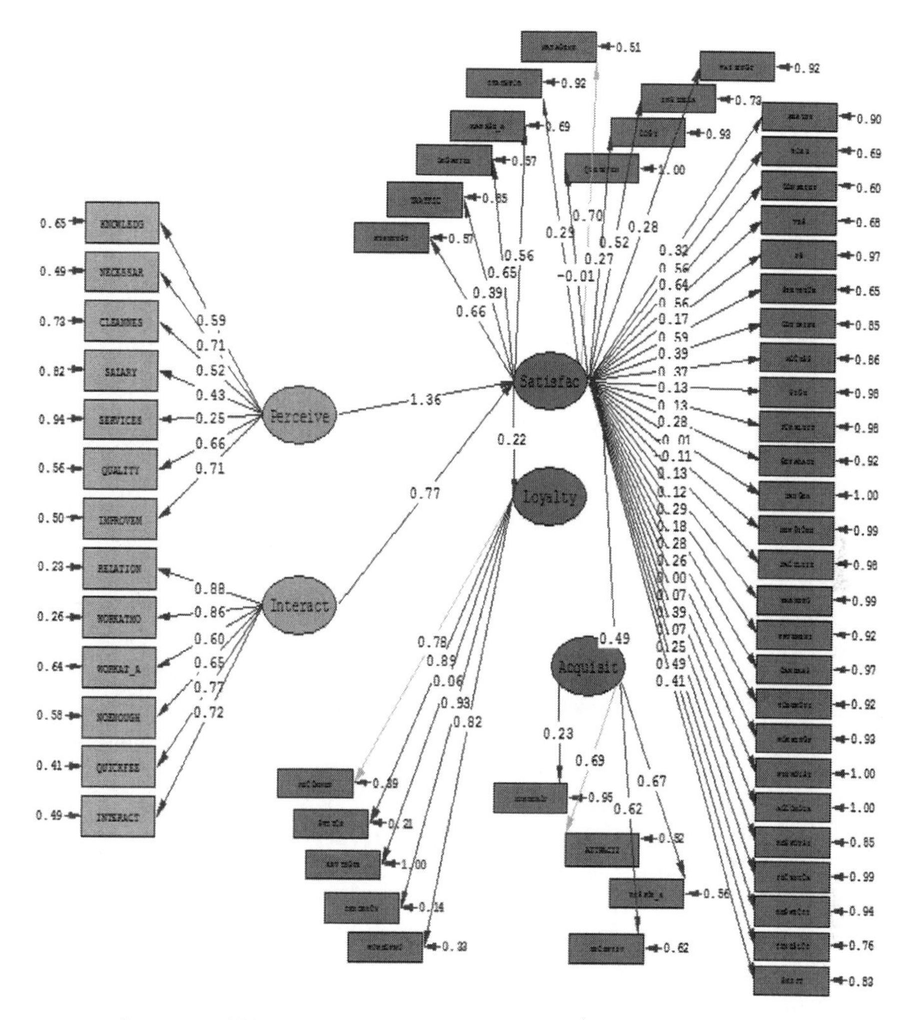

Fig. 6.2 The causal relationship between the variables derived from structural equation modelling in standard solution state (See Appendix 6 for detail)

Figure 6.2 shows the relationship between the major I-CRM factors (latent variables) and their observed variables with the derived standardized value from the relationships of the variables. In this diagram, the boxes ☐ contain latent variables and the ellipses ⬭ depict observed variables. The arrow → shows the one-way relationship between one variable and another.

Figure 6.2 depicts the relationship between latent variables which are perceived value, interactivity, customer satisfaction, loyalty and customer acquisition. The coefficients in the above modelling have been standardised. Structural equation modelling also generates standardized formats of the parameter estimates. In standardized estimates, only the latent covariance has been standardized and correlated. Also, all estimates are illustrated as a standardized metric.

To be able to ascertain whether or not our hypotheses have been proven, we must check the t-value coefficient which will be discussed in the next figure.

6.7.4 Significance of Coefficient Derived from Structural Equation Modelling

The t-value is the likelihood of observing a statistical sample as significant as the test statistic. It also shows the significance and meaningfulness of the relationships. According to the diagram below, perceived value and interactivity have a significant relationship with customer satisfaction. Also, the influence of customer satisfaction on customer acquisition is significant. However, customer satisfaction does not have a significant relationship with loyalty according to all the data collected from the questionnaire survey. T-value proposes that there should be adequate evidence that the model and its estimations have valid variance within the relationship between the variables.

Figure 6.3 delineates the relationship between the significant variables (latent variables) and their observed variables with the acquired significant values from the relationships between variables. In this diagram, the boxes ▭ contain latent variables and the ellipses ⬭ depict observed variables. The arrow → shows a one-way relationship between one variable and another.

As can be seen in the figure, some of the relationships between latent variables and observed variables are depicted in red which means that these must be rectified or omitted. The figure depicts the t-value derived from SEM software. Hence, an organization needs to provide strategies to improve the system. The majority of the observed variables for customer satisfaction have a positive and direct association with it. However, there were observed variables that have a negative impact and indirect relationship with their latent variable which is customer satisfaction, such as queue time, intelligent system, sign, forklift, danger, new store, facilities, parking, cameras, work operations at the Fremantle port, petrol station, accommodation, technical problems and respect. This implies negligence on the part of the port authorities or some weaknesses in providing the mentioned services to the customers which must be enhanced. Having said that, loyalty has observed variables which have a negative relationship with one latent variable, that is, the number of years the client has been working with his/her current employer. Customer acquisition has an indirect relationship with one of its observed variables which is a decrease in the number of drivers in comparison with the previous year.

Figure 6.3 clearly shows that customer satisfaction does not lead to loyalty. However, customer satisfaction will lead to customer acquisition. Next, these two hypotheses will be discussed.

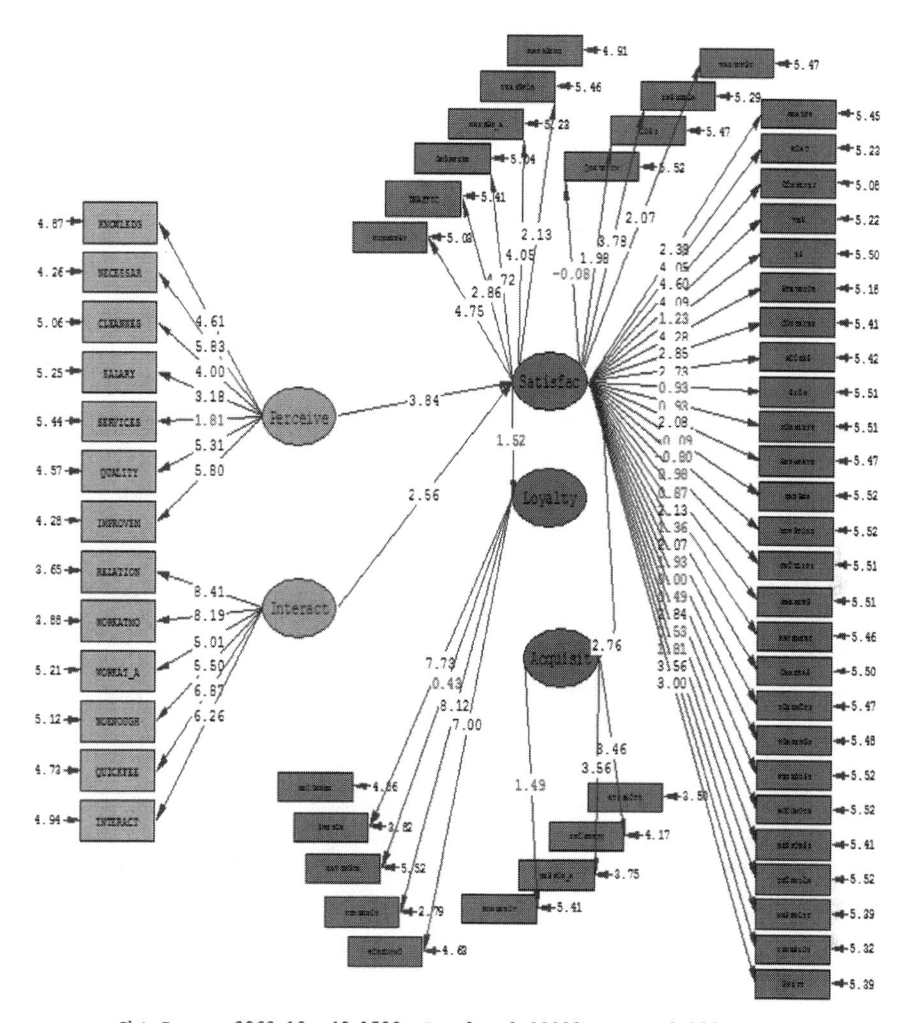

Fig. 6.3 The causal relationship between the variables, derived from structural equation modelling in T-value situation (See Appendix 6 for detail)

Number of Y-Variables = 45
"Number of Y" is defined as the number of observed variables or the questions related to the dependent variables.

Number of X-Variables = 13
"Number of X" is defined as the number of observed variables or the question related to the independent (Latent) variables.

Number of ETA-Variables = 3
"Number of ETA" is defined as the combination of latent or major variables which are independent.

Number of KSI-Variables = 2
Number of KSI is defined as the combination of latent variables which are dependent.

Number of Observations = 60
In this research, "number of observation" is defined as the overall number of respondents who participated in the survey.

Degrees of Freedom = 1,589
"Degrees of Freedom" is defined as the number of values in the final estimation of a statistic.

Full Information ML Chi Square = 3,363.19 (P = 0.0).

Root Mean Square Error of Approximation (RMSEA)

RMSEA is an index utilised to estimate the error of approximation in the population and has a range from 0 to 1. This measurement is according to the non-centrality parameter. Those models which have an RMSEA of 0.05 or less are very good models. However, models in which RMSEA is 0.10 or more have an acceptable fit.

Based on Lavee [4], RMSEA is an index used to show the average difference between observed data; if the value is less than 0.1, this would be more effective.

- Root Mean Square Error of Approximation (RMSEA) = 0.14
- 90 Percent Confidence Interval for RMSEA = (0.13; 0.14)
- P-Value for Test of Close Fit (RMSEA < 0.05) = 0.00.

Figures 6.2 and 6.3 show that based obtained parameters and significant coefficients in our structural equation modelling (SEM) for the patent variables, perceived value has a significant link with customer satisfaction. Likewise, interactivity has a significant link with customer satisfaction. This implies that perceived value and interactivity have a positive causal relationship with customer satisfaction (y = 3.84 and 2.56 respectively). Hence, both hypotheses are supported.

On the other hand, customer satisfaction does have a direct causal relationship with customer acquisition, but no relationship with loyalty (y = 2.76, 1.52 respectively). We also, must view subsidiary or secondary hypotheses which are the relationship between the latent variables and the observed variables.

Based on Fig. 6.3, all observed variables related to perceived value have a positive bond with it, apart from services in which y = 1.81. Therefore, the secondary hypothesis for that is refuted.

Also, all observed variables related to interactivity have a positive bond with it and all of the secondary hypotheses related to interactivity are supported.

Based on Fig. 6.3, among the observed variables related to customer satisfaction, there were several which have a positive or negative relation with it. Based on the analysis, and all of the raw data, we cannot establish a relationship between customer satisfaction, with queue time, intelligent system, sign, forklift, danger, new store, facilities, parking, cameras, working procedure, accommodation, technical support, petrol station and respect (y = −0.08, 1.23, 0.93, 0.93, −0.09, −0.80, 0.98, 0.87, 1.36, 1,93, 0.49, 0.53, 0.00 and 1.81).

6.8 Validation Mapping of Hypotheses Related to the Fremantle Port and I-CRM

Based on Fig. 6.3 and the information collected from the software, we must look at the t-value coefficients. If they are within the range of 1.96 to −1.96, our hypotheses will be rejected. If the coefficients are not within 1.96 to −1.96, our hypotheses will be supported. To be able to prove whether or not our model has an appropriate and valid reliability, there must be the following relationships among indices:

1. The portion of Chi square to degree of freedom (df) must be less than 3, which is true for our result.
2. RMSEA which is Root Mean Square Error of Approximation must be less than 0.1 and according to our analysis AGFI, it is about that at 0.14.
3. GFI and AGFI must be higher than 0.9 which is true.

Goodness-of-fit index (GFI) is influenced by sample size and can be large for models that are weakly characterised. The goodness-of-fit index (GFI) indicates the degree of discrepancy [5]. GFI can range from 0 to 1, but theoretically might produce negative values. A large sample size enhances GFI. As GFI increases in comparison with other fit models, researchers propose that utilising 0.95 produces an interruption in activities. Finally, GFI should be equal to or greater than 0.90 in order for the model to be accepted.

Adjusted Goodness-of-fit Index (AGFI) adjusts the GFI using a degree of freedom ratio that has been employed in the model. Here, the values range from 0 to 1 where higher values indicate better fit and but simultaneously their values are less than GFI values. However, our model has acceptable reliability. The next section discusses the relationship between the factors. The hypotheses will be revisited followed by recommendations which are intended to improve the system.

H1 Perceived value has a direct relationship with customer satisfaction.

As indicated by Fig. 6.3, perceived value has a positive relationship with customer satisfaction, as the t-value is 3.4, and as discussed, it is above 1.96.

Perceived value has several observed variables. According to our analysis, all observed variables have a proper correlation with the latent variable, but the services variable has been rejected as it is 1.1 (less than 1.96). Hence, companies need to provide better services in different sections of the Fremantle port.

H2 Interactivity has a positive relationship with customer satisfaction.

According to our results, interactivity has a direct relationship with customer satisfaction, as the result is 2.6 which is greater than 1.96 and therefore the hypothesis is accepted. All observed variables have a logical relationship with the latent variable.

H3 Customer satisfaction has a positive impact on customer acquisition.

Customer satisfaction has a direct and positive relationship with customer acquisition and, according to our analysis; the system will be able to generate more customers for the organization. T-value for the relationship between customer satisfaction and customer acquisition is 2.76 (greater than 1.96). So, the hypothesis is accepted and these two factors have a positive influence and impact on each other.

H4 Customer satisfaction has a positive relationship with customer loyalty.

Customer satisfaction has a negative relationship with loyalty as the analysis produced 1.52 (less than 1.96), and so the hypothesis is rejected. There may be many reasons for this and the observed variables in customer satisfaction must be explored further in order to confirm the rejection of the hypothesis. In doing so, we will provide new recommendations to improve the system.

According to Table 6.6, the rejection of the mentioned hypothesis indicates the higher expectations that individuals have of Fremantle port and the companies for whom they are working. The latter needs to make some changes and improvements if they are to retain the customers (drivers).

Table 6.6 The outcomes derived from SEM in research hypotheses testing

Result of the test	Significant number	Standard coefficient	Hypothesis
The hypothesis is supported	3.84	1.36	H1: Perceived value has a direct relationship with customer satisfaction
The hypothesis is supported	2.56	0.77	H2: Interactivity has a positive relationship with customer satisfaction
The hypothesis is supported	2.76	0.49	H3: Customer satisfaction has a positive impact on customer acquisition
The hypothesis is refuted	1.52	0.22	H4: Customer satisfaction has a positive relationship with customer loyalty

6.9 Recommendations

One could say that the current customers (drivers) are loyal since many of them have been working for Fremantle port for many years. Nevertheless, we provide the following recommendations intended to improve the system.

1. The waiting time in the queue must be decreased for the drivers.
2. The intelligence system and devices used at the Fremantle port need to be upgraded as they are not up-to-date and fall short of current world standards.
3. There must be enough signs and arrows to direct drivers and other customers to various locations. The lack of adequate road signage creates unnecessary inconvenience and angst for the customers.
4. The performance of the forklift drivers is not satisfactory, according to the responses to questionnaires and during interviews. There should be adequate incentives for forklift drivers as there are for other clients at the Fremantle port. In the majority of cases, forklift drivers do not want to finish the work before taking a lunch or smoke break. Given enough motivation, they might want to finish the task at hand before taking a break.
5. Danger created in the roads is another factor that makes the drivers disloyal to the Fremantle port. The risk of road accidents and lack of adequate security are two complaints in this regard. Hence, the port authority must recruit new staff to monitor all movement at the Freemantle port. Furthermore, dangers in the roads must be monitored by the port authority. Road rules and regulations must be complied with to reduce risks to the drivers.
6. Drivers who are dealing with the Fremantle port need a new store or supermarket. The lack of a store within the port area forces some drivers to drive miles away to buy necessities for themselves or their families. A new store would mean that drivers can remain at the port without having to go out of their way to shop, and will increase their loyalty to the port. Also, Fremantle port must provide adequate overnight accommodation for its drivers, particularly those from interstate. Furthermore, another petrol station is needed since at present there is only one, which is inadequate in meeting driver needs.
7. Lack of facilities at the Fremantle port is another issue that adds to driver dissatisfaction. Like any other customers elsewhere, drivers and various customers at the port should have their needs acknowledged and respected. Adequate facilities must be provided for all individuals, particularly since they are giving their valuable time to the Fremantle port.
8. The issue of having enough parking spaces is another problem facing the companies at the port. Fremantle port has insufficient parking areas; this impacts on the port itself as well as on surrounding residential areas.
9. Cameras need to be located to better monitor the various areas and to help drivers get to their specific destination. Also, with enough cameras, the Port authority will be able to better control various areas.

10. Technical support is another issue that may affect driver loyalty or disloyalty. If the drivers receive appropriate support when they are in difficulty, they are more likely to remain loyal. A decrease in technical assistance will decrease driver loyalty. So, the Fremantle port authority must provide better services at reasonable prices.
11. Respect is another very outstanding issue that has already made some drivers disloyal. Each individual must be considered as an entity and respected separately irrespective of his/her nationality, ethnicity and religion.

In terms of implementing the recommendations above, the existing CRM system and current strategies do not have the ability to handle customer dissatisfaction and are not intelligent. To become aware of the issues and develop efficient solutions to them, we need to do the modelling for I-CRM and dissatisfaction modelling and to formulate the relationships using lab tools which will be discussed in the next chapter.

All the data was taken into account to test and analyse our hypotheses without grouping our customers. If we categorize our customers into different groups, we might obtain support for this hypothesis as well. However, as customer satisfaction has many observed variables, it will be difficult to analyse. Likewise, our loyalty criteria are few, and this alone is another issue in testing the H4 hypothesis. As explained in this chapter, we cannot prove hypothesis number 4 with all of the data from 60 customers. Hence, we intend to prove it using our categorized customers. This has not yet been fully explained. In the next chapter, we intend to categorize our customers as 'ordinary', 'important' and 'very important' to find key customers and establish the relationship of all the variables including customer satisfaction and loyalty. To achieve, another method and other tools will be used for the analysis.

6.10 Significance of Validation of Mapping and Impact Factors Through Hypotheses Formulation and Proof of Concept

In Chap. 2, a review was conducted of the existing literature pertaining to customer satisfaction, its antecedents and consequences. To the best of our knowledge, many researchers have undertaken a vast amount of research on the majority of variables. However, for the first time, in this research we have included interactivity as one of the main variables in the CRM process and one of the most significant antecedents of customer satisfaction. Researchers [6] have presented new surveys on CRM and customer satisfaction in which customer satisfaction was measured by proposing hypotheses and testing them using SEM and structural equation modelling. In Gallarza and Gil Saura [7], the authors evaluate perceived value, customer satisfaction and loyalty and their impact on customer behavior, using structural equation modelling. Also, in another study conducted by Slevitch and Oh [8], analysis of variance was performed to examine whether control may create various levels of responses to produce customer satisfaction. Also, Carvajal et al. [9] used structure equation modelling to handle complaints and generate customer loyalty.

In this chapter, the hypotheses were examined by utilizing pivotal approaches. Although, in previous studies, researchers have tested and ascertained various hypotheses, the content and variables included in this research, and their implications, are absolutely original and unique. Also, none of the previous works has utilized a similar approach and methodology. In Chap. 5, we first transcribed our collected interviews and used text mining analysis in order to prioritize the customer complaints. Using that, we categorized and analysed customer complaints and followed this with in-depth information regarding each of the issues at Fremantle port. Based upon the customer complaints identified and prioritized in Chap. 5, the new hypotheses were established in Chap. 6. In this chapter, we propose two sets of hypotheses in accordance with our conceptual framework and the results derived in Chap. 5.

Firstly, based on our model, thirteen hypotheses pertaining to customer dissatisfaction were formulated. Also, we consider drivers as customers and regard them as our main sample and aim to address the issues which emerged from their complaints. The hypotheses that we aim to address are dissatisfaction due to the lack poor management performance, poor health and safety, weak infrastructure, poor stevedore performance, lack of proper time management, lack of proper rules and regulations, poor container sorting, poor booking system and cost issues at Fremantle port.

In the second part of our conceptual framework, we explore variables including interactivity, perceived value, customer satisfaction, loyalty and customer acquisition in order to ultimately provide recommendations to improve the effectiveness at Fremantle port. Customer responses are analysed in order to ascertain the extent to which they are satisfied with the current situation and what is needed to bring about improvements at Fremantle port. For this part, four hypotheses are proposed which relate to interactivity and perceived t-value to customer satisfaction, since ultimately, customer satisfaction is linked to loyalty and customer acquisition.

To test our hypotheses, and for the purposes of quantitative research, a questionnaire was designed and subsequently tested for validity and reliability. Questionnaire results are presented in Chap. 9 of this thesis. The results produced by questionnaire analysis of each of the nine customer satisfaction groups are compared by analysis of variance and F-test to determine whether there is a significant difference in the level of customer satisfaction between categories. The hypotheses were tested using analysis of variance.

Structural equation modelling and SEM software was used to test the second set of hypotheses. Utilising this methodology, we support and accept the relationship between interactivity and perceived value to customer satisfaction. Additionally, we support the notion that customer acquisition is positively associated with customer satisfaction. However, our hypothesis regarding the relationship between customer satisfaction and loyalty is refuted because all the collected data was imported into the software and the results were based on the entire body of data.

To be able to test and analyse the relationships, researchers sometimes stick with a standard solution; however, to be able to support the hypotheses and rectify and modify the models, we need to look at the t-value model derived from the software.

In this chapter, we contribute to knowledge and science. Researchers have already undertaken an enormous amount of investigation into customer relationship management and customer complaint management systems. However, since this study is concerned with CRM and a customer complaint management system in the logistics and transport system, it is a new contribution. Moreover, as the data was collected in one of Australia's most significant ports, this study is innovative and therefore both the analysis and the recommendations should make a valuable contribution to this field.

Researchers use SEM and analysis of variance for numerous reasons and in many studies. However, we combine these two tools and utilize them to test our hypotheses in order to arrive at recommendations to improve the conditions for the customers working at Fremantle port. We tested and confirmed the support for our hypotheses to ensure that our conceptual framework is legitimate and that our proposed method is scientifically based. Having proven our hypotheses, we are better placed to propose recommendations to address each of the shortcomings at Fremantle port.

In the majority of cases, researchers propose a model after collecting a comprehensive amount of data. However, there is no methodical approach for analysing the data, and rarely is anyone interested in undertaking research that specifically targets customer complaints.

In this thesis, not only do we collect enough data through both interview and questionnaire, we proffer a new methodology and conceptual framework, in that we utilize new variables such as interactivity and perceived value to analyse customer complaints. Also, we propose relevant hypotheses which are subsequently tested and legitimized. Additionally, a contribution is made from the perspective of science as a combination of software and appropriate methodologies is used.

Previously, no researchers have used our methodologies to test and establish hypotheses. Even the variables that were utilized in this study are new in terms of having them in CRM and a customer complaint management system. In the proposed conceptual framework, interactivity and perceived value are introduced for the first time and linked to customer satisfaction. In this chapter, their relationship was supported. However, there was no adequate support for the relationship between satisfaction and loyalty which will be tested and validated using another approach in subsequent chapters.

In traditional studies, researchers have employed either one or other of our proposed methods to test and prove hypotheses in service-oriented environments, but we propose a new challenge using anomalies regarding hypotheses proposition. Our methodology in this chapter clarifies that the key to resolving complaint issues and increasing customer satisfaction is to utilize both these methods together with the proposed conceptual framework.

We affirm that our approach is new and, as it is a combination of two methods, companies will be able to simply adopt our conceptual framework and handle their customer complaints using CRM and our additional elements. Using the current approach, the research platform is made more competitive as companies today are mostly service-based and customer-oriented. In the following, we introduce new guidelines which should be followed when implementing the proposed methodology. These will assist researchers to use our conceptual framework and propose recommendations in their own studies:

1. Collect real-world data.
2. Data must be in the formats of interviews and questionnaire.
3. Researchers must initiate the procedure after categorizing the data and applying text mining (quantitative) analysis to the collected interviews.
4. The data must be collected physically to ensure the integrity of the data.
5. Seek expert assistance to analyse the collected data.
6. Adopt the conceptual framework proposed in Chap. 4 of this thesis.
7. Company managers control the data processing.
8. Propose various hypotheses based on this chapter.
9. Test the hypotheses to determine the relationship between variables.
10. Use analysis of variance to see whether there is significant difference in the level of customer satisfaction between variables.
11. Use structural equation modelling to test the relationship between variables.
12. Propose recommendations to rectify the system based on supported hypotheses.
13. Reconsider refuted hypothesis and try to re-validate the relationship and ascertain the refuted one using new methodologies.
14. Using a combination of Analysis of variance and SEM, researchers can propose more effective recommendations, thereby increasing the relative satisfaction of customers.
15. SEM can also generate another model allowing researchers to interpret other results.
16. While these measurements may be adequate for testing and analysing the hypotheses and providing new recommendations, another methodology may need to be adopted in order to formulate the relationships in this chapter. This is discussed in subsequent chapters.
17. While some research has not been able to determine the bond between perceived value, interactivity, customer satisfaction, loyalty and customer acquisition, the results of this study provide an appropriate perspective and model which may assist researchers in future.

6.11 Cronbach Alphas Estimation for Each Variable

"Cronbach's alpha" is defined as a coefficient of reliability. In this thesis, we must estimate this for each section of our questionnaire data to measure reliability.

6.11.1 Customer Satisfaction Section

Case Processing Summary

		N	%
Cases	Valid	58	96.7
	Excluded(a)	2	3.3
	Total	60	100

Reliability Statistics

Cronbach's alpha	Number of questions
0.809	36

Among all the customer dissatisfaction points, we excluded two of the questions which were related to the feedbacks and willingness of the drivers (customer) to remain in a queue. These two questions were not answered using the Likert scale. We used the willingness of the customers in our demographic section. The result of reliability derived from software shows 0.809 for customer satisfaction and illustrates good reliability with acceptable consistency.

6.11.2 Customer Loyalty Section

Reliability Statistics

Cronbach's alpha	Number of questions
0.911	4

We chose five questions for the loyalty section, from which we excluded one question which asked for the employment history of the driver. Also, the responses were not based on the Likert scale and as the question was not consistent with other questions, it was excluded. However, we used it in our demographic section to find key customers.

The analysis of Cronbach Alpha shows 0.911 for customer loyalty and indicates strong consistency and reliability.

6.11.3 Perceived Value Section

Reliability Statistics

Cronbach's alpha	Number of questions
0.740	7

There were seven questions in this section related to customer perceived value and all were used. The analysis of customer perceived value shows 0.740 and implies acceptable reliability and stability of the data in this section. Also, it shows the correlation of this section with other sections.

6.11.4 Interactivity Section

Reliability Statistics

Cronbach's alpha	Number of questions
0.884	6

In this section, we had six questions to estimate Cronbach's Alpha for interactivity. The result was 0.884 which indicates good reliability and consistency. Also, it illustrates the high correlation of interactivity with the rest of the sections in our questionnaire.

6.11.5 Customer Acquisition Section

Reliability Statistics

Cronbach's alpha	Number of questions
0.700	3

In this section, we excluded one of the questions due to the lack of correlation with other questions; however, we analyse Cronbach's alpha using three related questions and obtained 0.70 which indicates reliability and consistency.

6.12 Measurement in I-CRM

For measurement purposes, we used the Likert scale [10]. To estimate the validity of the data, the experts may give the score based on any form of a Likert scale ranging from 1 (Very irrelevant) to 7 (Highly relevant). The experts can also comment on each of the issues in the questionnaire. The experts can provide comments, feedback and suggestions regarding the questionnaire [11]. A 7-point Likert scale (1 = totally unacceptable, 7 = Perfectly Acceptable) and (1 = strongly dissatisfied, 7 = extremely satisfied) were utilized to measure the variables of the conceptual framework. Except for the demographic questions, current numbers of drivers in comparison with previous year which we put the multiple answers based on (1 = Yes, 2 = No and 3 = I do not know) and employment history of the drivers which we make the question and answer using (1 = less than 1 year, 2 = less than 1 year, 3 = 1–4 years, 4 = 5–10 years, and 5 = More than 10 years). The questionnaire was cleansed again after a pre-test conducted with supervisors and following input from ten researchers who contributed their expertise [12].

6.13 Conclusion

In this chapter, the hypotheses regarding dissatisfaction and various issues at the Fremantle port were subjected to testing, validation and verification. Statistical testing was used for the first set of hypotheses to validate the existence and significance of issues which emerged in Chap. 5 utilising text analysis as well as ascertaining the hypotheses.

Our second set of hypotheses regarding the relationship between customer satisfaction, perceived value, interactivity, loyalty and customer acquisition were also tested. However, of the four hypotheses, three were verified but the hypothesis suggesting a relationship between customer satisfaction and loyalty was rejected due to the fact that there were quite a lot of customers with many complaints. There should be another way to test and prove this hypothesis and validate it irrespective of the number of variables. Ultimately, we cannot claim that our customer relationship management system is intelligent so far.

Also, in this chapter, the numerous variables, whether latent or observed, created intricacy and complexity. Also, the existing CRM is complex and based on the literature reviewed in Chap. 2, it does not have the intelligence and capability to handle the issues which have been identified. Hence, we need to provide intelligence in CRM systems that require non-linear programming and a non-linear tool to be able to evaluate the variables which are investigated in Chap. 7. To address the mentioned issues and provide more effective recommendations, we need to have an Intelligent CRM. Hence, the relationships need to be formulated using particular lab tools which will be discussed in Chap. 7. Also, relationship modelling will be carried out in the following chapter, as well as a revalidation of the hypotheses using another tool and methodology. In subsequent chapters, the methodology for making the system intelligent will be discussed.

References

1. Montgomery, D. C. (2001). *Design and analysis of experiments*. New York: Wiley.
2. Joo, Y. G., & Sohn, S. Y. (2008). Structural equation model for effective CRM of digital content industry. *Expert Systems with Applications, 34*, 63–71.
3. SSIcentral (2011, 2012). Structural equation models. Available http://www.ssicentral.com/lisrel/complexdocs/chapter5_web.pdf.
4. Lavee, Y. (1988). Linear structural relationship in family research. *Journal of marriage and the family*, 937.
5. Barrett, P. (2007). Structural equation modelling: Adjudging model fit. *Personality and Individual Differences, 42*, 815–824.
6. Hsu, S. H., Chen, W., & Hsieh, M. (2006). Robustness testing of PLS, LISREL, EQS and ANN-based SEM for measuring customer satisfaction. *Total Quality Management and Business Excellence, 17*, 355–372.
7. Gallarza, M. G., & Gil Saura, I. (2006). Value dimensions, perceived value, satisfaction and loyalty: An investigation of university students' travel behaviour. *Tourism Management, 27*, 437–452.
8. Slevitch, L., & Oh, H. (2010). Asymmetric relationship between attribute performance and customer satisfaction: A new perspective. *International Journal of Hospitality Management, 29*, 559–569.
9. Carvajal, S. A., Ruzzi, A. L., Nogales, Á. F., & Moreno, V. M. (2011). The impact of personalization and complaint handling on customer loyalty. *African Journal of Business Management, 5*, 13187–13196.
10. Vagias, W. M. (2006). Likert-type scale response anchors. In *Clemson International Institute for Tourism and Research Development* (p. 1). Department of Parks, Recreation and Tourism Management, Clemson University.
11. Uggioni, P. L., & Salay, E. (2012). Reliability and validity of a scale to measure consumer attitudes regarding the private food safety certification of restaurants. *Appetite, 58*, 470–477.
12. Garrido-Moreno, A., & Padilla-Meléndez, A. (2011). Analyzing the impact of knowledge management on CRM success: The mediating effects of organizational factors. *International Journal of Information Management, 31*, 437–444.

Chapter 7
Improving Customer Satisfaction Through Customer Type Mapping and I-CRM Strategies

7.1 Introduction

In this chapter, we carry out the modelling of dissatisfaction and satisfaction through the five steps of customer screening, data reduction, principal component analysis, data envelopment analysis, achieving potential improvement and providing recommendations in order to address the issues at the Fremantle port. We also re-validate our hypotheses for the major variables using a complete methodology. This process will help us to group our customers into clusters and we map the issues and the impact factors to these clusters. This in turn helps us to identify and prioritise the significant issues that should be addressed first in order to improve customer satisfaction within the shortest time with maximum results. This concept can be proven by the processes of analytical hierarchy and sensitivity analysis which provide mapping between customer type and satisfaction level.

7.2 A Framework for I-CRM Development and Customer Satisfaction

In order to convert customer dissatisfaction issues to customer satisfaction indicators, we developed the following five-step framework including consideration of customer type mapping and strategy development. Figure 7.1 illustrates how we define and evaluate the customer satisfaction. Also, we carry our correlation analysis and clustering of customer types. Finally, we map dissatisfaction issues and turn them into satisfaction indicators followed by the development of recommendations.

In the following framework, we summarize the five steps which are discussed next together with sensitivity analysis.

A. R. Faed, *An Intelligent Customer Complaint Management System with Application to the Transport and Logistics Industry*, Springer Theses, DOI: 10.1007/978-3-319-00324-5_7,
© Springer International Publishing Switzerland 2013

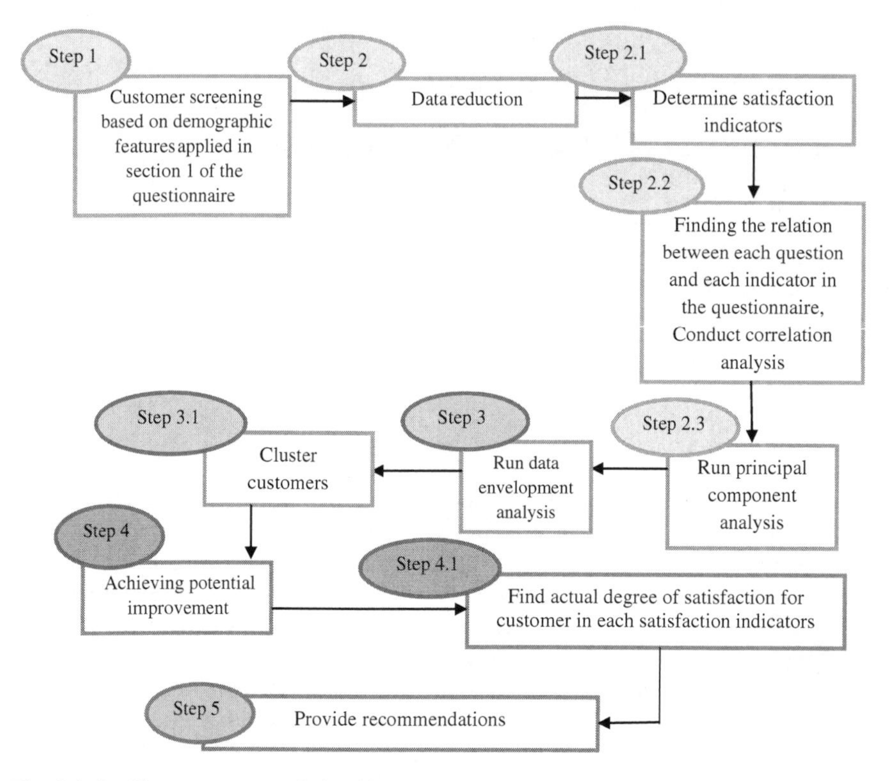

Fig. 7.1 Intelligent customer relationship management module for satisfaction

Step 1 Customer screening based on demographic features is applied in Section 1 of the questionnaire.

Step 2 Data reduction, determination of satisfaction indicators and finding the relation between each question and each indicator in the questionnaire, conduct correlation analysis. Next, we need to conduct principal component analysis to transform satisfaction indicators values to one value.

The principal component analysis process is performed for all categories of indicators and calculates a unique value for each satisfaction indicator. The output of the principal component analysis is used for further analysis and procedure of the proposed approach by data envelopment analysis.

Step 3 Run DEA in order to calculate the overall degree of satisfaction for each customer. Simultaneously, we need to cluster customers under various categories.

Step 4 Determine potential improvement and obtain the real degree of customer satisfaction.

Step 5 Propose recommendations to improve both the system and the organization.

In the following section, we give a detailed description of the five steps and the algorithms and methodology and the results of the algorithm on questionnaire data. This part is provided to improve customer satisfaction through customer type mapping and strategies for I-CRM development.

7.3 Step 1: Customer Screening

Output: Key customers

There are three categories of customers, which are defined below:

7.3.1 Category One Customers

Category one customers or key customers, in this thesis, have been type mapped (ranked) based on the following priority.

Customers would like to feel important and being called by their first names is one step which will make a difference. Meanwhile, each individual likes to be differentiated from the other customers by receiving extra services. They are the people or entities who climb the ladder of the company and show their agreement with services within the time frame. They need to be treated by the company in a way that exceeds their expectations. They have the highest level of expectations; hence, it is difficult to fully satisfy category one customers and this is the main focus of this research. By their feedback, they can easily leverage the decision-making process regarding services and products. We have grouped the customers under category one, category two and category three in terms of their education level, position, working condition, knowledge of the culture of Western Australia, character of the customers and working experience of the customers irrespective of gender and age.

7.3.2 Category Two Customers

Category two customers are the second important priority in our sample. Their presence is vital for analysis purposes. They have the potential to become category one customers. A company needs to make them feel very special in order to turn them into category one customers. We grouped our customers into category two on the basis of their education level, position, working condition, knowledge of Western Australian culture, character of the customers and working experience of the customers irrespective of their gender and age.

7.3.3 Category Three Customers

Category three customers are those with low decision-making impact on the company. Like the previous customer categories, category three customers will be grouped based on education level, position, condition, knowledge of Western Australian culture, character of the customers and working experience of the customers irrespective of gender and age.

In the screening procedure, the sixty customers (drivers) who were interviewed and responded to the questionnaire are screened and examined based on six criteria. These criteria are extracted from questions 3, 4, 5, 7 and 8 in section one and question number 3 in section three of the questionnaire.

The extracted criteria are:

7.3.4 Level of Education

"Education" is defined as a means of elevation and improvement in life. Education can be either academic or non-academic, or both. Either of the alternatives can present a new avenue for the individual for development, confidence and self-actualisation. Education is globally believed to be one of the critical stepping stones for individuals' enhancement, and alleviates poverty. In this study, level of education is one of the significant criteria in the search for key customers and to find category one customers or the most important ones.

Also, it has a direct relationship with experience. Education has four levels which are:

(1) Some high school
(2) High school graduate
(3) Some college/technical
(4) College graduate

We consider levels 3 and 4 as category two customers (important) and category one customers (most important customers) respectively. High school graduates are considered as category three customers (ordinary customers).

7.3.5 Position

"Position" is defined as the current working status of an individual in a work setting. Position is very important when selecting the key customers. We categorized our drivers (customers) as novices, drivers or operators, experienced drivers, or driver owners. Respondents respond to this item according to their position in the company. Each customer has one of the following positions:

7.3.5.1 A Novice

A novice customer is the driver who has started this work recently and does not have any previous experience in this type of job.

7.3.5.2 Driver and Operator

There are some drivers who are operators as well and have no connection with any logistics companies. They are all self-employed.

7.3.5.3 Experienced Driver

As described in the section above.

7.3.5.4 Driver and Owner

These are drivers who own their own trucks; however, they are contracted to and under the control of a logistics company. We consider experienced drivers and owners as category one customers. Customers who are both driver and operator are category two customers and finally, novice drivers are category three customers.

7.3.6 Work Status

"Work status" is defined as the existing status of an employee's contract. Based on this study, working status is one of the most significant factors that contribute to finding key customers. Customers may have a full-time, part-time and casual working contract with the company. Those who are employed as full-time workers for the company can be considered as loyal and important customers and accordingly, part-time and casuals can be considered as less important customers of the company because they allocate less time to the company. The basis on which the customer is employed determines the status of his/her work. Work status is one of the following:

(1) Full-time
(2) Part-time
(3) Casual

A full-time driver can be considered as a category one customer. A part-time driver is considered as a category two customer and a casual driver is a category three customer.

7.3.7 Knowing the Culture of WA

"Knowing the culture" is defined as the familiarity with the culture of a particular region. Knowing the culture is alike knowing the language of a region. Culture is that concealed relationship which provides a bond between the individual and a society. The behaviour of people also reflects culture. Various cultures in different regions of the world create diversity in the ways people live.

Culture is essentially important as it gives a unique identity to every area of the globe. Each region may be well known by that particular culture and characteristics and culture constructs the personality of individuals in that community. In order to be able to undertake and complete all tasks successfully, people need to have a good knowledge of a particular culture with its beliefs, customs, traditions and celebrations.

We chose this question to make a new contribution to this field and obtain a more accurate result in the selection of category one and two customers. In Australia, like other countries and regions around the world, each state or region has a separate culture, rules and regulations. Likewise, customers or the drivers of the company must be familiar with the culture of Western Australian in order to perform better and communicate more effectively with people of that region. Based on this study and to find key customers, we claim that those who know the Western Australian culture may have better performance in comparison with individuals who know little or nothing about the culture.

The level of familiarity with the working culture in Western Australia (WA) also determines the importance of a customer. Customers are familiar with the working culture in WA at one of the following levels:

(1) Poor
(2) Fair
(3) Good
(4) Very good.
(5) Excellent

Customers who have good, very good, or excellent familiarity with the working culture are considered as category one customers. Fair familiarity delineates them as category two customers and those who have little knowledge of the culture are category three customers.

7.3.8 Character of the Customer

"Character" is defined as the representation of traits and moral features of an individual. Character of the customers and in our study, the character of drivers, may be responsible for word of mouth. There are some psychological aspects of drivers that separate them from the others. We included the psychological aspect to extend the horizon in this field and find our key customers.

Character of a customer (driver) that is demonstrated by the customer (driver) is another criterion for screening. These characteristics are:

(1) Relaxed and calm
(2) Easy-going and friendly
(3) Easily distracted
(4) Easily frustrated

Relaxed, calm, easy-going and friendly customers are the category one customers. Category two customers may have characteristics 1 and 2; however, due to stress in the environment, familial issues or problems occurring at the port, they may get distracted. As, category three customers are novice, and unfamiliar with the system and working environment, they may become easily distracted, frustrated and annoyed. Despite this fact, they can be very courteous and friendly people.

7.3.9 Work Experience

"Work experience" is defined as the knowledge and capabilities that an individual has acquired since the beginning of his working journey. The number of years that the driver has had experience on the job is critically important when we want to search for key customers. It is clear that a person with several years of experience can be very helpful in our study.

Work experience is the last criterion for screening customers. Experience levels are:

(1) less than 1 year
(2) 1–4 years
(3) 5–10 years
(4) More than 10 years

Customers with more than five year experience are category one customers. Customers who have between one and for an experience of four years are category two customers and customers with less than one year experience are considered as category three customers.

Based on the aforementioned procedure, category one customers are determined as presented in Table 7.1.

Category two customers are presented in Table 7.2.

The remainder are category three customers or 'ordinary customers'.

Table 7.1 Category one customers and their levels in each criterion

#	Customer	Level of education	Position	Work status	Knowing the culture	Character	Experience
1	Customer 15	2	3	1	4	2	3
2	Customer 18	2	4	1	5	2	3
3	Customer 21	3	3	1	5	2	3
4	Customer 22	3	4	1	4	2	4
5	Customer 23	3	4	1	5	2	4
6	Customer 25	2	4	1	5	2	3
7	Customer 41	2	3	1	3	1	3

Table 7.2 Category two customers and their levels in each criterion

#	Customer	Level of education	Position	Work status	Knowing the culture	Character	Experience
1	Customer 4	3	4	1	5	2	1
2	Customer 8	2	3	1	5	2	2
3	Customer 11	2	3	1	5	1	2
4	Customer 12	2	2	1	3	2	1
5	Customer 13	3	3	1	5	2	2
6	Customer 20	3	3	1	3	2	2
7	Customer 24	4	3	1	4	2	1
8	Customer 27	2	4	1	3	3	2
9	Customer 29	2	3	1	4	2	2
10	Customer 30	3	3	1	4	1	2
11	Customer 35	3	3	1	3	1	2
12	Customer 44	3	3	1	3	1	2
13	Customer 47	2	2	1	4	1	2
14	Customer 48	2	3	1	5	1	1
15	Customer 49	3	3	1	4	1	2
16	Customer 50	4	3	1	4	2	2
17	Customer 51	2	3	1	5	2	1
18	Customer 52	3	4	1	3	2	2
19	Customer 54	2	3	1	3	2	2
20	Customer 57	3	3	1	5	2	2
21	Customer 58	2	3	1	3	4	2

7.4 Step 2: Data Reduction

Output: key satisfaction indicators

7.4.1 Step 2.1: Indicators Categorization

In this step, nine categories of indicators are established according to the sub-main questions of this research. These indicator categories are:

(1) Management
(2) Booking system
(3) Infrastructure
(4) Health and safety
(5) Container sorting
(6) Cost
(7) Jurisdiction
(8) Stevedores performance
(9) Time management

In each category, there are related questions in Section 2 of the questionnaire. Table 7.3 shows the relationship between questions and indicator categories. In Table 7.3, numbers after Q indicate the section and question number in the questionnaire, respectively. For example, Q2-1 is the first question in Section 2 of the questionnaire which is related to the management issues at the Fremantle port.

Table 7.3 The relation between questions and indicator categories

Indicator category	Questions
(1) Management	Q2-1, Q2-2, Q2-3, Q2-4, Q2-5, Q2-6, Q2-10, Q2-12, Q2-15, Q2-16, Q2-19, Q2-20, Q2-21, Q2-28, Q2-29, Q2-31, Q2-32, Q2-33, Q2-34, Q2-35, Q2-36
(2) Booking system	Q2-5, Q2-6, Q2-7, Q2-8, Q2-9, Q2-14, Q2-17, Q2-18
(3) Infrastructure	Q2-8, Q2-10, Q2-15, Q2-19, Q2-23, Q2-24, Q2-25, Q2-26, Q2-27, Q2-29, Q2-30, Q2-31, Q2-33, Q2-34
(4) Health and safety	Q2-11, Q2-21, Q2-22, Q2-28
(5) Container sorting	Q2-17, Q2-18
(6) Cost	Q2-10, Q2-23, Q2-24, Q2-25, Q2-30
(7) Jurisdiction	Q2-4, Q2-12, Q2-13
(8) Stevedore performance	Q2-13, Q2-16, Q2-20
(9) Time management	Q2-6, Q2-7, Q2-20, Q2-21, Q2-23, Q2-35, Q2-36

7.4.2 *Step 2.2: Correlation Analysis*

In this step, a correlation analysis is performed for the questions in each category of indicators. The purpose of correlation analysis is to understand the relationship between the questions in each category. High correlation between questions in a category reveals a high potential for data reduction. Furthermore, the correlation matrix in each category is the input for PCA which will be presented in the next step.

The results of correlation analysis in all categories are presented in Appendix 2. Table 7.4 presents a summary of results. In this table, the number and percentage of the significant correlations are specified. Significant correlations in 90 % confidence level are those correlations whose p value is less than 0.1.

Table 7.4 The correlation summary results

Indicator category	Total number of questions	Total possible paired correlations	Number of significant correlations	Percentage of significant correlations (%)
(1) Management	21	210	63	30
(2) Booking system	8	28	13	46
(3) Infrastructure	14	91	14	15
4) Health and safety	4	6	1	17
5) Container sorting	2	1	0	0
6) Cost	5	10	1	10
7) Jurisdiction	3	3	3	100
8) Stevedore performance	3	3	0	0
9) Time management	7	21	4	19

As seen in Table 7.4, the percentage of significant correlations in each indicator category varies between 0 and 100 %. However, in the categories with total number of questions more than 5, at least 15 % of correlations are significant. Since, as we prefer to have as few indicators as possible in data envelopment analysis, we perform data reduction with principal component analysis for all 9 categories of indicators.

7.4.3 Step 2.3: Principal Component Analysis (PCA)

Principal component analysis is adopted to assign a number to each category of questions related to each satisfaction variable. In this step, the precision of data has been decreased and turned into one number.

7.4.3.1 Principal Component Analysis Process

Principal component analysis (PCA) is extensively used in multivariate statistics such as factor analysis. It is utilised to diminish the number of variables under study and for the analysis of decision-making units (DMUs) such as industries, universities, hospitals, cities, etc. The goal of PCA is to determine a combination of elements and variables such that each new variable, called a principal component, is a linear combination of original variables. The first new variable y_1 accounts for the maximum variance in the sample data. The new variables (principal components) are uncorrelated. Principal component analysis is performed by identifying the eigenstructure of the covariance or singular value decomposition of the original data.

Principal component analysis is performed by identifying the eigenstructure of the correlation matrix of the original data. Suppose X is a m × n matrix composed by new x_{ij}'s defined as the value of jth index (indicator) for ith customer. Furthermore, suppose $\hat{X}$ is the standardized matrix of X with x_{ij}'s defined as the value of jth standardized indicator for ith customer. PCA is performed to identify new independent variables or principal components (defined as Y_j for $j = 1,\ldots,n$), which are respectively different linear combination of $\hat{x}_1 \ldots \hat{x}_n$. As mentioned, this is achieved by identifying the eigenstructure of the correlation matrix of the original data. The principal components are defined by a m × n matrix $Y = (y_1 \ldots y_n)_{m \times n}$ composed of y_{ij}'s. we have:

To obtain the L and consequently vectors Y and PCA scores, the following steps are performed:

PCA Step 1: Standardize data X to $\hat{X}$

PCA Step 2: Calculate the sample spearman correlation matrix R for $\hat{X}$.

PCA Step 3: Calculate eigen values (λ) and eigen vectors (L) of correlation matrix R Those eigen vectors comprise the principal components Y_i. The components in eigenvectors are respectively the coefficients in each corresponding *Yi* (Eq. 1):

$$Y = L * \hat{X} \tag{7.1}$$

PCA Step 4: Calculate the weights w_j of the principal components and PCA scores Z (the value of each indicator category).

$$w_j = \lambda_j / \sum_{j=1}^{n} \lambda_j = \lambda_j / n \quad ,j = 1, \ldots, n \tag{7.2}$$

$$z_i = \sum_{j=1}^{n} w_j Y_{ij} \quad i = 1, \ldots, m \tag{7.3}$$

7.4.3.2 PCA Results of Indicator Categories

The principal component analysis process in Sect. 7.4.3.1 is performed for all nine categories of indicators. The complete result of Principal component analysis in each indicator category by MINITAB® Release 14.1 is presented in Appendix 3. The overall indicator values calculated using Eq. 7.3 for all the customers is presented in Table 7.5.

In Table 7.5, abbreviations are:

Category indicator	Abbreviation
Management	MNG
Booking system	BS
Infrastructure	INF
Health and safety	HS
Container sorting	CONS
Cost	CO
Jurisdiction	JUR
Stevedore performance	STP
Time management	TMNG

Table 7.5 Indicator actual values obtained from PCA

DMU	Actual indicator values obtained from PCA								
	MNG	BS	INF	HS	CONS	CO	JUR	STP	TMNG
Customer 1	3.3	4.0	3.9	4.4	2.7	4.7	3.0	3.7	2.6
Customer 2	4.5	4.8	3.4	3.4	2.7	3.3	5.0	4.0	7.0
Customer 3	3.8	4.6	2.6	4.2	3.9	2.9	4.2	4.1	4.6
Customer 4	3.0	3.4	4.2	4.9	4.6	3.3	4.0	4.2	2.5
Customer 5	4.3	4.5	2.8	2.2	3.9	1.0	4.0	4.3	4.4
Customer 6	2.5	5.2	4.5	5.1	6.3	4.9	4.3	3.3	6.0
Customer 7	6.5	2.5	5.5	1.2	1.2	3.3	6.1	5.0	5.6
Customer 8	4.7	3.9	3.6	3.2	2.4	6.4	4.7	4.9	5.3
Customer 9	2.8	3.3	1.1	2.3	2.1	3.4	3.8	3.7	3.9
Customer 10	4.4	3.6	4.0	4.6	3.6	4.6	4.7	5.0	3.9
Customer 11	1.9	4.8	5.0	1.6	6.5	2.3	1.8	2.3	5.5
Customer 12	3.4	6.3	3.0	3.1	2.2	1.2	1.1	1.8	4.5
Customer 13	3.1	5.2	5.0	4.7	4.8	4.4	3.3	4.2	4.8
Customer 14	5.6	2.7	3.3	3.4	4.6	4.1	4.1	4.8	2.4
Customer 15	7.0	1.0	3.8	2.0	1.2	4.2	7.0	7.0	1.9
Customer 16	2.6	6.3	1.6	3.5	4.4	2.0	4.9	3.3	5.5
Customer 17	2.4	5.0	4.3	5.6	4.8	2.4	3.2	2.3	4.0
Customer 18	3.8	4.8	3.3	2.7	3.9	3.6	2.5	2.3	6.3
Customer 19	4.9	3.2	1.0	2.9	3.6	3.3	5.2	3.9	5.2
Customer 20	2.9	5.3	3.2	3.4	2.7	4.4	4.0	4.3	5.3
Customer 21	2.2	5.0	3.9	6.0	4.6	2.0	2.3	3.1	6.1
Customer 22	1.4	5.7	2.7	4.4	3.9	1.4	3.3	3.5	4.9
Customer 23	1.0	5.9	3.7	5.8	7.0	3.7	1.5	1.7	6.4
Customer 24	6.8	2.9	4.1	1.0	1.4	1.1	6.0	6.5	5.3
Customer 25	2.0	7.0	5.0	7.0	4.7	6.3	1.0	1.0	6.9
Customer 26	3.3	5.5	4.1	4.9	4.4	2.9	3.3	3.0	4.0
Customer 27	4.3	4.3	3.0	4.0	3.9	4.6	2.3	3.8	4.7
Customer 28	4.2	5.3	7.0	3.4	5.1	7.0	4.0	4.3	4.9
Customer 29	2.9	4.8	4.7	2.7	3.9	3.9	4.0	2.8	6.3
Customer 30	2.9	3.8	2.9	4.8	3.9	3.0	3.0	3.8	2.5
Customer 31	4.2	3.7	4.0	4.6	3.6	4.6	4.6	5.0	4.5
Customer 32	2.5	3.8	3.2	2.4	2.7	1.3	2.5	3.6	4.8
Customer 33	4.5	4.4	4.3	3.0	2.7	2.2	3.3	3.6	6.1
Customer 34	4.4	3.7	4.6	4.2	3.3	3.4	4.5	3.8	4.7
Customer 35	4.0	4.3	3.5	4.9	3.9	2.6	3.6	4.7	2.4
Customer 36	2.3	3.9	2.7	3.2	4.8	4.5	3.3	3.5	1.9
Customer 37	1.9	3.7	4.1	4.9	3.6	4.8	5.3	3.9	2.5
Customer 38	2.7	4.0	4.3	3.7	4.8	4.4	3.0	2.4	3.4
Customer 39	1.9	3.7	4.0	5.4	5.3	3.9	3.9	3.6	5.5
Customer 40	2.6	5.3	1.8	5.2	2.0	2.0	2.8	3.1	5.6
Customer 41	2.5	4.0	3.1	4.7	2.1	3.3	3.6	3.1	5.8

(continued)

Table 7.5 (continued)

DMU	Actual indicator values obtained from PCA								
	MNG	BS	INF	HS	CONS	CO	JUR	STP	TMNG
Customer 42	3.9	4.7	4.7	1.4	1.3	3.8	4.0	4.1	4.2
Customer 43	3.9	4.0	3.3	2.4	1.0	2.6	3.5	4.6	4.1
Customer 44	4.2	4.1	6.4	3.0	3.4	6.0	4.0	4.3	1.0
Customer 45	3.0	5.9	5.6	2.5	4.1	6.1	2.2	2.0	4.8
Customer 46	4.3	4.8	5.5	2.6	6.0	6.1	3.7	3.4	3.8
Customer 47	4.9	3.9	4.2	2.3	2.0	3.7	3.0	3.6	2.5
Customer 48	5.4	4.0	4.1	3.7	3.9	4.3	4.6	5.6	1.2
Customer 49	4.9	3.8	3.9	4.3	6.0	3.7	2.8	4.6	4.3
Customer 50	4.0	2.7	4.4	4.5	2.4	3.5	3.6	3.3	2.5
Customer 51	4.2	5.1	4.7	3.8	4.1	4.2	4.1	2.3	6.3
Customer 52	2.8	6.3	3.6	2.0	4.4	3.1	2.3	2.5	4.5
Customer 53	4.5	4.7	2.4	2.8	3.9	3.6	3.5	4.7	2.5
Customer 54	3.6	3.1	2.9	3.0	5.3	3.9	3.0	4.6	4.1
Customer 55	3.0	3.6	1.9	1.5	3.3	3.4	1.9	4.1	4.8
Customer 56	3.9	4.3	4.0	3.7	4.4	2.9	5.0	3.3	5.6
Customer 57	3.0	4.9	4.2	3.3	1.0	1.4	4.2	3.5	3.5
Customer 58	3.5	5.6	5.9	3.4	2.5	3.4	2.4	2.3	5.8
Customer 59	5.4	5.0	4.8	2.1	5.1	4.7	3.6	3.9	5.5
Customer 60	4.0	4.5	3.5	3.6	4.8	2.0	3.7	4.2	3.3

7.5 Step 3: Data Analysis with Data Envelopment Analysis

In this step, we aim to define and reach the level of optimum precision; data envelopment analysis is applied to the three categories of customers (category one, category two and category three) to identify the strengths and weaknesses of customer services. Essentially, we decrease data precision by clustering customers into three groups and then increase our understanding of the customer viewpoint by utilizing data envelopment analysis. In other words, we enhance our precision in a particular part of collected data. The optimization ability of Data Envelopment Analysis (DEA) is used to generate improvement at the Fremantle port.

Output: Calculating an overall degree of satisfaction for each customer including the key customers.

7.5.1 Data Envelopment Analysis: An Overview

In accordance with data envelopment analysis terminology, a decision-making unit (DMU) must be assigned to the customers in the assessment categorization. Data envelopment analysis generates a level which is the boundary in which it pursuits the maximum performers and envelops the remainder [1]. Therefore, the empirical frontier predicated on real DMU is utilised. The empirical frontier is generated throughout connecting thorough 'relatively best' DMUs in the recognised population. Current observations develop a non-parametric empirical frontier so there is no need to explicitly achieve the function that relates inputs to outputs. Accordingly, the effectiveness score of each DMU is compared with the rest of the DMUs, in which the distance between a specific DMU and the frontier can be measured and achieved. It must be noted that if the action of noticed DMUs is weak, then the empirical frontier produces inaccurate outcomes.

Based on Luo and Homburg [2], data envelopment analysis is a mathematical programming tool that estimates and assesses the effectiveness of resource application and exercise. Data envelopment analysis measures the comparative effectiveness of the companies [2]. When we create the observed effectiveness of individual DMUs, DEA assists to recognise attainable measurements within which performance might be addressed. The competent and systematic DMUs are acknowledged as the 'peer' group for the ineffective DMU. Sometimes, it is essential to estimate the data on the peer DMUs to have an impartial evaluation of the inefficient DMU with the peer DMUs. The scaled and evaluated consolidations of peers might create a standard measurement for relatively less effective DMUs. The genuine levels of input utilisation or output creation of efficient DMUs can supply as a specific goal for less useful DMUs, whilst the procedures of measurement and benchmarking DMUs can be clarified and communicated for the information of managers of DMUs who have previously appointed to intensify the performance. The ability of data envelopment analysis to recognise possible peers or role models and simple efficiency scores makes it a very valuable method which many researchers use and rely upon. Also, this approach is a trustworthy one and distinguished in comparison with other performance measurement approaches.

The ultimate objective of Data Envelopment Analysis (DEA) is for achieving the weights that intensify the effectiveness of the DMU under assessment, when restricting the efficiency of all DMUs to less than or equal to 1.0 (for input-oriented models). Charnes et al. [1], recognised the hardship in searching the standard set of weights to formulate the comparative effectiveness, which may create DMUs to value inputs and outputs variously and therefore employ various weights. The authors propose that each DMU needs to employ weights which place it in the most favourable light in relation to the rest of the DMUs. The adaptability in the weights can be a deficiency and stability of this method. It can be considered as a weakness because the thoughtful option of weights by a DMU does not pertain to the value of any input or output that might permit a DMU to expose efficiency. It may have more to do with the choice of weights than operational efficiency. This adaptability and flexibility might be regarded as strength due to the fact that an inefficient DMU with even the most favourable weights cannot claim that the weights are unjust. Also, relationships may be included into the model to limit the weights as considered appropriate.

It must be stated that the efficiency values created by DEA are considered valid only within that specific team of peers. A DMU that is adequate and has the effectiveness for one group might be considered inadequate and inefficient when compared with another group. Likewise, if a group of very poor DMUs were estimated using DEA, we would still have efficient DMUs. As suggested, building a generic frontier is one way of addressing this problem. Additionally, if the set of DMUs is small, there is little discrepancy between elements.

7.5.2 Applied Data Envelopment Analysis Model

In the applied DEA model, we have 60 DMUs where each DMU indicates a customer. The only input for all the DMUs is considered as a unity vector. The reason is that there is no input in the customer satisfaction modelling. All the nine categories of indicators are considered as outputs. The value of these indicators for each customer is obtained from PCA analysis.

The applied DEA model for DMU_0 is shown in Model 1.

$$\begin{aligned} &\text{Max } \theta \\ &s.t. \\ &\sum_{j=1}^{60} \lambda_j \leq 1 \\ &\sum_{j=1}^{60} \lambda_j y_{rj} - s_r = \theta y_{ro} \quad r = 1{:}9 \\ &\lambda_j, s_r \geq 0 \end{aligned} \tag{7.4}$$

In Model 1, θ is the technical efficiency or in our case, the overall degree of customer satisfaction (ODCS). This model is for the customer with index 0. In this model, y_{rj} is the value of rth indicator for customer j. s_r is the surplus variable for the rth constraint.

7.5.3 Data Envelopment Analysis Results

Table 7.5 shows the raw output data for DEA. Each summary indicator in the columns of Table 7.5 is considered as an output for DEA. As mentioned earlier, each indicator in Table 7.5 shows the degree of customer satisfaction regarding a particular aspect of organizational performance such as management, booking system, infrastructure, health and safety, container sorting, cost, jurisdiction, stevedores performance, and time management.

Table 7.6 presents the overall degree of customer satisfaction (ODCS) in the first column based on Model 1 for each customer. The surplus variables are presented in the remaining columns. The efficiency score is between zero and one. Hence, those branches with efficiency of one are efficient. The weak efficient DMUs have non-zero surplus and the optimum objective value of one, while the strong efficient DMUs have zero surplus and the optimum objective value of one.

From the results of Model 1, an optimization procedure is presented. This procedure utilizes the values of surpluses and efficiency scores (ODCS) in the optimal solution of Model 1. Let us define a virtual DMU as a target. The output of this virtual target DMU is obtained from Eq. (2).

$$\hat{y}_{r0} = \frac{1}{\theta} y_{r0} + s_r \tag{2}$$

where $\hat{y}$ is target value, s_r is surplus in Model 1 and θ is optimum objective value (ODCS) of Model 1. From efficiency improvement point of view the target values of output in Eq. (2) are the optimal values and an optimization procedure can be described as follows: An inefficient DMU should increase its outputs to target outputs. A week efficient DMU should increase its outputs by surpluses.

The results in Table 7.6 indicate the values of surpluses. For instance, consider DMU_1 (Customer 1). According to the efficiency scores, this customer is not efficient. In other words, this customer is not satisfied compared to the other customers. If we (as decision makers) want to have customer 1 as a satisfied customer, we should increase the level of satisfaction in all categories of indicators by 11 % as $\frac{1}{\theta} = \frac{1}{0.9} = 1.11$.

In addition, because of non-zero surpluses for customer 1 in indicators MNG, INF, CONS, JUR, and TMNG, the target values of each output (indicator) for customer 1 should be calculated using Eq. 2. For example, the target values of MNG (management) and BS (booking system) for customer 1 are calculated as follows:

$$M\hat{N}G_1 = \frac{1}{0.9} MNG_1 + s_1^{MNG} = \frac{1}{0.9} 3.3 + 0.2 = 3.9$$

$$\hat{BS}_1 = \frac{1}{0.9} BS_1 + s_1^{BS} = \frac{1}{0.9} 4 + 0.0 = 4.5$$

If we decide to change the zero customer (new customer) to a satisfied customer, or if we intend to decrease the level of complaints, we need to change the actual value (yr0) to target value ($\hat{y}$r0).

The actual degree of satisfaction for customer 1 regarding management is calculated as 3.3 while the target value is 3.9. This target value is the minimum (optimum) level of satisfaction and if realized, will make customer 1 a satisfied customer. It should be noted that this improvement in satisfaction should occur in all aspects of organizational performance according to the target values in Table 7.7.

7.5.4 Customer Type Mapping

The DEA model 1 is not capable of ranking efficient DMUs as it assigns a common index of one to all the efficient DMUs. Andersen and Petersen [3] modified this

Table 7.6 Overall degree of customer satisfaction and model surpluses

DMU	ODCS (θ)	Indicator surplus								
		MNG	BS	INF	HS	CONS	CO	JUR	STP	TMNG
Customer 1	0.90	0.2	0.0	0.2	0.0	1.0	0.0	0.5	0.0	1.7
Customer 2	1.00	0.0	0.0	0.0	0.0	0.0	0.0	0.0	0.0	0.0
Customer 3	0.97	0.0	0.0	1.1	0.0	0.0	1.2	0.0	0.0	0.0
Customer 4	0.99	0.1	0.6	0.0	0.0	0.0	0.9	0.0	0.0	2.3
Customer 5	0.92	0.0	0.0	3.4	0.6	0.2	4.8	0.0	0.0	0.2
Customer 6	1.00	0.0	0.0	0.0	0.0	0.0	0.0	0.0	0.0	0.0
Customer 7	1.00	0.0	0.0	0.0	0.0	0.0	0.0	0.0	0.0	0.0
Customer 8	1.00	0.0	0.0	0.0	0.0	0.0	0.0	0.0	0.0	0.0
Customer 9	0.77	1.2	0.0	3.3	0.0	0.7	0.0	0.0	0.0	0.0
Customer 10	1.00	0.0	0.0	0.0	0.0	0.0	0.0	0.0	0.0	0.0
Customer 11	1.00	0.0	0.0	0.0	0.0	0.0	0.0	0.0	0.0	0.0
Customer 12	1.00	0.0	0.0	0.0	0.0	0.0	0.0	0.0	0.0	0.0
Customer 13	1.00	0.0	0.0	0.0	0.0	0.0	0.0	0.0	0.0	0.0
Customer 14	1.00	0.0	0.0	0.0	0.0	0.0	0.0	0.0	0.0	0.0
Customer 15	1.00	0.0	0.0	0.0	0.0	0.0	0.0	0.0	0.0	0.0
Customer 16	1.00	0.0	0.0	0.0	0.0	0.0	0.0	0.0	0.0	0.0
Customer 17	0.98	0.0	0.3	0.2	0.0	0.0	2.9	0.0	0.4	1.4
Customer 18	0.96	0.0	0.1	0.4	0.9	0.0	0.0	1.3	1.1	0.0
Customer 19	0.99	0.0	0.5	3.3	0.0	0.0	0.0	0.0	1.1	0.0
Customer 20	1.00	0.0	0.0	0.0	0.0	0.0	0.0	0.0	0.0	0.0
Customer 21	1.00	0.0	0.0	0.0	0.0	0.0	0.0	0.0	0.0	0.0
Customer 22	0.98	1.4	0.0	1.0	0.0	0.7	2.3	0.3	0.0	0.3
Customer 23	1.00	0.0	0.0	0.0	0.0	0.0	0.0	0.0	0.0	0.0
Customer 24	1.00	0.0	0.0	0.0	0.0	0.0	0.0	0.0	0.0	0.0
Customer 25	1.00	0.0	0.0	0.0	0.0	0.0	0.0	0.0	0.0	0.0
Customer 26	0.98	0.0	0.0	0.0	0.0	0.0	1.8	0.0	0.2	1.0
Customer 27	0.96	0.0	0.0	1.3	0.0	0.9	0.0	0.7	0.0	0.0
Customer 28	1.00	0.0	0.0	0.0	0.0	0.0	0.0	0.0	0.0	0.0
Customer 29	0.99	1.0	0.5	0.0	1.4	0.0	1.0	0.0	0.7	0.0
Customer 30	0.92	0.0	0.1	0.7	0.0	0.0	0.0	0.3	0.0	2.0
Customer 31	1.00	0.0	0.0	0.0	0.0	0.0	0.0	0.0	0.0	0.0
Customer 32	0.80	1.6	0.0	0.6	0.0	0.0	2.5	1.7	0.0	0.0
Customer 33	0.96	0.0	0.0	0.0	0.1	0.2	1.8	1.0	0.0	0.0
Customer 34	1.00	0.0	0.0	0.0	0.0	0.0	0.0	0.0	0.0	0.0
Customer 35	1.00	0.0	0.0	0.0	0.0	0.0	0.0	0.0	0.0	0.0
Customer 36	0.85	1.2	0.1	2.0	0.4	0.0	0.0	0.0	0.0	2.8
Customer 37	1.00	0.0	0.0	0.0	0.0	0.0	0.0	0.0	0.0	0.0
Customer 38	0.82	0.0	0.2	0.0	0.0	0.0	0.1	0.3	0.8	1.3
Customer 39	1.00	0.0	0.0	0.0	0.0	0.0	0.0	0.0	0.0	0.0
Customer 40	0.97	0.0	0.0	2.7	0.0	2.7	2.2	0.0	0.0	0.0
Customer 41	0.93	0.1	0.8	0.8	0.0	2.9	0.4	0.0	0.0	0.0

(continued)

Table 7.6 (continued)

DMU	ODCS (θ)	Indicator surplus								
		MNG	BS	INF	HS	CONS	CO	JUR	STP	TMNG
Customer 42	0.92	0.0	0.0	0.4	1.8	3.1	1.2	0.0	0.0	0.0
Customer 43	0.90	0.1	0.0	0.0	0.2	1.5	0.7	0.7	0.0	0.0
Customer 44	0.95	0.0	0.6	0.0	0.1	1.2	0.5	0.0	0.0	3.6
Customer 45	0.96	0.0	0.0	0.0	2.6	0.6	0.1	0.1	0.5	1.0
Customer 46	1.00	0.0	0.0	0.0	0.0	0.0	0.0	0.0	0.0	0.0
Customer 47	0.89	0.0	0.0	0.0	0.0	2.0	0.3	0.9	0.6	1.2
Customer 48	1.00	0.0	0.0	0.0	0.0	0.0	0.0	0.0	0.0	0.0
Customer 49	1.00	0.0	0.0	0.0	0.0	0.0	0.0	0.0	0.0	0.0
Customer 50	0.96	0.0	1.5	0.0	0.0	0.9	1.7	0.0	0.4	2.0
Customer 51	1.00	0.0	0.0	0.0	0.0	0.0	0.0	0.0	0.0	0.0
Customer 52	0.99	0.0	0.0	0.0	2.7	0.2	1.2	0.8	0.0	1.5
Customer 53	0.98	0.0	0.0	2.7	0.7	0.5	1.6	0.8	0.0	0.7
Customer 54	0.97	1.3	0.6	1.2	0.9	0.0	0.0	0.2	0.0	0.0
Customer 55	0.86	1.2	0.2	2.7	1.3	0.0	0.0	2.7	0.0	0.0
Customer 56	1.00	0.0	0.0	0.4	0.0	0.0	1.2	0.0	0.7	0.0
Customer 57	0.92	0.1	0.0	0.0	0.0	3.4	3.2	0.0	0.1	1.0
Customer 58	0.99	0.0	0.0	0.0	1.3	1.9	2.8	0.5	0.6	0.0
Customer 59	1.00	0.0	0.0	0.0	0.0	0.0	0.0	0.0	0.0	0.0
Customer 60	0.94	0.0	0.0	1.9	0.0	0.0	3.5	0.0	0.0	0.7

DEA model (known as AP DEA model) for DEA-based ranking purposes as shown in Model 3:

$$
\begin{aligned}
&\text{Min } \theta \\
&\quad s.t. \\
&\quad \sum_{\substack{j=1 \\ j\neq 0}}^{n} \lambda_j x_{ij} \leq \theta x_{i0} \quad i = 1, 2, \ldots m \\
&\quad \sum_{\substack{j=1 \\ j\neq 0}}^{n} \lambda_j y_{rj} \geq y_{ro} \quad r = 1, 2, \ldots s \\
&\quad \lambda_j \geq 0, \forall j j \neq 0 \text{ and } \theta \, is\, free
\end{aligned} \tag{3}
$$

We use Model 3 for the full ranking of the customers. In Model 3, a DMU is efficient if the efficiency score of that DMU is equal to or greater than one. All inefficient DMUs have an efficiency score less than one. The ranking results for Model 3 are presented in the third and fourth columns of Table 7.8.

Table 7.7 Target values for each indicator

DMU	Indicator target values								
	MNG	BS	INF	HS	CONS	CO	JUR	STP	TMNG
Customer 1	3.9	4.5	4.6	4.9	4.0	5.2	3.9	4.1	4.6
Customer 2	4.5	4.8	3.4	3.4	2.7	3.3	5.0	4.0	7.0
Customer 3	3.9	4.7	3.8	4.3	4.0	4.2	4.3	4.2	4.7
Customer 4	3.1	4.0	4.2	4.9	4.6	4.3	4.0	4.2	4.8
Customer 5	4.7	4.9	6.4	3.0	4.5	5.9	4.3	4.7	5.0
Customer 6	2.5	5.2	4.5	5.1	6.3	4.9	4.3	3.3	6.0
Customer 7	6.5	2.5	5.5	1.2	1.2	3.3	6.1	5.0	5.6
Customer 8	4.7	3.9	3.6	3.2	2.4	6.4	4.7	4.9	5.3
Customer 9	4.8	4.3	4.8	3.0	3.4	4.4	5.0	4.8	5.1
Customer 10	4.4	3.6	4.0	4.6	3.6	4.6	4.7	5.0	3.9
Customer 11	1.9	4.8	5.0	1.6	6.5	2.3	1.8	2.3	5.5
Customer 12	3.4	6.3	3.0	3.1	2.2	1.2	1.1	1.8	4.5
Customer 13	3.1	5.2	5.0	4.7	4.8	4.4	3.3	4.2	4.8
Customer 14	5.6	2.7	3.3	3.4	4.6	4.1	4.1	4.8	2.4
Customer 15	7.0	1.0	3.8	2.0	1.2	4.2	7.0	7.0	1.9
Customer 16	2.6	6.3	1.6	3.5	4.4	2.0	4.9	3.3	5.5
Customer 17	2.5	5.4	4.6	5.7	4.9	5.4	3.3	2.7	5.5
Customer 18	4.0	5.1	3.8	3.7	4.1	3.8	3.9	3.5	6.6
Customer 19	4.9	3.7	4.4	2.9	3.6	3.3	5.2	5.0	5.2
Customer 20	2.9	5.3	3.2	3.4	2.7	4.4	4.0	4.3	5.3
Customer 21	2.2	5.0	3.9	6.0	4.6	2.0	2.3	3.1	6.1
Customer 22	2.8	5.8	3.7	4.5	4.6	3.7	3.7	3.6	5.3
Customer 23	1.0	5.9	3.7	5.8	7.0	3.7	1.5	1.7	6.4
Customer 24	6.8	2.9	4.1	1.0	1.4	1.1	6.0	6.5	5.3
Customer 25	2.0	7.0	5.0	7.0	4.7	6.3	1.0	1.0	6.9
Customer 26	3.4	5.6	4.2	5.0	4.5	4.8	3.4	3.2	5.0
Customer 27	4.5	4.5	4.4	4.2	4.9	4.8	3.1	4.0	4.9
Customer 28	4.2	5.3	7.0	3.4	5.1	7.0	4.0	4.3	4.9
Customer 29	3.9	5.3	4.7	4.1	3.9	4.9	4.0	3.5	6.3
Customer 30	3.2	4.2	3.9	5.2	4.2	3.3	3.6	4.1	4.7
Customer 31	4.2	3.7	4.0	4.6	3.6	4.6	4.6	5.0	4.5
Customer 32	4.7	4.7	4.6	3.0	3.4	4.1	4.8	4.5	6.0
Customer 33	4.7	4.6	4.5	3.3	3.0	4.1	4.4	3.8	6.4
Customer 34	4.4	3.7	4.6	4.2	3.3	3.4	4.5	3.8	4.7
Customer 35	4.0	4.3	3.5	4.9	3.9	2.6	3.6	4.7	2.4
Customer 36	3.9	4.7	5.2	4.2	5.6	5.3	3.9	4.1	5.0
Customer 37	1.9	3.7	4.1	4.9	3.6	4.8	5.3	3.9	2.5
Customer 38	3.3	5.1	5.2	4.5	5.9	5.5	4.0	3.7	5.5
Customer 39	1.9	3.7	4.0	5.4	5.3	3.9	3.9	3.6	5.5
Customer 40	2.7	5.5	4.6	5.4	4.8	4.3	2.9	3.2	5.8
Customer 41	2.8	5.1	4.2	5.0	5.2	3.9	3.9	3.3	6.2

(continued)

Table 7.7 (continued)

DMU	Indicator target values								
	MNG	BS	INF	HS	CONS	CO	JUR	STP	TMNG
Customer 42	4.2	5.1	5.5	3.3	4.5	5.4	4.4	4.5	4.6
Customer 43	4.4	4.4	3.7	2.8	2.6	3.6	4.6	5.1	4.6
Customer 44	4.4	4.9	6.7	3.3	4.8	6.8	4.2	4.5	4.7
Customer 45	3.1	6.2	5.8	5.2	4.9	6.5	2.4	2.6	6.0
Customer 46	4.3	4.8	5.5	2.6	6.0	6.1	3.7	3.4	3.8
Customer 47	5.5	4.4	4.7	2.6	4.2	4.5	4.2	4.6	4.0
Customer 48	5.4	4.0	4.1	3.7	3.9	4.3	4.6	5.6	1.2
Customer 49	4.9	3.8	3.9	4.3	6.0	3.7	2.8	4.6	4.3
Customer 50	4.2	4.4	4.6	4.7	3.4	5.4	3.8	3.8	4.6
Customer 51	4.2	5.1	4.7	3.8	4.1	4.2	4.1	2.3	6.3
Customer 52	2.8	6.4	3.6	4.7	4.6	4.3	3.1	2.5	6.0
Customer 53	4.6	4.8	5.1	3.5	4.5	5.3	4.4	4.8	3.3
Customer 54	5.0	3.8	4.2	4.0	5.5	4.0	3.3	4.8	4.2
Customer 55	4.7	4.4	4.9	3.0	3.8	4.0	4.9	4.8	5.6
Customer 56	3.9	4.3	4.4	3.7	4.4	4.1	5.0	4.0	5.6
Customer 57	3.3	5.3	4.6	3.6	4.5	4.8	4.6	3.9	4.8
Customer 58	3.5	5.7	6.0	4.7	4.4	6.2	3.0	3.0	5.9
Customer 59	5.4	5.0	4.8	2.1	5.1	4.7	3.6	3.9	5.5
Customer 60	4.3	4.8	5.6	3.8	5.1	5.6	3.9	4.5	4.2

7.6 Step 4: Organization Strengths and Weaknesses

Output: areas in organizational practices which are excellent/poor; the potential improvements in each practice will be determined.

The difference between target values in Table 7.7 obtained from DEA and actual values in Table 7.5 are the potential satisfaction improvement that exists in different aspects of organizational performance (Table 7.8). Indeed, these potentials are not the same for all the customers. However, we know that some customers are considered to be more important than the others. Therefore, with reference to the key customers specified in Step 1, we can discuss the strengths and weaknesses of the organizational performance in terms of satisfying its customers.

Table 7.8 Differences between target values and real values for each indicator

DMU	Customer importance	Full rank degree of satisfaction (θof model 3)	Rank	Indicator difference values								
				MNG	BS	INF	HS	CONS	CO	JUR	STP	TMNG
Customer 1	C3	0.958	54	0.6	0.5	0.7	0.6	1.4	0.6	0.9	0.4	2.0
Customer 2	C3	1.116	6	0.0	0.0	0.0	0.0	0.0	0.0	0.0	0.0	0.0
Customer 3	C3	0.989	38	0.1	0.1	1.2	0.1	0.1	1.3	0.1	0.1	0.1
Customer 4	C2	0.997	30	0.1	0.7	0.0	0.0	0.1	1.0	0.0	0.0	2.2
Customer 5	C3	0.968	49	0.4	0.4	3.7	0.7	0.6	4.9	0.3	0.3	0.6
Customer 6	C3	1.099	8	0.0	0.0	0.0	0.0	0.0	0.0	0.0	0.0	0.0
Customer 7	C3	1.086	9	0.0	0.0	0.0	0.0	0.0	0.0	0.0	0.0	0.0
Customer 8	C2	1.045	15	0.0	0.0	0.0	0.0	0.0	0.0	0.0	0.0	0.0
Customer 9	C3	0.908	60	2.0	1.0	3.7	0.7	1.3	1.1	1.1	1.2	1.2
Customer 10	C3	1.012	22	0.0	0.0	0.0	0.0	0.0	0.0	0.0	0.0	0.0
Customer 11	C2	1.022	19	0.0	0.0	0.0	0.0	0.0	0.0	0.0	0.0	0.0
Customer 12	C2	1.011	23	0.0	0.0	0.0	0.0	0.0	0.0	0.0	0.0	0.0
Customer 13	C2	1.026	17	0.0	0.0	0.0	0.0	0.0	0.0	0.0	0.0	0.0
Customer 14	C3	1.009	24	0.0	0.0	0.0	0.0	0.0	0.0	0.0	0.0	0.0
Customer 15	**C1**	1.182	2	**0.0**	**0.0**	**0.0**	**0.0**	**0.0**	**0.0**	**0.0**	**0.0**	**0.0**
Customer 16	C3	1.132	4	0.0	0.0	0.0	0.0	0.0	0.0	0.0	0.0	0.0
Customer 17	C3	0.992	36	0.0	0.4	0.3	0.1	0.1	3.0	0.1	0.4	1.5
Customer 18	**C1**	0.982	43	**0.1**	**0.3**	**0.5**	**1.0**	**0.2**	**0.1**	**1.5**	**1.2**	**0.2**
Customer 19	C3	0.997	31	0.0	0.5	3.4	0.0	0.0	0.0	0.0	1.1	0.0
Customer 20	C2	1.008	25	0.0	0.0	0.0	0.0	0.0	0.0	0.0	0.0	0.0
Customer 21	**C1**	1.029	16	**0.0**	**0.0**	**0.0**	**0.0**	**0.0**	**0.0**	**0.0**	**0.0**	**0.0**
Customer 22	**C1**	0.994	34	**1.4**	**0.1**	**1.1**	**0.1**	**0.8**	**2.3**	**0.4**	**0.0**	**0.4**
Customer 23	**C1**	1.100	7	**0.0**	**0.0**	**0.0**	**0.0**	**0.0**	**0.0**	**0.0**	**0.0**	**0.0**
Customer 24	C2	1.124	5	0.0	0.0	0.0	0.0	0.0	0.0	0.0	0.0	0.0
Customer 25	**C1**	1.245	1	**0.0**	**0.0**	**0.0**	**0.0**	**0.0**	**0.0**	**0.0**	**0.0**	**0.0**
Customer 26	C3	0.993	35	0.1	0.1	0.1	0.1	0.1	1.9	0.0	0.3	1.0
Customer 27	C2	0.983	42	0.2	0.2	1.4	0.2	1.1	0.2	0.8	0.2	0.2
Customer 28	C3	1.172	3	0.0	0.0	0.0	0.0	0.0	0.0	0.0	0.0	0.0
Customer 29	C2	0.997	29	1.0	0.5	0.0	1.4	0.1	1.0	0.1	0.7	0.1
Customer 30	C2	0.968	50	0.3	0.5	0.9	0.4	0.4	0.2	0.5	0.3	2.2
Customer 31	C3	1.022	18	0.0	0.0	0.0	0.0	0.0	0.0	0.0	0.0	0.0
Customer 32	C3	0.922	59	2.2	0.9	1.5	0.6	0.7	2.8	2.3	0.9	1.2
Customer 33	C3	0.982	45	0.2	0.2	0.2	0.3	0.3	1.9	1.2	0.2	0.2
Customer 34	C3	1.000	27	0.0	0.0	0.0	0.0	0.0	0.0	0.0	0.0	0.0
Customer 35	C2	1.017	20	0.0	0.0	0.0	0.0	0.0	0.0	0.0	0.0	0.0
Customer 36	C3	0.940	57	1.6	0.8	2.5	1.0	0.8	0.8	0.6	0.6	3.1
Customer 37	C3	1.063	12	0.0	0.0	0.0	0.0	0.0	0.0	0.0	0.0	0.0
Customer 38	C3	0.928	58	0.6	1.1	0.9	0.8	1.0	1.0	1.0	1.3	2.1
Customer 39	C3	1.014	21	0.0	0.0	0.0	0.0	0.0	0.0	0.0	0.0	0.0
Customer 40	C3	0.987	39	0.1	0.2	2.8	0.2	2.8	2.3	0.1	0.1	0.2
Customer 41	**C1**	0.972	48	**0.3**	**1.1**	**1.0**	**0.4**	**3.0**	**0.6**	**0.3**	**0.2**	**0.5**
Customer 42	C3	0.967	52	0.4	0.4	0.8	1.9	3.2	1.6	0.3	0.3	0.4
Customer 43	C3	0.960	53	0.5	0.5	0.3	0.4	1.6	1.0	1.1	0.5	0.5
Customer 44	C2	0.978	46	0.2	0.8	0.4	0.3	1.3	0.7	0.2	0.2	3.7
Customer 45	C3	0.982	44	0.2	0.3	0.3	2.6	0.8	0.4	0.2	0.6	1.2
Customer 46	C3	1.055	13	0.0	0.0	0.0	0.0	0.0	0.0	0.0	0.0	0.0
Customer 47	C2	0.957	55	0.6	0.5	0.5	0.3	2.3	0.8	1.2	1.1	1.5
Customer 48	C2	1.047	14	0.0	0.0	0.0	0.0	0.0	0.0	0.0	0.0	0.0
Customer 49	C2	1.081	10	0.0	0.0	0.0	0.0	0.0	0.0	0.0	0.0	0.0
Customer 50	C2	0.983	41	0.2	1.7	0.2	0.2	1.0	1.9	0.2	0.5	2.2
Customer 51	C2	1.008	26	0.0	0.0	0.0	0.0	0.0	0.0	0.0	0.0	0.0
Customer 52	C2	0.997	32	0.0	0.0	0.0	2.7	0.3	1.2	0.8	0.0	1.5
Customer 53	C3	0.992	37	0.1	0.1	2.7	0.8	0.6	1.6	0.9	0.1	0.7
Customer 54	C2	0.986	40	1.4	0.7	1.4	1.0	0.2	0.2	0.3	0.2	0.2
Customer 55	C3	0.943	56	1.7	0.8	3.0	1.5	0.5	0.6	3.0	0.7	0.8
Customer 56	C3	0.999	28	0.0	0.0	0.5	0.0	0.0	1.2	0.0	0.8	0.0
Customer 57	C2	0.967	51	0.3	0.4	0.3	0.3	3.5	3.3	0.4	0.4	1.3
Customer 58	C2	0.996	33	0.0	0.1	0.1	1.3	2.0	2.9	0.6	0.7	0.1
Customer 59	C3	1.078	11	0.0	0.0	0.0	0.0	0.0	0.0	0.0	0.0	0.0
Customer 60	C3	0.975	47	0.3	0.3	2.1	0.3	0.3	3.6	0.2	0.2	0.9
C1 sum				1.8	1.5	2.6	1.5	4	3	2.2	1.4	1.1
C2 sum				4.3	6	5.1	6.8	10.3	10.5	4.5	3.6	15.1
Sum				11.1	8.7	30.8	14	18.2	34.5	14	10.8	17.8

C1: Category one; C2: Category two; C3: Category three

As can be noted from the above table, we have highlighted category one and two customers, these being 'most important' and 'important' customers respectively. We also obtain the sum for all categories which will be utilised in the next section [4].

7.7 Step 5: Submit Recommendations to Decrease Weaknesses and Strengthen Good Practices

We have specified seven key (most important) customers of which customers 15, 21, 23, and 25 are satisfied customers. In the full ranking DEA model, they have got ranks 2, 16, 7, and 1, respectively. But five key customers are not satisfied with the current performance of organization. For instance, Customer 41 has got the rank of 48 between 60 customers so this customer is not satisfied. The difference between actual and target values for this customer show that for booking system and infrastructure, there is a high potential for this customer to be satisfied. The last three rows of Table 7.8 present the sum of all potential improvements in each indicator for category 1 customers (C1), Category two customers (C2) and category three customers, respectively.

According to all the key customers, a huge potential for improvement exists in container sorting and cost. Key customers are not satisfied with the way their containers are sorted or with the cost of food, accommodation, parking, etc. These two aspects may be considered as the weaknesses of the current organization. Furthermore, the same two aspects can be considered as weaknesses when category one and two customers are considered as the reference population. Another aspect that may be considered as a weakness is the management system. According to category one and two customers, management is a poor organizational practice and there is great potential for its improvement in this system.

According to the overall sum of potential improvement for the key customers, the four aspects of booking system, health and safety, stevedore performance, and time management are the strong points in organizational practices since the potential for improvement of these aspects is the least. Stevedore performance is also a strength with respect to category two customers (important customers).

7.8 Analytical Hierarchal Process for Strategic I-CRM Development

In this section, we address the sensitivity of the issues using our different customers who were identified in the previous sections. Here, we utilize the analytical hierarchal process to create a new model after our objectives have been structured, and to measure our objectives and alternatives. Lastly, we synthesise our results and obtain the sensitivity analysis. The analytical hierarchal process returns the ratio scales from paired comparisons.

We benefitted from the results derived from data envelopment analysis regarding key customers. In this section, using the analytical hierarchal process, we attempt to produce a decision-making process with respect to the customers who constitute most of the system. The following table clearly highlights the sum of category one customers followed by category two and three customers. We also obtain the ratios and after AHP analysis, we subtract the numbers and put them in different rows. The analytic hierarchy process creates the necessary objective to process the preferences and priorities in order to carry out optimum decision-making. It also enhances priorities for our alternatives and the criteria that we need to have for our evaluations. Furthermore, our priorities are achieved according to pairwise comparison. The analytical hierarchal process has some advantages because it provides ratio scale, normalising the results and creating sensitivity diagrams [5].

Additionally, the analytical hierarchal process is a tool and Eigen value method for decision-making utilising several criteria. It generates a methodical approach to rectify and balance the numbers in quantitative and qualitative analysis [6]. The analytical hierarchal process can assess different alternatives and provide reasonable feedback. Pair- wise comparison is conducted between the main factors in relation to the significance of the objectives. Furthermore, using AHP, we can normalise our results and as it is allied to quantitative research, it has the potential to obtain sensitivity and certainty of the results [7].

7.8.1 Satisfaction Indicators on a Pairwise Comparison Approach for Customers

In this section, we use an analytical hierarchy process to find which category of customer has the lowest satisfaction and the input is the rate of potential improvement derived from data envelopment analysis. We need to choose a group from one of the categories of customers. Then, considering priorities and the amount of improvement, we will do the decision-making for the one which needs the most improvement. (Table 7.9)

Table 7.9 Category Indicators and abbreviations

Category indicator	Abbreviation
Management	MNG
Booking System	BS
Infrastructure	INF
Health and Safety	HS
Container Sorting	CONS
Cost	CO
Jurisdiction	JUR
Stevedore Performance	STP
Time Management	TMNG

Table 7.10 Calculation of target values and real values for each indicator

	MNG	BS	INF	HS	CONS	CO	JUR	STP	TMNG
C1 sum	1.8	1.5	2.6	1.5	4	3	2.2	1.4	1.1
C2 sum	4.3	6	5.1	6.8	10.3	10.5	4.5	3.6	15.1
C3 sum	11.1	8.7	30.8	14	18.2	34.5	14	10.8	17.8
Total sum	17.2	16.2	38.5	22.3	32.5	48	20.7	15.8	34
C2-C1	2.5	4.5	2.5	5.3	6.3	7.5	2.3	2.2	14
C3-C1	9.3	7.2	28.2	12.5	14.2	31.5	11.8	9.4	16.7
C3-C2	6.8	2.7	25.7	7.2	7.9	24	9.5	7.2	2.7
C2/C1	2.3889	4	1.9615	4.5333	32.5	3.5	2.0455	2.5714	13.727
C3/C1	6.1667	5.8	11.846	9.3333	4.55	11.5	6.3636	7.7143	16.182
C3/C2	2.5814	1.45	6.0392	2.0588	1.767	3.2857	3.1111	3	1.1788

Fig. 7.2 Envelopment surface and orientation of the nodes (objectives of satisfaction)

According to Table 7.10, cost has the highest number which illustrates its potential for improvement. It does not reflect customer satisfaction and it illustrates that the company as well as Fremantle port authorities are able to control the current rules, regulations and new technologies. However, it appears that it is difficult to change the booking system and stevedore performance. This may require the assistance of an expert and the consideration of other important factors in order to rectify the shortcomings and to provide a solution. In this section, we consider each of the issues that were discussed in Chaps. 5 and 6 as objectives of our analysis.

From Fig. 7.2, cost is our first priority, followed by time management. Container sorting and infrastructure have similar type mapping. Then, we have health and safety followed by booking system, jurisdiction issues, management problems and stevedore performance. Likewise, our inconsistency is 0.03 which is reasonable. Also, our proportions and ratios have the same logical order. As shown in Fig. 7.3 below, if we normalize our result as for the previous figure, we can say that if cost is 100 percent of our objective, it is twice as significant as booking system. However, container sorting and infrastructure have a similar trend.

In order to create the following figure, we conducted a pairwise comparison between each objective and three categories. Figure 7.4 below shows the sorting and synthesis of results. It illustrates that category three customers are the best

Fig. 7.3 Normalized and synthesized objectives of Fig. 7.2

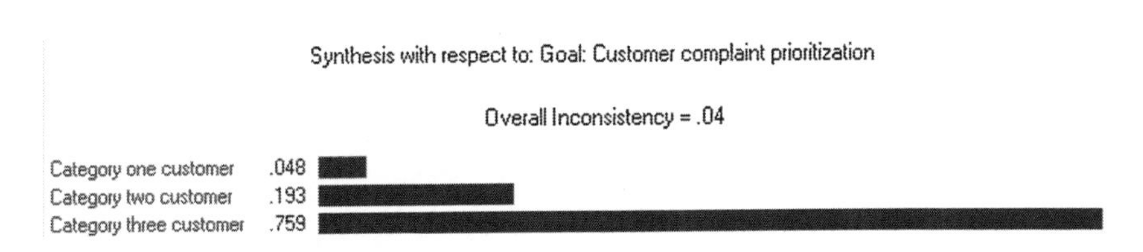

Fig. 7.4 Synthesizing customers

contributors in our project, followed by category two customers and category one customers. Previously, we identified our category one customers using other methods; however, with this approach we endeavour to determine the sensitivity of each group. Based on previous analysis, we find our key customers and name them ‘category one’ and ‘category two’ customers.

7.8.2 Dynamic Sensitivity Analysis

Dynamic sensitivity analysis is a horizontal bar graph that enables the user to change the priority settings of each of the variables and make them visible [8].

‘Node’ in the analytical hierarchy process is defined as an objective which we need to address. It also highlights the type and the rate of customer satisfaction as well as the importance of each of the nine indicators. In Fig. 7.5 below, using sensitivity dynamic analysis for each of the nodes, we consider ‘what-if’ scenarios to validate our results. According to this figure, cost is the most important one followed by time management, container sorting and infrastructure issues. Then we have health and safety, booking system, jurisdiction, management, and stevedore performance respectively. These results indicate that ordinary customers are the preferable result derived by the software. Also, if we increase each of the factors, category three customers who are the ordinary ones, are still considered as the most preferable.

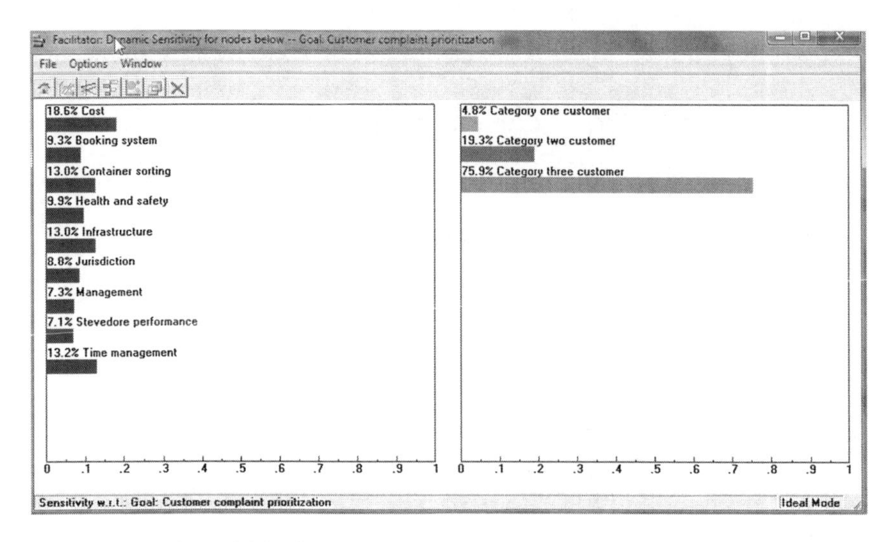

Fig. 7.5 Dynamic sensitivity for the objectives

7.8.3 Performance Sensitivity Analysis

Performance sensitivity represents real-time and interactive type mapping of the choices as the decision makers make comparisons between various possibilities. It also provides a graphical display of the various types of ranks [9]. In the following figure, we discuss sensitivity analysis and its influence on shifts in priorities. The sensitivity analysis is pivotally significant because any change in the significance of features produces various levels of cost, infrastructure, management, jurisdiction, time management, health and safety, booking system, stevedore performance and container sorting.

Figure 7.6, shows a performance sensitivity diagram. The X axis represents the objectives and Y axis is the objective percentage and our alternatives are on the right of the figure. As seen in the figure below, cost is the most important objective and can be seen in the Y axis on the left. Overall, ordinary customers are more preferable; however, they are worse on time management and booking system. Also, they are by far the best on infrastructure and cost as in most of the cases they do not have any expectations from their companies. Although it is vital for the companies to maintain their key customers, it is really critical to pay attention to category three customers by providing new strategies to retain them. They are the individuals who can meet the requirements of the companies as well as those of the Fremantle port authorities. What is really interesting is that ordinary customers do not complain about infrastructure as do the key customers. This implies that if companies improve their booking system, management and time management, they will be more likely to retain the ordinary customers (drivers) for a long time. Also, category two are more preferable than category one customers. Important customers' objectives are high for time management and booking system;

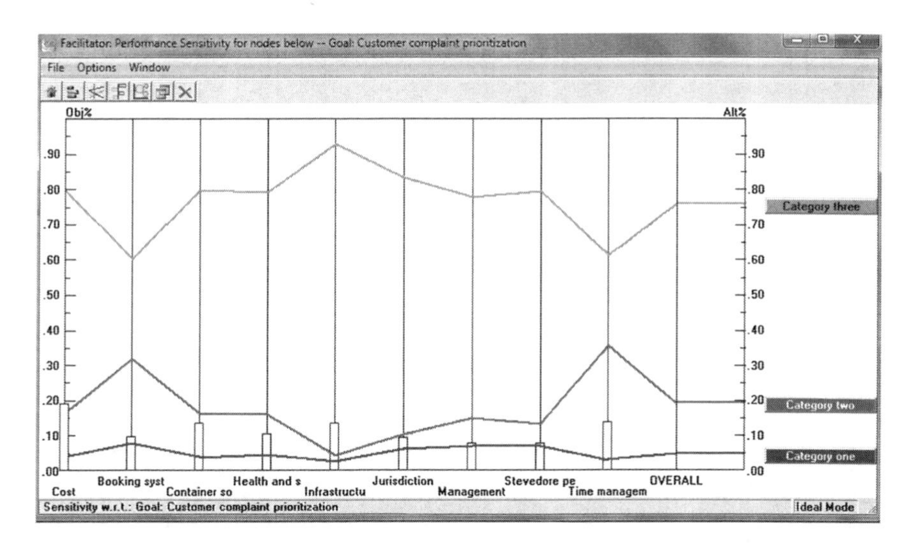

Fig. 7.6 Performance sensitivity for the objectives

however, category one customers, due to their expectations must have all the objectives met in order to be satisfied and companies are greatly challenged to meet the needs of key customers. Likewise, to strike a balance, authorities must have the same vision for all customers regardless of who they are, and should control situations so that all customer needs are met and customers are content.

7.8.4 Gradient Sensitivity Analysis

The gradient sensitivity analysis allocates each objective to a different gradient graph. The vertical line shows the existing priority of the chosen objective. The straight lines highlight the alternatives. The existing priority of an alternative is the place where the alternative line crosses the vertical objective line [8].

The gradient sensitivity analysis of expert choice as illustrated in Fig. 7.7 indicates the ranking to changes in cost. The vertical line depicts the priority weight of cost and can be seen from the x axis intersection. As the vertical line moves within the x axis, it automatically changes the weights, the horizontal lines shows new priority of customers which can be noted from the y axis. Also, as can be noted, category three customers have an increasing trend which needs companies' attention. Category two customers start to have a decreasing trend; however, category one customers is flattened and implies lack of attention to category one (truck drivers). It indicates that, if cost as one of the objectives increases from 0.05 to 0.18, the customers' ranking does change and in the second scenario, if the cost increases to 0.8, the ranking is changed. While the vertical line meets the intersection of three alternatives, the ranking of two alternatives starts to change.

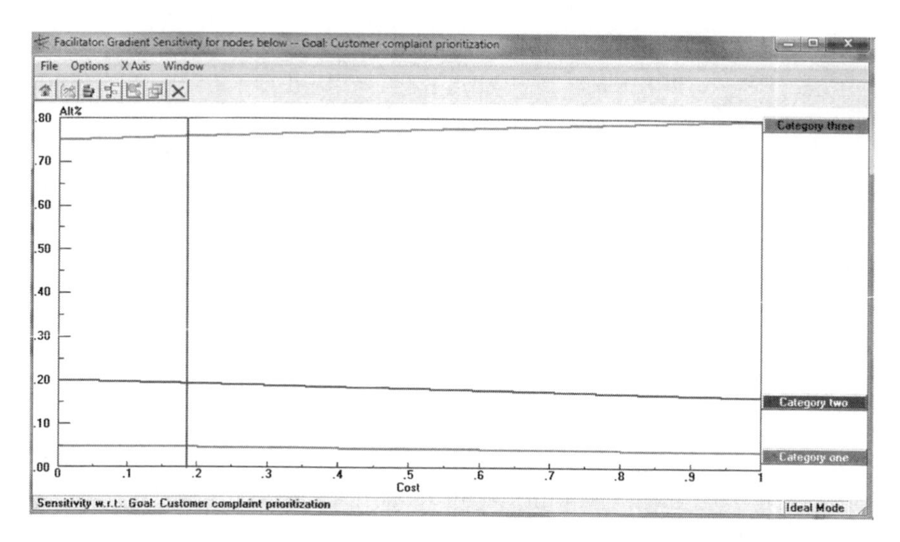

Fig. 7.7 Gradient sensitivity of customers on cost

Since the lines of alternatives do not cross each other, it implies that the ranking of the alternatives never changes, unless the company or the Fremantle port authorities provide new changes to the system and strategies. Also, they may wish to adopt new technologies and employ experts to handle the issues. Moreover, as the business needs change, the priorities may also change due to necessities that the company may confront.

Based on AHP findings and our previous findings using DEA, as the majority of the customers are 'ordinary' rather than 'key' customers, if the company manages to fully meet their expectations, it is highly unlikely that any other port or organization will be able to compete with the Fremantle port and they will dominate other industries.

7.9 Determining and Addressing Significant Issues to Achieve Maximum Timely Satisfaction

In Chap. 2, we performed an extensive survey of the extant research in customer relationship management and its components. Also, we carried out extensive research on customer complaints. To the best of our knowledge, we take a state-of-the art approach in this thesis. In the following, we present works that have been done previously that are relevant to the content of this chapter.

In studies undertaken by researchers, separate approaches have been used to explore customer relationship management, loyalty and customer satisfaction. For example, [10] uses DEA to evaluate the effectiveness of customer satisfaction and loyalty. Also, [11] conducted a study of acquisition in the insurance industry and utilized DEA for their formulation and analysis. In addition, in some studies such as [12], researchers used an analytic hierarchy process and propose a performance measurement of CRM utilizing AHP. For example, in [13], the authors used the AHP approach to access the relative significance of customer needs and plan for its attributes. Also, [14] used AHP to analyse customer requirements in logistics companies. Additionally, to assess customer satisfaction and to estimate the loyalty level of customers, researchers tend to use fuzzy AHP.

This chapter makes a major contribution to the existing literature in that it proposes a complete methodology for performance measurement, customer complaint management and customer ranking. Likewise, using our methodology, we optimize our model to reach our results and recommendations.

We initiated this chapter by introducing the five-step algorithm for analysis of customer satisfaction. We introduce different customers such as category one customers, category two and category three customers, each of them in detail. In the next part, we extract the main criteria from the questionnaire and discuss them in detail. Then, we differentiate our customers. We conduct a correlation analysis for the questions in each category related to customer satisfaction, loyalty, perceived value, interactivity and customer acquisition respectively. When the correlation is high, principal component analysis produces better results.

In the next section, as there is a large amount of data, it is reduced utilising principal component analysis (PCA). To this aim, we refer to our questionnaires and determine the relationship between each issue (complaint) and frequency of the question for that particular issue. PCA is also applied to analyse decision making units (DMU). We formulate PCA for all nine categories.

After reducing the data, we conduct data envelopment analysis to find key customers who influence the decision-making process. DEA also helps to identify possible benchmarks towards which performance can be targeted. We also rank our customers using DEA. The ultimate results of DEA can generate information for improvement potential of the indicators. Utilising DEA, we propose a second set of recommendations for Fremantle port.

In the final stage of this chapter, we use the analytic hierarchy process to analyse improvement potential and sensitivity. It also assists in creating a total ranking of alternatives which are our various customers. As it is a decision-making tool, it has a positive relationship with DEA and provides a holistic view of the customers.

To the best of our knowledge, in none of the previous academic endeavours have researchers utilised the same approach to find key customers and explore the potential for improvement in order to create optimization. However, in this thesis, we optimise the results of our customers. Also, this is the first study that uses a combination of principal component analysis, data envelopment analysis and analytic hierarchy processes. Also, we are quite sure that such research has not previously been undertaken in the field of logistics with key customers being identified based on the stated attributes.

This chapter implicitly contributed to the existing body of knowledge and science, by acquiring and refining the data from questionnaire. Also, apart from demographic questions, we included new psychological questions in order to find key customers. We also formulated the relationship which may help researchers learn from the points we mention. Also, sensitivity analysis in this thesis is conducted to manipulate the strength of the ranking among the alternative customers. Additionally, we benefitted from combining methodologies and tools to analyse and find key customers using the optimisation process and we introduced new convincing recommendations to increase the efficiency and productivity of the system.

In the following section, we provide guidelines and advice for researchers who may wish to use this chapter as a platform for their analysis. Our methodology can assist researchers to have a better understanding of their problems and better recognise their customers and the environment in which they may wish to work. The steps that we introduce are sequential and by following these steps, researchers can effectively find their way and appropriately analyse their systems. For the purpose of implementing our model and methodology in similar works, various steps are taken as follows:

1. Screen customers
2. Define extracted variables from questionnaire in detail
3. Consider each indicator as a hypothesis
4. Categorize customers according to the categories of one, two or three.
5. Initially reduce data by relating major complaints to the relevant questions
6. Reduce the number of variables and analysis of decision-making units (DMUs), utilising principal component analysis.
7. Analyse data using data envelopment analysis
8. Use applied DEA model to identify key customers
9. Rank the customers
10. If we want to change an unhappy customer to a satisfied and happy customer and decrease the level of complaints, we must act according to the formula proposed in Chap. 7 and change yro which is the actual value to ∧yro which is the target value.
11. Based on DEA, each customer must be considered as a decision-making unit (DMU)
12. Based on customer categorization and the results from DEA, we can address the reasons for each customer's dissatisfaction.
13. Propose recommendations and solutions to the customer complaints.
14. Use the analytic hierarchy process to create a decision-making process with respect to the customers who make the major contribution to the system.
15. Also, future researchers may wish to use the fuzzy analytic hierarchy process accompanied by DEA for their evaluation of sensitivity.

7.10 Conclusion

In this chapter, we proposed a methodology for customer satisfaction by generating a five-step algorithm. Using PCA, we reduced the number of variables in order to work with the data more easily. Then, using the survey data, we subsequently proposed data envelopment analysis to find key customers. To eliminate the undesirable effects of the current regulations, we need to provide recommendations and solutions to an organization. Hence, we present and submit our recommendations to reduce weaknesses and strengthen the system by suggesting better or different practices.

As aforementioned, the issues that we discussed and for which we provided recommendations with the existing approaches were all optimized. In the last section of Chap. 7, utilizing an analytic hierarchy process as one of the pivotal decision-making techniques, we analyse sensitivity and create an overall ranking of our alternatives which are our customers. Now, it is time to examine the second part of our conceptual framework and find the legitimacy of the relationship between them. In the next chapter, we will discuss and analyse the relationship between satisfaction, perceived value, interactivity, loyalty and customer acquisition using the fuzzy logic approach.

References

1. Charnes, A., Cooper, W. W., & Rhodes, E. (1978). Measuring the efficiency of decision making units. *European Journal of Operational Research, 2*, 429–444.
2. Luo, X., & Homburg, C. (2007). Neglected outcomes of customer satisfaction. *Journal of Marketing, 71*, 133–149.
3. Andersen, P., & Petersen, N. C. (1993) A procedure for ranking efficient units in data envelopment analysis. *Management science,* 1261–1264.
4. Faed, A., Hussain, O. K., Faed, M., Saberi, Z. (2012) Linear modelling and optimization to evaluate customer satisfaction and loyalty. *Presented at the the 9th IEEE International Conference on e-Business Engineering.*
5. Saaty, T.L., & Vargas, L.G. (2001) The seven pillars of the analytic hierarchy process. *Models, Methods, Concepts & Applications of the Analytic Hierarchy Process,* 27–46.
6. Vaidya, O. S., & Kumar, S. (2006). Analytic hierarchy process: an overview of applications. *European Journal of Operational Research, 169*, 1–29.
7. Zhu, L., Aurum, A., Gorton, I., & Jeffery, R. (2005). Tradeoff and sensitivity analysis in software architecture evaluation using analytic hierarchy process. *Software Quality Journal, 13*, 357–375.
8. Shafiq N., & Khamidi, M. F. (2009) Performing sensitivity analysis of pipeline risk failure using analytical hierarchy process.
9. Udo, G. G. (2000). Using analytic hierarchy process to analyze the information technology outsourcing decision. *Industrial Management & Data Systems, 100*, 421–429.
10. Bayraktar, E., Tatoglu, E., Turkyilmaz, A., Delen, D., & Zaim, S. (2012). Measuring the efficiency of customer satisfaction and loyalty for mobile phone brands with DEA. *Expert Systems with Applications, 39*, 99–106.

11. Cummins, J. D., & Xie, X. (2008). Mergers and acquisitions in the US property-liability insurance industry: Productivity and efficiency effects. *Journal of Banking & Finance, 32*, 30–55.
12. Öztaysi, B., Kaya, T., & Kahraman, C. (2011). Performance comparison based on customer relationship management using analytic network process. *Expert Systems with Applications, 38*, 9788–9798.
13. Lin, M.-C., Wang, C.-C., Chen, M.-S., & Chang, C. A. (2008). Using AHP and TOPSIS approaches in customer-driven product design process. *Computers in Industry, 59*, 17–31.
14. Wu, M., & Cai, Y. (2011) Research of evaluation of customer relationship of logistics enterprise based on AHP, 1–4.

Chapter 8
Linear and Non-linear Analytics and Opportunity Development in I-CRM

8.1 Introduction

In this chapter, the relationship between customer satisfaction and factors related to I-CRM, namely: perceived value, interactivity, customer loyalty, and customer acquisition, are analysed. As previously suggested, customer satisfaction plays a significant role in I-CRM. In order to verify the relationship between our proposed variables based on the conceptual framework, we utilise linear modelling to formulate the relationship between customer satisfaction with its antecedents and consequences. Also, utilizing a formula, we test and prove the relationships. Likewise, we adopt non-linear modelling to structure and evaluate the impact of perceived value and interactivity on customer satisfaction issues and then we estimate how customer satisfaction issues influence customer loyalty and customer acquisition, using fuzzy inference systems. Based on the Takagi-Sugeno type fuzzy inference system, we define fuzzy rules; we also, define three aspects of presentation. Firstly, we provide rules for interactivity, perceived value and customer satisfaction. The second aspect includes rules about the relationship between nine issues and loyalty. Thirdly, we obtain rules for the relationship between those nine satisfaction issues and customer acquisition.

Previously, we tested and analysed our hypotheses in Chap. 6, and encountered challenges regarding the utilization of all data for all 60 customers. Despite this, three of our hypotheses were supported. However, the relationship between customer satisfaction and loyalty remained obscure and not validated. To address this with the aim of establishing a legitimate relationship between these two variables and proving our hypothesis, a new methodology was chosen.

This relationship is assessed at two levels:

Level 1: Here, the impacts of perceived value and interactivity on customer satisfaction are assessed.

A. R. Faed, *An Intelligent Customer Complaint Management System with Application to the Transport and Logistics Industry*, Springer Theses, DOI: 10.1007/978-3-319-00324-5_8,
© Springer International Publishing Switzerland 2013

8.1.1 Research Questions

(a) Does perceived value affect customer satisfaction?
(b) Does interactivity affect customer satisfaction?

Level 2: In the second level, the impacts of customer satisfaction on customer loyalty and customer acquisition are assessed. Specifically, the research questions are:

(c) Does customer satisfaction affect loyalty?
(d) Does customer satisfaction affect customer acquisition?

Based on Chap. 4, we have developed the following conceptual framework in which perceived value and interactivity are positively associated with customer satisfaction and are considered as antecedents of customer satisfaction. After generating customer satisfaction, it will lead to loyalty and customer acquisition which are considered to be the consequences of customer satisfaction. The five aforementioned variables have been identified in Chap. 4 and validated in Chap. 6. Figure 8.1 depicts the main aspects of the conceptual framework.

Fig. 8.1 Conceptual framework

The structure of the analysis is shown in Fig. 8.2. Customer issues have been previously identified in Chap. 6 and validated in Chap. 7.

Figure 8.2, illustrates the relationships between customer issues at the Fremantle port with the antecedents and consequences of customer satisfaction. The above figure shows three various levels. As mentioned, in level one, using linear and non-linear modelling, we need to obtain the relationship between perceived value and interactivity with customer issues. In the second level, we need to ascertain the relationship between customer issues and customer loyalty; and in the third level, we ascertain the relationship between customer issues and customer acquisition. The primary objective of this relationship analysis is to quantify the relationship between factors by choosing linear or non-linear models. At first, linear models are used to describe the relationship between factors. Although linear models have the advantage of being simple and easy to understand, in the case of complex and non-linear relations, they fail to accurately capture the relationships.

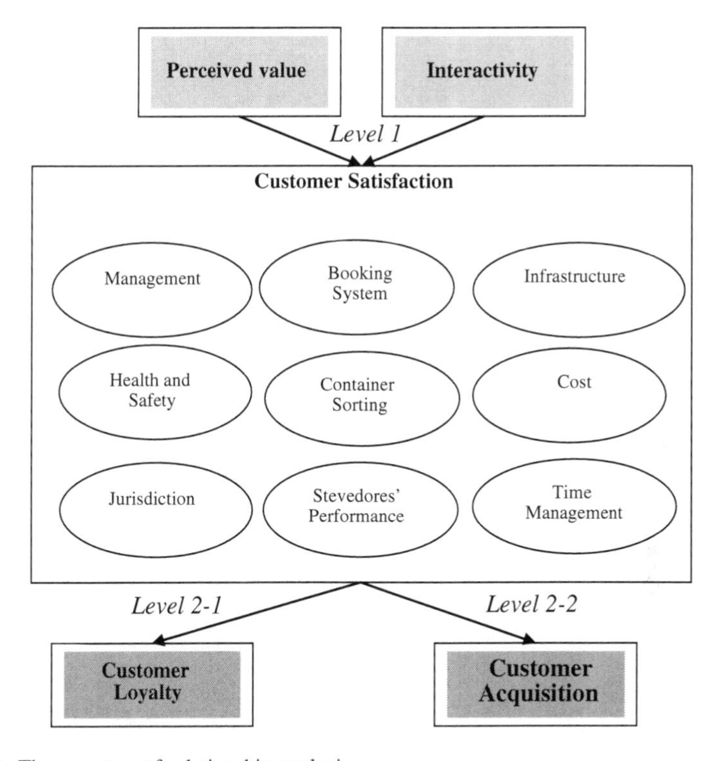

Fig. 8.2 The structure of relationship analysis

Several quantification methods and models are available for the study of complex relationships between different factors. Artificial Neural Networks (ANNs) and fuzzy inference systems are two such methods. Artificial neural networks can properly handle data complexity and non-linearity and have the advantage of intelligent modelling. Fuzzy inference systems deal with uncertainty and ambiguity in non-crisp data sets and can be used for fuzzy data modelling. Furthermore, fuzzy inference systems can be integrated with ANNs to exploit the advantages of both ANN and fuzzy inference systems. The adaptive network-based fuzzy inference system (ANFIS) is the resulting modelling technique that can not only handle data complexity and non-linearity, but is also able to deal with fuzzy uncertainty in the data and can perform fuzzy data modelling.

Since the data for this study were collected through a questionnaire survey, the linguistic expressions used to answer the questions carry a fuzzy uncertainty which affects any calculations pertaining to the factors under study, i.e. customer satisfaction, perceived value, interactivity, customer loyalty, and customer acquisition.

This fuzzy uncertainty in the data can be handled well by using an adaptive network-based fuzzy inference system as a modelling technique to assess the relationship between customer satisfaction and other factors. The output of ANFIS is one or more fuzzy rules that describe the relationship between factors.

Table 8.1 Value finding table using DEA

	Importance	Loyalty	Perceived	Interactivity	Acquisition	Satisfaction PCA full	Satisfaction DEA	Satisfaction DEA full
Customer 4	C2	3.583452	3.567845	4.689414728	4.72744157	3.2	6.8	3.5
Customer 8	C2	3.595932	5.333061	4.286772075	5.16946498	4.6	7.0	4.7
Customer 11	C2	1.289567	1	1	1	3.1	7.0	2.3
Customer 12	C2	4.710696	2.697317	4.151763524	3.89779059	4.2	7.0	1.7
Customer 13	C2	1.704526	3.847747	3.329243565	4.16858095	2.9	7.0	3.3
Customer 15	**C1**	7	6.786258	4.702191764	6.20353479	6.7	7.0	7.0
Customer 18	**C1**	6.393103	4.609712	4.922430401	7	4.0	5.9	2.8
Customer 20	C2	3.844326	5.824782	5.066673814	3.71458185	4.1	7.0	3.4
Customer 21	**C!**	3.280003	3.354238	4.299866846	2.97251248	2.5	7.0	2.0
Customer 22	**C1**	7	2.743089	4.377245367	2.71369781	3.3	6.6	2.0
Customer 23	**C1**	1	2.694523	1.838274093	2.30486017	1.2	7.0	1.0
Customer 24	C2	7	5.850227	7	5.22386083	7.0	7.0	6.2
Customer 25	**C1**	3.540877	2.91365	3.435610285	3.69337177	1.0	7.0	1.1
Customer 27	C2	6.235839	2.60418	3.139760328	3.01374376	3.8	5.9	3.3
Customer 29	C2	2.515842	3.441616	4.080193478	5.99911597	3.7	6.8	3.3
Customer 30	C2	6.393103	3.508303	4.576744743	3.89779059	3.4	4.9	2.8
Customer 35	C2	1.593016	3.791686	4.476818001	2.9513024	3.7	7.0	3.5
Customer 41	**C1**	1.434351	3.655965	4.209956243	3.93097636	3.8	5.2	2.7
Customer 44	C2	3.844326	4.456909	6.126222089	5.9659302	3.9	5.7	4.6
Customer 47	C2	1	2.726421	3.380781492	4.339814	5.0	4.2	4.0
Customer 48	C2	7	2.848926	4.229036544	4.98625624	4.4	7.0	4.9
Customer 49	C2	4.97489	4.910797	3.352321875	3.98537221	3.4	7.0	3.9
Customer 50	C2	2.589358	4.557926	4.653206467	3.32695428	4.1	5.9	3.6
Customer 51	C2	6.235839	5.545405	4.59564799	2.74688358	3.6	7.0	3.4
Customer 52	C2	3.583452	1.411823	2.577978497	2.77613917	3.8	6.8	2.4
Customer 54	C2	5.317593	3.850358	4.573383396	6.35355776	3.9	6.1	3.7
Customer 57	C2	7	2.229323	2.797754945	4.96504616	4.7	4.9	3.2
Customer 58	C2	3.499704	1.845924	2.730498063	5.25311642	3.7	6.8	2.9

The remainder of this section is organized as follows. In Sect. 8.2, linear modelling of the relations between factors is described and its results are presented. In Sect. 8.3, ANFIS is used to specify the fuzzy relationship of the factors in terms of fuzzy if–then rules.

An adaptive network-based fuzzy inference system has the supportive ability to generate customer satisfaction using the neuro-fuzzy logic method [1].

By using DEA in the previous chapter, we were able to establish the key customers' priorities in their order of importance. Here, in this chapter, we use DEA to find the value of each variable (Table 8.1).

8.2 Linear Modelling Between Customer Satisfaction and I-CRM

In this section, an analysis is performed to determine the linear relationship between customer satisfaction and other factors. The structure for the analysis is shown in Fig. 8.2. Considering Fig. 8.3, five of the nine decision-making units (DMUs) are on the efficient frontier. A general rule is that the DMUs should appear there at least three times, as there are variables (inputs + outputs) in the model.

Here, a 2-step procedure to analyse these relationships is proposed.

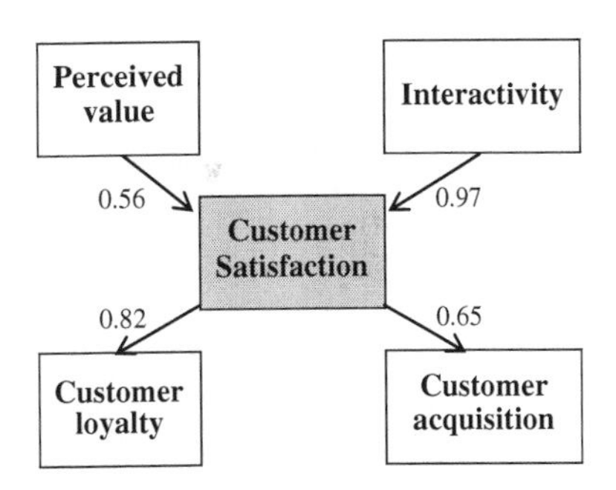

Fig. 8.3 The structure of relation analysis

8.2.1 Factor Quantification: Calculate an Overall Degree of Agreement for Each Customer in the Four Sections of Perceived Value, Interactivity, Customer Loyalty, and Customer Acquisition

Similar to the principal component analysis (PCA) process performed for indicator categories in the customer satisfaction section, the overall degree of agreement is calculated by principal component analysis. The principal component analysis is performed for Sects. 8.3, 8.4, 8.5, and 8.6 in the questionnaire. The number of questions in each section is more than one and in each section, principal component analysis will aggregate all the questions and yield an overall degree of agreement.

The overall degree of satisfaction is derived from principal component analysis of 9 categories of indicators in Sect. 8.2 of the questionnaire. We develop a code in MATLAB to conduct the principal component analysis in a more convenient way (see Appendix 3).

8.2.2 Relationship Analysis Using Linear Model Estimation Between Customer Satisfaction, Perceived Value and Interactivity

In this step, we must estimate the best linear model between customer satisfaction as a dependent variable and perceived value and interactivity as independent variables (level 1).

In this step, with reference to the most important customers, a linear model indicating the relationship between customer satisfaction as a dependent variable and perceived value and interactivity as independent variables is estimated. The result of model estimation is as follows:

$$\begin{aligned} CS &= \alpha + \beta \times Per.val + \gamma \times Interactivity \\ CS &= -2.7 + 0.56 \times Per.val + 0.97 \times Interactivity \\ \mathrm{R}^2 &= 87\,\% \end{aligned} \tag{8.1}$$

The formula 8.1 displays new coefficients as α, β, and γ. "α" is defined as a constant coefficient and an independent variable of the equation.

In formula 8.1, "β" is defined as a coefficient that outlines the positive impact that perceived value has on customer satisfaction. Also, it is the regression coefficient for variables perceived value and interactivity. A positive value of beta shows a direct relationship between *P* value and customer satisfaction.

Likewise, "γ" is defined as a coefficient indicating the influence that interactivity has on customer satisfaction. "R^2" is defined as the coefficient to determine regression. It is also the regression coefficient for variables 'perceived value' and 'interactivity'.

This linear estimated model has a very good fitness for the data related to the key customers with a regression coefficient of determination being $R^2 = 87\,\%$. In statistics, the coefficient of determination R^2, with values between 0 and 1, is used mostly in the context of a statistical model to show goodness of fit. It indicates the proportion of variance in a chosen dependent variable of a data set that is accounted for by independent variables. It provides a measure of how well one or more variables explain another's behaviour, within a multi-linear regression model.

As seen in model 8.1, both perceived value (per.val) and interactivity have a positive impact on customer satisfaction (CS) as both of the regression coefficients are positive.

It should be noted that all data used in model 8.1 are scaled between 1 and 7 (as in a Likert scale). The proportional impact of perceived value is estimated to be 0.56 on CS. In other words, a unit change in perceived value will increase the customer satisfaction by 0.56.

For interactivity, the impact on CS is estimated as 0.97 which is a relatively larger positive impact when compared with perceived value.

8.2.3 Relationship Analysis Using Linear Model Estimation for Loyalty, Acquisition and Customer Satisfaction

In the second step, we must estimate the best linear model between loyalty and customer acquisition as dependent variables, and customer satisfaction as an independent variable (level 2).

The aim of this step is to ascertain whether customer satisfaction will improve loyalty and customer acquisition. With reference to key customers and a linear regression analysis, we will provide an answer to this question.

First, we consider customer loyalty. The estimated model is presented in 8.2.

$$\begin{aligned} loyalty &= \alpha + \beta \times CS \\ loyalty &= 1.59 + 0.82 \times CS \\ \mathrm{R}^2 &= 38\,\% \end{aligned} \tag{8.2}$$

As seen in model 8.2, for key customers, the impact of customer satisfaction on loyalty cannot really be estimated with a linear model as the regression coefficient of determination $\mathrm{R}^2 = 38\ \%$ is not good enough to rely on for making a judgment. However, this model shows a negative impact of customer satisfaction on loyalty for the key customers.

The linear relationship between customer satisfaction and customer acquisition is estimated in model 8.3.

$$\begin{aligned} Acquisition &= \alpha + \beta \times CS \\ Acquisition &= 2 + 0.65 \times CS \\ \mathrm{R}^2 &= 49\,\% \end{aligned} \tag{8.3}$$

In model 8.3, the impact of customer satisfaction on customer acquisition is estimated to be a positive number of 0.65 indicating that customer satisfaction will increase customer acquisition. Again, it should be noted that this linear model is not a well-fitting model due to the low coefficient of determination: $\mathrm{R}^2 = 49\ \%$ [2, 3].

8.2.4 Summary of Customer Satisfaction and I-CRM

We have evaluated the relationship between customer satisfaction and its antecedents and consequences which have been considered as I-CRM variables. As can be noted from Fig. 8.3, we formulated the relationships using linear modelling. We also revalidated the relationship between the variables. In Chap. 6, we tested the hypotheses using structural equation modelling and we ascertained several relationships using all collected data from the questionnaire survey. However, the hypothesis suggesting a relationship between customer satisfaction and loyalty was not supported. In this chapter, we use categorised data and categorised customers to ascertain the relationship, based on what we achieved in Chap. 7. We use categories one and two customers and create a formula to test the hypotheses. Using category one and two customers, perceived value has a positive relationship with customer satisfaction with 0.56 as an outcome. Interactivity has a positive relationship with customer satisfaction with 0.97 as an outcome which shows the strong relationship with customer satisfaction. Customer satisfaction is directly associated with customer acquisition with the outcome of 0.65 which shows a definite link between the two variables. Finally, customer satisfaction is positively associated with customer loyalty with the result of 0.82 that shows a robust relationship between two variables.

8.3 Fuzzy Approach Using Adaptive Neuro-Fuzzy Inference System Modelling

8.3.1 Methodology: Takagi-Sugeno Type Fuzzy Inference System

After considering numerous tools and methodologies to test and validate all of our hypotheses, Adaptive Neuro-Fuzzy Inference System (ANFIS) was chosen as a legitimate solution to prove the relationship. At the same time, it can produce rules. Various types of FIS are reported in the literature [4, 5], and [6] and each is characterized only by its consequent parameters. The Takagi-Sugeno fuzzy inference system (TS-FIS) (also known as TSK fuzzy model) was proposed by [6]. TS-FIS is a systematic approach to infer an output from a set of inputs that the relation between output and inputs is defined by fuzzy IF–THEN rules. A typical fuzzy rule in TS-FIS has the following form:

$$IF\, x_1\; is\, A_1\; and\,(or)\, x_2\; is\, A_2\; THEN\, x_3 = f(x_1, x_2),$$

where A_1 and A_2 are fuzzy sets in the antecedent (IF part of fuzzy rule) while $x_3 = f(x_1,x_2)$ is a crisp function in the consequent (THEN part of fuzzy rule). Usually $f(x_1,x_2)$ is a polynomial in the input variables x_1 and x_2. When $f(x_1,x_2)$ is a first-order polynomial, the resulting TS-FIS is called a first-order TS-FIS.

Figure 8.4 shows the fuzzy reasoning procedure for a first order TS-FIS model with one output and two inputs, having two membership functions for each input. This TS-FIS comprises two fuzzy rules:

$$Rule\ 1: \ IF\ x\ is\ A_1\ and\ y\ is\ B_1\ THEN\ f_1 = p_1x + q_1y + r_1$$

$$Rule\ 2: \ IF\ x\ is\ A_2\ and\ y\ is\ B_2\ THEN\ f_2 = p_2x + q_2y + r_2$$

where x and y are arbitrary values in the discourse sets of the inputs X and Y. A_1 and A_2 are the fuzzy values (sets) of input X. B_1 and B_2 are the fuzzy values (sets) of input Y. p_1, q_1, and r_1 are the coefficients of consequent outputs f_1 and p_2, q_2, and r_2 are the coefficients of consequent outputs f_2.

Since each rule has a crisp output, the overall output f is obtained via weighted average, thereby avoiding the time-consuming process of defuzzification required in a Mamdani model [4]. w_1 and w_2 are the firing strength of rules 1 and 2, respectively. w_1 is calculated as the degree of membership x to A_1 ($\mu_{A1}(x)$) multiplied by the degree of membership y to B_1 ($\mu_{B1}(y)$). Similarly, w_2 is calculated as the degree of membership x to A_2 ($\mu_{A2}(x)$) multiplied by the degree of membership y to B_2 ($\mu_{B2}(y)$).

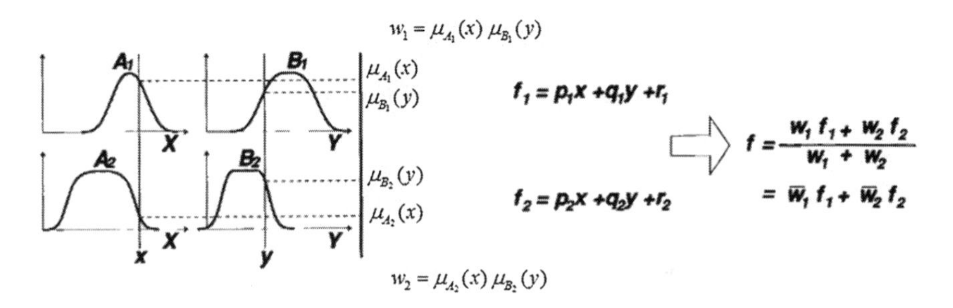

Fig. 8.4 A two-input first-order Takagi-Sugeno fuzzy model with two rules [7]

8.3.2 Adaptive Network-Based Fuzzy Inference System

Neuro-fuzzy modelling, developed in the neural network literature [8], refers to the application of various learning techniques to fuzzy modelling or a fuzzy inference system. Neuro-fuzzy systems, which combine neural networks and fuzzy logic, have recently attracted great interest in the areas of research and application. The neuro-fuzzy approach has the added advantage of reducing training time not only due to its smaller dimensions, but also because the network can be initialized with parameters relating to the problem domain. Such results emphasize the benefits of the fusion of fuzzy and neural network technologies as it facilitates an accurate initialization of the network in terms of the parameters of the fuzzy reasoning system.

A specific approach in neuro-fuzzy development is the adaptive neuro-fuzzy inference system (ANFIS), which has shown significant results in modelling nonlinear functions [9]. ANFIS uses a feed forward network to search for fuzzy decision rules that perform well on a given task. Using a given input–output data set, ANFIS creates a FIS whose membership function parameters are adjusted using a back propagation algorithm alone or a combination of a back propagation algorithm with a least squares method. This allows the fuzzy systems to learn from the data being modelled.

Consider the first order TS-FIS presented in Fig. 8.5. Here, the functioning of ANFIS is a five-layered feed-forward neural structure, and the functionality of the nodes in these layers can be summarized as follows. Note that in each layer, the nodes are numbered from up to down in descending order. At the first layer, for each input, the membership grades in the corresponding fuzzy sets are estimated as output. Output from each node i in layer 1 ($O_{1,i}$) is calculated in Eq. (8.4).

$$\begin{aligned} O_{1,i} &= \mu_{A_i}(x) \quad && \mathrm{i} = 1,2 \\ O_{1,i} &= \mu_{B_{i-2}}(y) \quad && \mathrm{i} = 3,4 \end{aligned} \tag{8.4}$$

where x or y is the input to the nodes, A_i or $B_{i\text{-}2}$ is a fuzzy set associated with node ith of layer 1.

At the second layer, all potential rules between the inputs are formulated by applying fuzzy intersection (AND). The product operation is utilised to evaluate the outstanding power of each formula. Output from each node i in layer 2 ($O_{2,i}$) is calculated with Eq. (8.5).

$$O_{2,i} = w_i = \mu_{A_i}(x) \times \mu_{B_i}(y), \quad \mathrm{i} = 1,2 \tag{8.5}$$

The third layer is used for estimation of the ratio of the ith rule's firing strength to the sum of all rule's firing strengths. Output from each node i in layer 3 ($O_{3,i}$) is calculated with Eq. (8.6).

$$O_{3,i} = \bar{w}_i = \frac{w_i}{w_1 + w_2} \quad \mathrm{i} = 1,2 \tag{8.6}$$

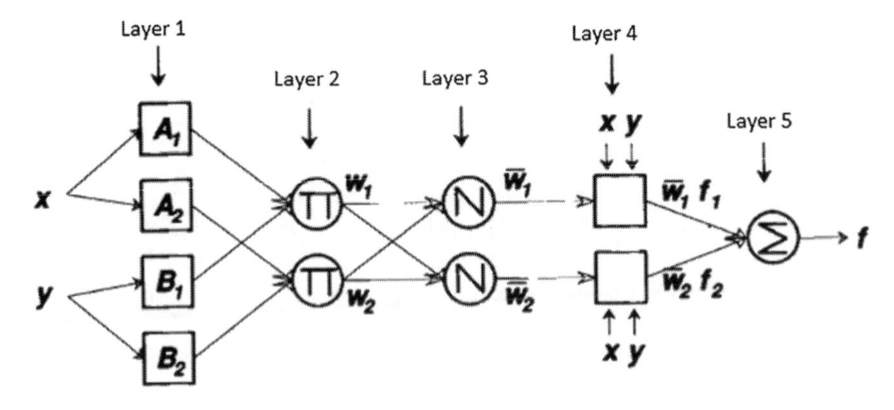

Fig. 8.5 Equivalent ANFIS architecture of FIS in Fig. 8.8 [7]

In layer 4 all nodes are adaptive and their outputs ($O_{4,i}$) are calculated with Eq. (8.7).

$$O_{4,i} = \bar{w}_i f_i = \bar{w}_i(p_i x + q_i y + r_i) \tag{8.7}$$

where $\bar{w}_i$ is the output of layer 3 and [10] is the parameter set. Parameters in this layer will be referred to as consequent parameters.

The final layer computes the overall output ($O_{5,i}$) as the summation of all incoming signals from layer 4 (Eq. 8.8).

$$\text{Overall output} = O_{5,i} = \sum_i \bar{w}_i f_i = \frac{\sum_i w_i f_i}{\sum_i w_i} \tag{8.8}$$

In this ANFIS architecture, there are two adaptive layers (1, 4). Layer 1 has three modifiable parameters (a_i, b_i and c_i) pertaining to the input MFs. These parameters are called *premise* parameters. Layer 4 has also three modifiable parameters (p_i, q_i and r_i) pertaining to the first order polynomial. These parameters are called *consequent* parameters. Optimizing the values of the adaptive parameters is of vital importance for the performance of the adaptive system. Reference [9] developed a hybrid learning algorithm for ANFIS which is faster than the classical back-propagation method to approximate the precise value of the model parameters. The hybrid learning algorithm of ANFIS consists of two alternating phases: (1) gradient descend which computes error signals recursively from the output layer backward to the input nodes, and (2) least squares method, which finds a feasible set of consequent parameters. To run the ANFIS, a reliable and easy-to-use software package is the fuzzy toolbox of MATLAB software.

8.3.3 Subtractive Clustering

Since, there is no optimal solution for the best number of fuzzy sets in ANFIS, we propose to use subtractive clustering to obtain the number of fuzzy sets for each input of ANFIS.

The subtractive clustering method assumes that each data point is a potential cluster centre and calculates a measure of the likelihood that each data point would define the cluster centre, based on the density of surrounding data points. The algorithm does the following:

1. Selects the data point with the highest potential to be the first cluster centre.
2. Removes all data points in the vicinity of the first cluster centre, in order to determine the next data cluster and its centre location.
3. Iterates this process until all of the data is within a pre-determined acceptable vicinity of a cluster centre.

The subtractive clustering method is an extension of the mountain clustering method proposed by Yager and Filev [11]. To perform subtractive clustering, *genfis2* function in fuzzy logic toolbox of MATLAB® software has been used. The output of *genfis2* is a Takagi-Sugeno type fuzzy inference system which should be trained with ANFIS to obtain the final fuzzy inference system. *genfis2* generates a Sugeno-type FIS structure using subtractive clustering and requires separate sets of input and output data as input arguments. When there is only one output, *genfis2* is used to generate an initial FIS for anfis training. *genfis2* accomplishes this by extracting a set of rules that models the data behavior.

The rule extraction method first uses the sub-cluster function to determine the number of rules and antecedent membership functions and then uses linear least squares estimation to determine each rule's consequent equations. This function returns a FIS structure that contains a set of fuzzy rules to cover the feature space.

8.4 Adaptive Neuro-Fuzzy Inference Systems Results

In this section, using the Adaptive Neuro-Fuzzy Inference Systems analysis which is a universal approximator and a useful applied technique, we create rules and provide recommendations regarding the issues. In this thesis, we need to validate the relationship between the variables, but, in this section, we evaluate perceived value and interactivity with customer satisfaction. Using this approach, we can achieve and interpret layers of perceived value and interactivity.

8.4.1 Adaptive Neuro-Fuzzy Inference Systems Results of Relation Analysis in Level 1

In level 1 of the analysis, the relation between perceived value, interactivity and customer satisfaction is analysed with ANFIS. The ultimate output of this analysis is a set of fuzzy rules which describe this relationship in terms of several linguistic

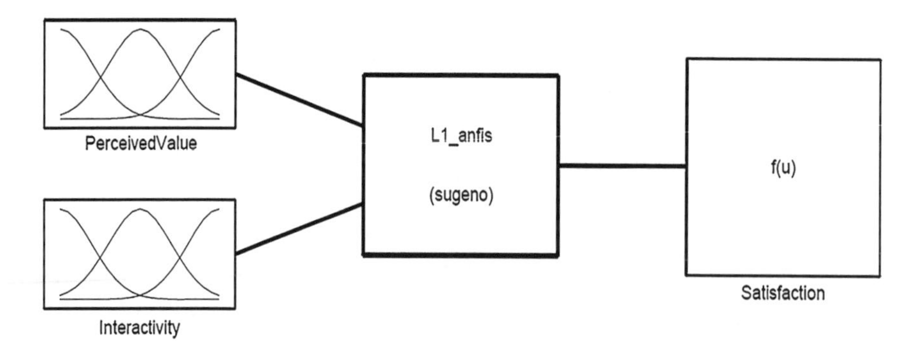

Fig. 8.6 Structure of input–output in ANFIS analysis of level 1 relationship

variables. Figure 8.6 shows the structure of input–output analysis in Level 1 which is the relationship between perceived value and interactivity with customer satisfaction issues at the Fremantle port (Fig. 8.7).

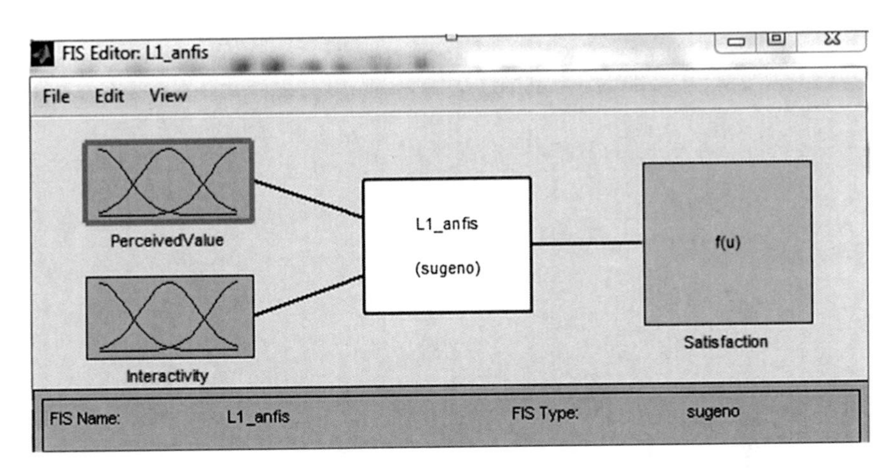

Fig. 8.7 Structure of input–output in adaptive neuro-fuzzy inference systems analysis of level 1 relationship taken from software

Prior to ANFIS analysis, the input variables should be represented in terms of fuzzy linguistic variables. Here, we adopt a data-driven approach to determine a set of linguistic variables for fuzzy modelling of the inputs. Here, the first question is how many linguistic variables should be assigned to each input. The answer to this question is provided by the subtractive clustering algorithm proposed in Sect. 8.4.3.

The subtractive clustering for perceived value and interactivity has returned 5 clusters ranging from Very Low to Very High in the discourse of 1–7 (Fig. 8.8).

The data available for inputs and output of ANFIS are used to run ANFIS. First, *genfis2* function of MATLAB® has generated an initial FIS and then this initial FIS is trained by *anfis* function to yield a final fuzzy inference system. The surface generated by the final FIS is shown in Fig. 8.9 (see Appendix 5 for MATLAB code).

The surface shown in Fig. 8.9 is one representation of the final fuzzy inference system resulting from ANFIS in level 1. Another result of this final FIS is its rule presentation. Actually, a FIS is a set of fuzzy IF–THEN rules. Final FIS of level 1 (L1_anfis) has 5 fuzzy rules as follows:

The rule-presentation of L-1_anfis is as follows:

Figure 8.10 generates the following rules:

1. If (PerceivedValue is VeryLow) and (Interactivity is VeryLow) then (Satisfaction is SatCluster1)
2. If (PerceivedValue is Low) and (Interactivity is Low) then (Satisfaction is SatCluster2)

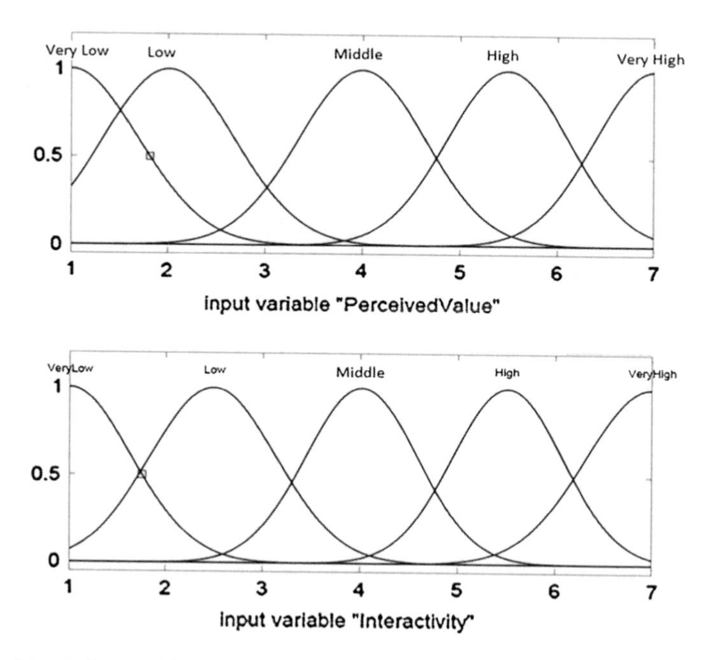

Fig. 8.8 Linguistic variables and membership functions of adaptive neuro-fuzzy inference systems inputs in level 1 which is the relationship between interactivity and perceived value with customer issues

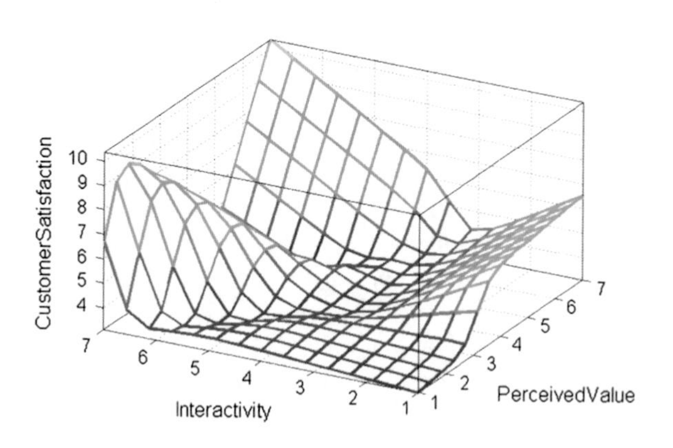

Fig. 8.9 Surface generated by the final fuzzy inference systems for the inputs-output of level 1 which is the relationship between interactivity and perceived value with customer issues

3. If (PerceivedValue is Middle) and (Interactivity is Middle) then (Satisfaction is SatCluster3)
4. If (PerceivedValue is High) and (Interactivity is High) then (Satisfaction is SatCluster4)
5. If (PerceivedValue is VeryHigh) and (Interactivity is VeryHigh) then (Satisfaction is SatCluster5)

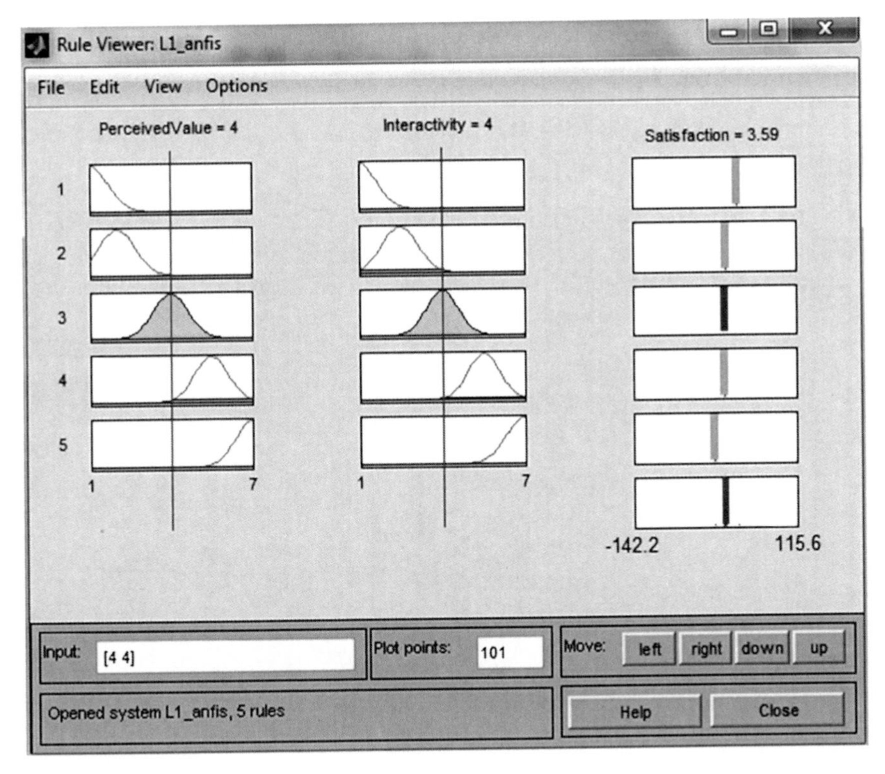

Fig. 8.10 Final fuzzy inference systems rules for the inputs–output of level 1 taken from software

where:

- SatCluster1 = [−7.905 * PerceivedValue + 15.11 * Interactivity − 4.03]
- SatCluster2 = [1.17 * PerceivedValue + 0.2539 * Interactivity + 0.2349]
- SatCluster3 = [0.2734 * PerceivedValue + 0.8356 * Interactivity − 0.8496]
- SatCluster4 = [−0.343 * PerceivedValue + 0.1342 * Interactivity + 5.045]
- SatCluster5 = [0.1597 * PerceivedValue − 1.286 * Interactivity − 6.706]

SatCluster 1–5 are all values of customer satisfaction in different clusters where linear functions show the relationship between customer satisfaction and perceived value, interactivity.

According to Fig. 8.10 and these 5 rules, the following conclusions can be drawn:

- When both perceived value and interactivity are very low or low, customer satisfaction is low.

- When both perceived value and interactivity are high or very high, a significant increase in customer satisfaction is observed.
- Increases in either perceived value or in interactivity will increase the satisfaction but this increase is not as big as the increase that concurrent synergy can generate.

Figure 8.11 which is obtained after running the codes and is based on level 1, illustrates the analysis of the relation between perceived value, interactivity and customer satisfaction. Figure 8.11 clearly shows that we generate an appropriate function for interactivity and perceived value relative to customer satisfaction [12].

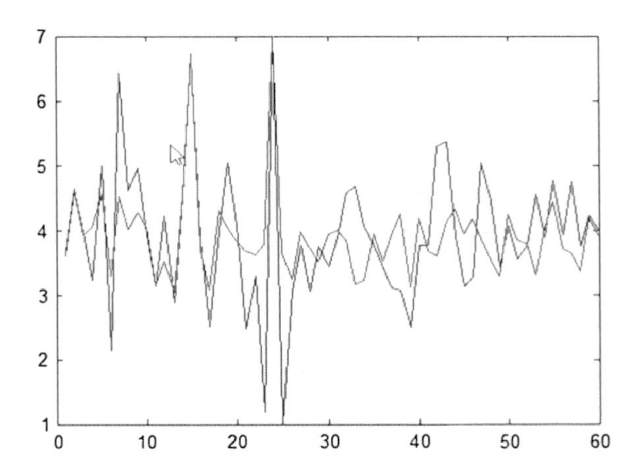

Fig. 8.11 Relation analysis for perceived value and interactivity layers

8.4.2 *Adaptive Neuro-Fuzzy Inference Systems Results of Relation Analysis in Level 2-1 Which is the Relationship Between the Issues and Loyalty*

In level 2 of the analysis, the relationship between customer loyalty, customer acquisition and customer satisfaction is analysed with ANFIS. First, we quantify the relationship between customer loyalty and customer satisfaction, hence the name Level 2-1. Again, the primary objective of ANFIS analysis is to derive a set of fuzzy rules which describe this relationship in terms of several linguistic variables. Figure 8.12 shows the structure of input–output analysis in Level 2-1. In this structure, customer satisfaction is represented by 9 variables each of which corresponds to a category in the questionnaire completed by customers.

Basically, customer satisfaction has been measured in terms of these 9 categories. The advantage of having these 9 satisfaction categories is that it allows analysis of relationship between loyalty and satisfaction in different managerial and operational areas of the organization (Fig. 8.13).

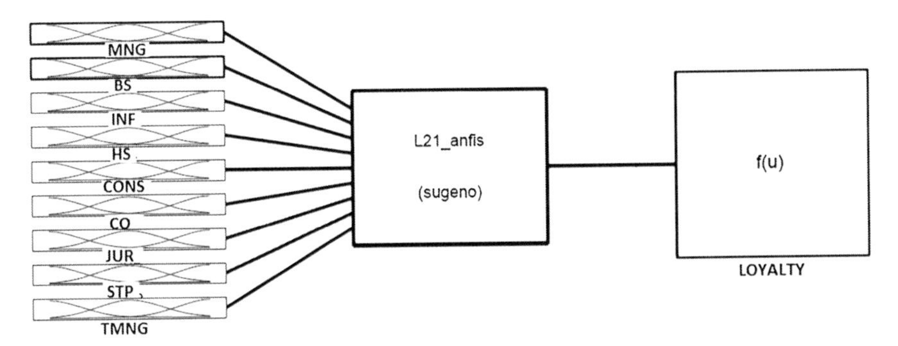

Fig. 8.12 Structure of input–output showing the relationship between interactivity and perceived value with customer issues analysis of level 2-1 relationship

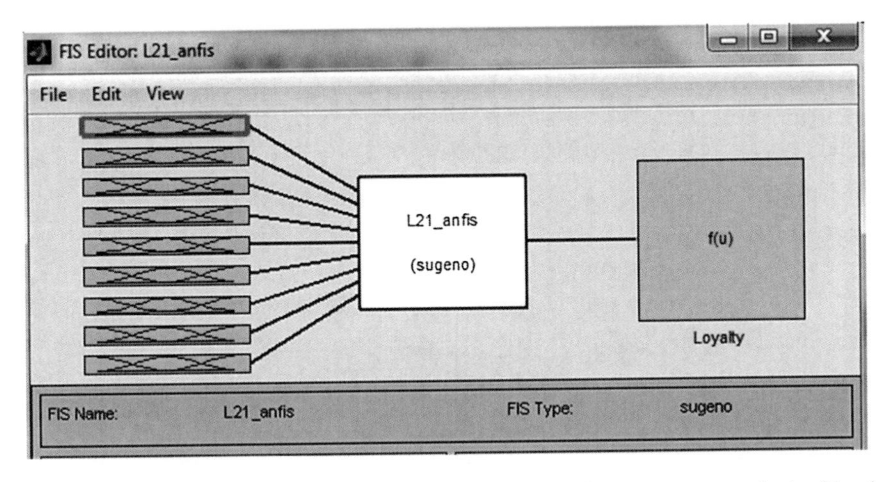

Fig. 8.13 Structure of input–output in adaptive neuro-fuzzy inference systems analysis of level 2-1 relationship between the issues and loyalty

To start ANFIS analysis, first the input variables should be represented in terms of fuzzy linguistic variables. Adopting a data-driven approach to determine the set of linguistic variables for fuzzy modelling of the inputs, the subtractive clustering algorithm is applied. The subtractive clustering for each of the inputs has returned two or three clusters ranging from Low to High in the discourse of 1–7 (Fig. 8.14). For two variables of management (MNG) and cost (CO) three clusters and consequently three fuzzy linguistic variables are derived, namely Low, Middle, and High. For the rest of the inputs, only two clusters of linguistic variables are derived which are named Low and High as demonstrated below.

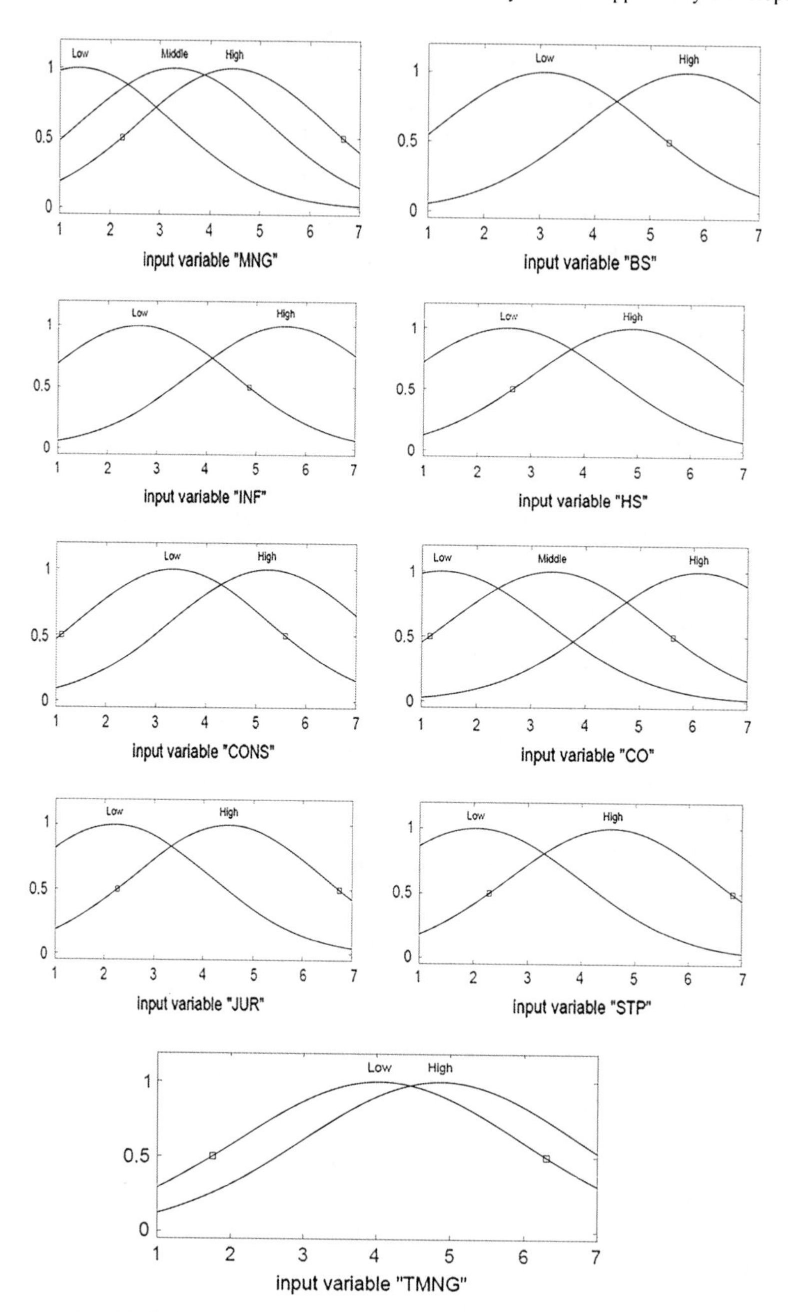

Fig. 8.14 Linguistic variables and membership functions of ANFIS inputs in level 2-1

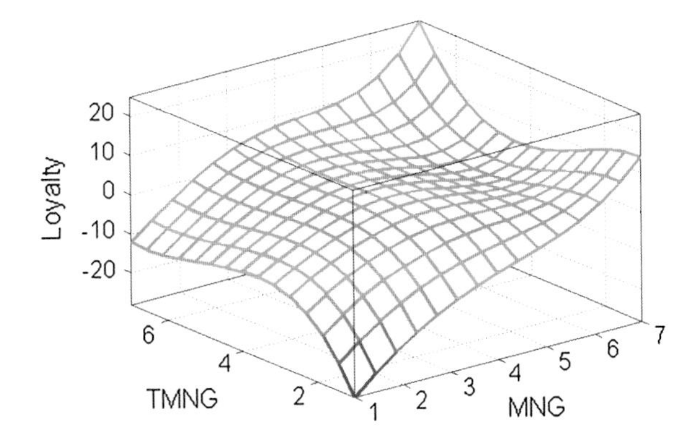

Fig. 8.15 Partial surface presentation of L2-1_adaptive neuro-fuzzy inference systems

The data available for inputs and output of ANFIS are used to run ANFIS code in MATLAB. First, genfis2 has generated an initial FIS and then this initial FIS is trained by anfis function to generate a final fuzzy inference system (named as **L2-1_anfis**) (see Appendix 5 for MATLAB code). The surface presentation of L2-1_anfis is not possible as it has more than two input variables (9 inputs). However, the relationship of loyalty with a pair of inputs is possible. For instance, satisfaction regarding management (MNG) and time management (TMNG) with loyalty have the relationship shown in Fig. 8.15.

As an example, we took the above outcome from the inference system to show the surface between loyalty, time management and management in accordance with the following rules:

The rule-presentation of L2-1_anfis is as follows:

Figure 8.16 represents and generates the following rules:

(Rule 1)

If (MNG is High) and (BS is High) and (INF is High) and (HS is High) and (CONS is High) and (CO is High) and (JUR is High) and (STP is High) and (TMNG is High)
Then Loyalty is LoyaltyCluster1 = 0.3 * MNG − 8.5 * BS − 1.7 * INF + 8.6 * HS + 3.9 * CONS + 5.8 * CO + 2.2 * JUR − 13.8 * STP − 13.4 * TMNG + 2.2

(Rule 2)

If (MNG is Low) and (BS is Low) and (INF is Low) and (HS is Low) and (CONS is Low) and (CO is Low) and (JUR is Low) and (STP is Low) and (TMNG is Low)
Then Loyalty is LoyaltyCluster2 = 21.8 * MNG + 20 * BS − 26.8 * INF + 6.9 * HS − 18.1 * CONS + 5.7 * CO + 25 * JUR + 0.3 * STP − 4 * TMNG + 1.3

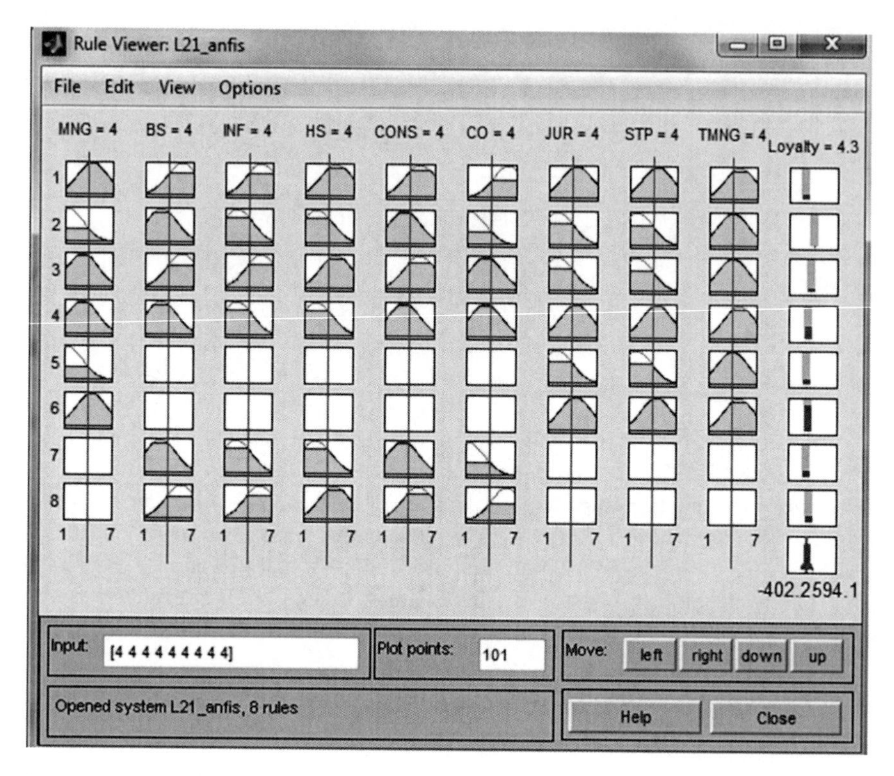

Fig. 8.16 Final fuzzy inference systems rules for the inputs–output of level 1 taken from software

(Rule 3)

If (MNG is Middle) and (BS is High) and (INF is High) and (HS is High) and (CONS is High) and (CO is Middle) and (JUR is Low) and (STP is Low) and (TMNG is Low)
Then Loyalty is LoyaltyCluster3 = 19.7 * MNG − 15.4 * BS + 12.5 * INF − 26.6 * HS + 0.6 * CONS − 14.4 * CO + 9.5 * JUR + 19.1 * STP + 13.8 * TMNG + 1.2

(Rule 4)

If (MNG is Middle) and (BS is Low) and (INF is Low) and (HS is Low) and (CONS is Low) and (CO is Middle) and (JUR is High) and (STP is High) and (TMNG is High)
Then Loyalty is LoyaltyCluster4 = 23.9 * MNG + 4.3 * BS − 15.9 * INF − 0.12 * HS + 1.3 * CONS − 10 * CO + 0.35 * JUR − 5.6 * STP − 2.1 * TMNG + 13.7

(Rule 5)

If (MNG is Low) and (JUR is Low) and (STP is Low) and (TMNG is Low)
Then Loyalty is LoyaltyCluster5 = −7.8 * MNG + 2.6 * BS + 1.9 * INF + 6.2 * HS + 4 * CONS − 0.7 * CO − 2.3 * JUR − 1.1 * STP − 9.2 * TMNG + 13.7

(Rule 6)

If (MNG is High) and (JUR is High) and (STP is High) and (TMNG is High)
Then Loyalty is LoyaltyCluster6 = −17.6 * MNG − 0.8 * BS + 7.5 * INF − 2.3 * HS + 1.9 * CONS − 1.1 * CO − 6 * JUR + 6.4 * STP + 2.6 * TMNG − 1

(Rule 7)

If (BS is Low) and (INF is Low) and (HS is Low) and (CONS is Low) and (CO is Low)
Then Loyalty is LoyaltyCluster7 = 17.1 * MNG − 8.9 * BS − 1.4 * INF + 2.2 * HS − 4.2 * CONS + 1.8 * CO − 15.7 * JUR + 2 * STP + 4.1 * TMNG − 14.6

(Rule 8)

If (BS is High) and (INF is High) and (HS is High) and (CONS is High) and (CO is High)
Then Loyalty is LoyaltyCluster8 = 7.7 * MNG − 1.2 * BS − 24.6 * INF − 8.1 * HS − 5.8 * CONS + 19.1 * CO − 0.7 * JUR + 11 * STP + 10.1 * TMNG + 4.4

The fuzzy rules above indicate the positive or negative linear relationship of loyalty with each of the satisfaction categories. The coefficient of each satisfaction category also indicates the magnitude of impact each satisfaction category has on loyalty. For example, in rule 1 where satisfaction in all categories is High, the LoyaltyCluster1 gives an estimate of loyalty. As can be seen, in LoyaltyCluster1 management (MNG), health and safety (HS), container sorting (CONS), cost (CO), and jurisdiction (JUR) have a positive impact on loyalty. In other words, we can say that loyal customers are happy with these managerial and operational performance aspects of the organization. On the other hand, as fuzzy rule 1 shows, loyalty has a negative relationship with satisfaction in booking system (BS), infrastructure (INF), stevedores' performance (STP), and time management (TMNG) indicating that loyal customers are not satisfied with these managerial and operational performance aspects of the organization. Similar conclusions can be drawn for other fuzzy values of satisfaction categories as indicated in the IF part of the rules.

An easier way to sum up the relationships is the graphical representation of loyalty changes versus each individual satisfaction category (Fig. 8.17). The following conclusions can be drawn from these relationships:

- MNG, INF, STP, CONS and TMNG share a similar relationship with loyalty as increases in these satisfaction categories parallels the average increase in loyalty.
- The reverse relationships can be seen for BS, HS, and JUR as increases in these satisfaction categories parallel the average decrease in loyalty.
- The relation between loyalty and cost satisfaction (CO) cannot be presented in a linear form. However, one could say that high satisfaction with cost has a positive impact on loyalty.

Each of the following diagrams represents the relationship between loyalty and one of the issues, the abbreviations of which are shown below:

Category indicator	Abbreviation
Management	MNG
Booking system	BS
Infrastructure	INF
Health and safety	HS
Container sorting	CONS
Cost	CO
Jurisdiction	JUR
Stevedores performance	STP
Time management	TMNG

Below, we provide a detailed interpretation of each diagram.

Figure 8.17 shows the relationship between loyalty and dissatisfaction issues which are dealt with in the section on customer satisfaction. As can be noted, management has an increasing trend in respect to loyalty; however, there are fluctuations as well. It starts from nearly 1 and increases dramatically to 4; however, it decreases from 4 to 6 before increasing again. The booking system has a negative trend compared to loyalty, and from 4 it descends sharply. Infrastructure issues with loyalty have a logical trend as it starts at 1 and with increasing fluctuations it peaks. Health and safety have an acceptable loyalty initiation, and despite having ups and downs it slumps. Container sorting has a positive and constant trend, and as discussed in the previous section, its loyalty is positively associated with customer satisfaction. At the end of the procedure, it reaches the climax.

Cost has a distinctive trend, though; it rises to the peak and achieves loyalty. It has an unexpected fall at 2 and then it develops from 5 onwards. Jurisdiction has a negative trend with respect to loyalty. This graph indicates that those in authority in Fremantle port cannot satisfy their customers using current rules and regulations and these may have to be changed. Stevedore performance has a positive

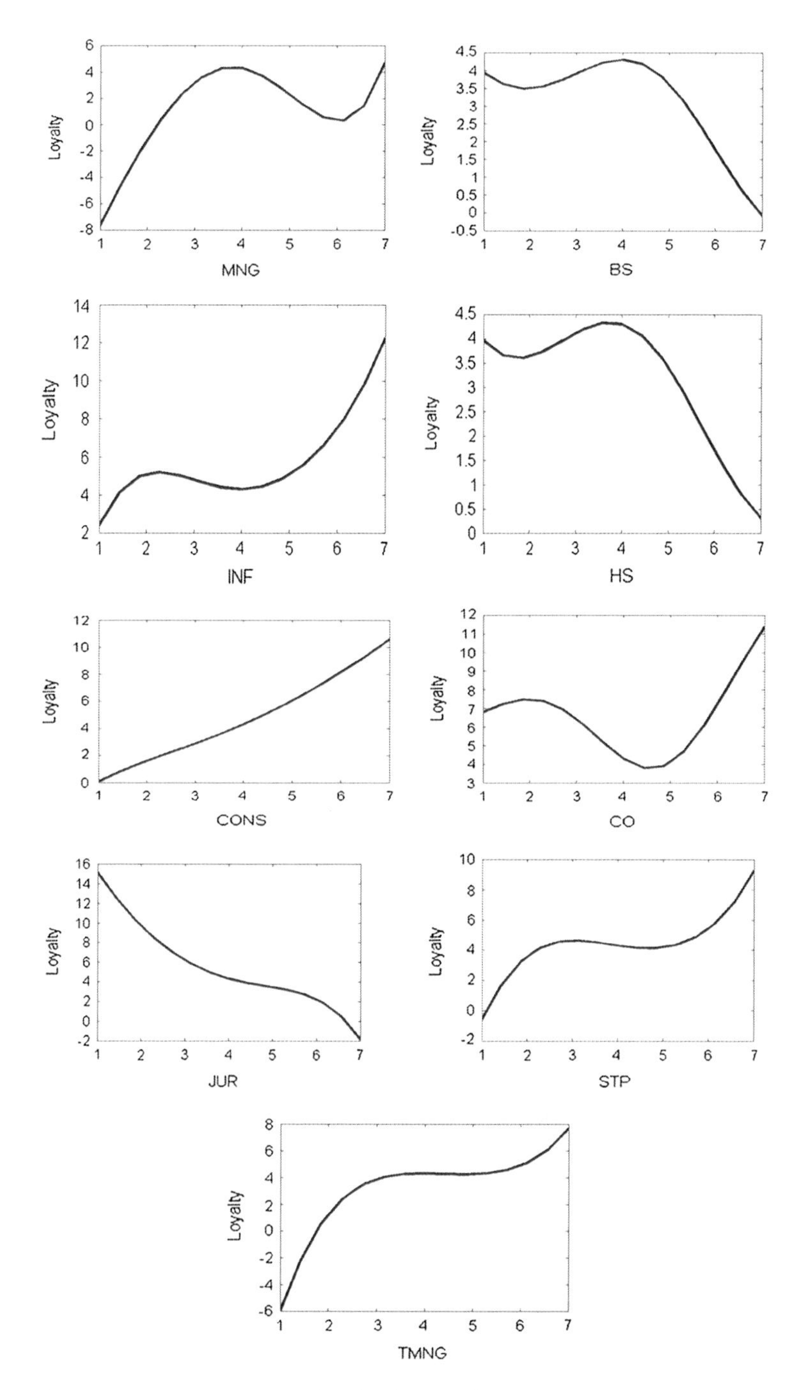

Fig. 8.17 Loyalty changes versus each individual satisfaction category

relationship with loyalty; however, from 3 to 5 it has a relative and constant decline, but from 5 it increases again. Time management has a somewhat similar trend to that of stevedore performance; however, from 3 to 5 it has a constant trend.

Figure 8.18 which is produced after running the program depicts level 2.1 which concerns the relationship between customer satisfaction and loyalty. In level 2.1, instead of considering customer satisfaction, we determine the relationship between our nine issues which we derived from the system and loyalty, in order to estimate the loyalty and its sensitivity. For each satisfaction in Fig. 8.18, we have a loyalty point and we rely upon the established ANFIS.

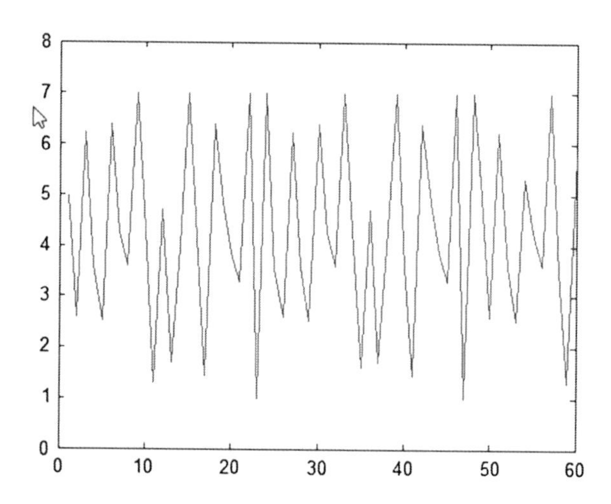

Fig. 8.18 Relationship analysis based on adaptive neuro-fuzzy inference systems

8.4.3 Summary

In level one, we ascertained the relationship between perceived value and interactivity with customer issues at the Fremantle port using non-linear modelling. To do this, we utilised ANFIS and generated models, graphs and rule to highlight the relations. Based upon the rules, we clarified how interactivity and perceived value can influence customer issues.

8.4.4 ANFIS Results of Relation Analysis in Level 2-2

In this section, the relation between customer acquisition and customer satisfaction is analysed with ANFIS (Level 2-2). We quantify the relationship between customer acquisition and customer satisfaction in terms of several linguistic variables. Figure 8.19 shows the structure of input–output analysis in Level 2-2. Similar to level 2-1, in this structure, customer satisfaction is represented by nine variables

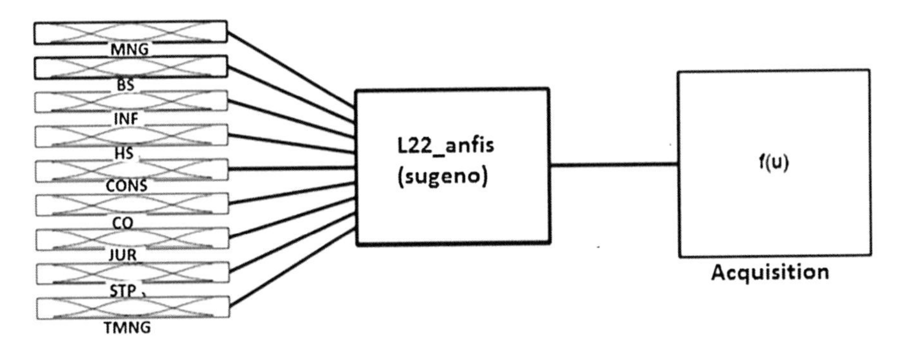

Fig. 8.19 Structure of input–output in adaptive neuro-fuzzy inference systems analysis of level 2-2 relationship

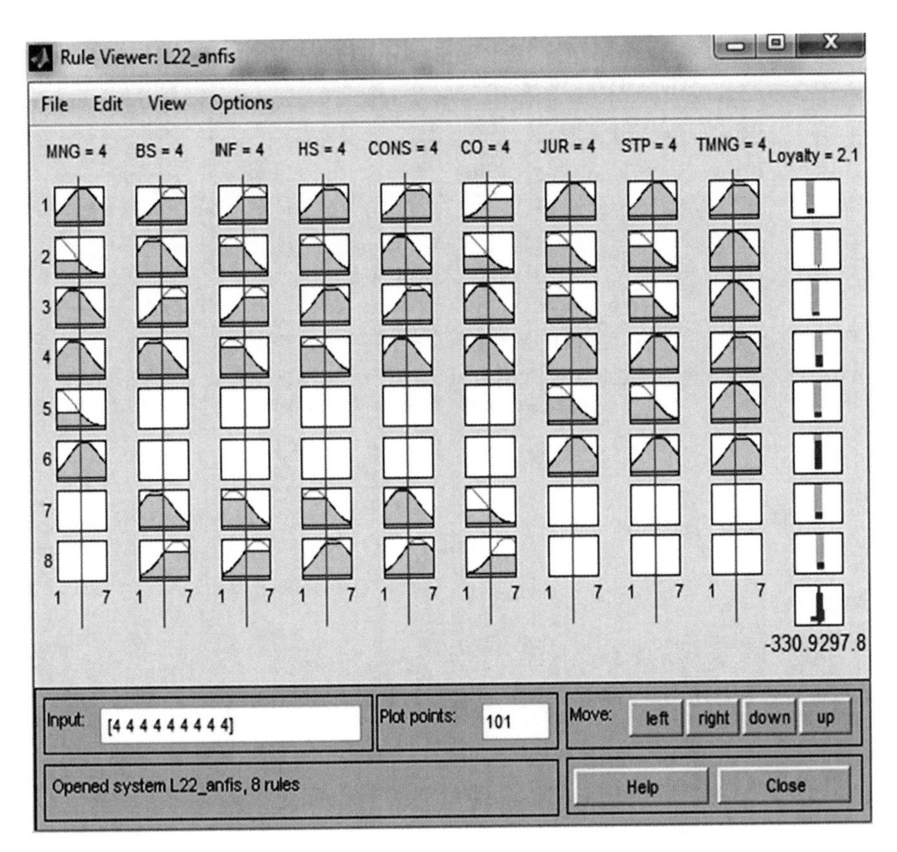

Fig. 8.20 Final fuzzy inference systems rules for the inputs-output of level 1 taken from software

each of which corresponds to a category of customer satisfaction. The advantage of having these nine satisfaction categories is that it allows us to ascertain the relationship between customer acquisition and customer satisfaction in terms of different managerial and operational aspects of the organization.

To start ANFIS analysis, first the input variables should be represented in terms of fuzzy linguistic variables. All input variables of the ANFIS in Level 2-2 have been previously fuzzified in Sect. 8.3.2.

The data available for inputs and output of ANFIS are used to run ANFIS code in MATLAB. First, *genfis2* has generated an initial FIS and then this initial FIS is trained by the *anfis* function to generate a final fuzzy inference system (named as **L2-2_anfis**) (see Appendix 5 for MATLAB code). The rule-presentation of L2-2_anfis which illustrates the relationship between customer issues and loyalty is as follows:

Rule (1)

If (MNG is High) and (BS is High) and (INF is High) and (HS is High) and (CONS is High) and (CO is High) and (JUR is High) and (STP is High) and (TMNG is High)
Then Acquisition is AcquisitionCluster1 = 0.7 * MNG + 2.5 * BS + 10.1 * INF − 29.3 * HS + 1.2 * CONS − 0.9 * CO − 3.6 * JUR −
5.3 * STP + 8.2 * TMNG − 2.4

Rule (2)

If (MNG is Low) and (BS is Low) and (INF is Low) and (HS is Low) and (CONS is Low) and (CO is Low) and (JUR is Low) and (STP is Low) and (TMNG is Low)
Then Acquisition is AcquisitionCluster2 = 2.4 * MNG − 4.1 * BS − 11.8 * INF + 2.3 * HS + 6.5 * CONS + 5 * CO + 4.6 * JUR + 7.8 * STP − 7.5 * TMNG + 1.6

Rule (3)

If (MNG is Middle) and (BS is High) and (INF is High) and (HS is High) and (CONS is High) and (CO is Middle) and (JUR is Low) and (STP is Low) and (TMNG is Low)
Then Acquisition is AcquisitionCluster3 = −12.5 * MNG + 4.5 * BS + 7.5 * INF + 13.1 * HS + 2.8 * CONS + 13.5 * CO − 10.1 * JUR − 18.6 * STP − 4.7 * TMNG + 1.2

Rule (4)

If (MNG is Middle) and (BS is Low) and (INF is Low) and (HS is Low) and (CONS is Low) and (CO is Middle) and (JUR is High) and (STP is High) and (TMNG is High)
Then Acquisition is AcquisitionCluster4 = −16.7 * MNG − 3.9 * BS − 11.9 * INF + 9.5 * HS + 7.1 * CONS −
10 * CO + 5.5 * JUR + 3.2 * STP − 1.6 * TMNG − 2.2

Rule (5)

If (MNG is Low) and (JUR is Low) and (STP is Low) and (TMNG is Low)
Then Acquisition is AcquisitionCluster5 = 2.8 * MNG + 1.1 * BS − 0.7 * INF − 4.9 * HS − 2.2 * CONS − 5.9 * CO + 0.8 * JUR + 5.7 * STP +2.4 * TMNG + 1.5

Rule (6)

If (MNG is High) and (JUR is High) and (STP is High) and (TMNG is High)
Then Acquisition is AcquisitionCluster6 = 4 * MNG + 0.4 * BS − 3 * INF + 3.3 * HS − 2.5 * CONS + 0.6 * CO − 1.3 * JUR + 5.3 * STP − 2.2 * TMNG − 14

Rule (7)

If (BS is Low) and (INF is Low) and (HS is Low) and (CONS is Low) and (CO is Low)
Then Acquisition is AcquisitionCluster7 = 3.4 * MNG + 2.2 * BS − 2.5 * INF − 5.1 * HS − 0.1 * CONS + 6.9 * CO + 2 * JUR − 7.6 * STP + 3 * TMNG − 1.7

Rule (8)

If (BS is High) and (INF is High) and (HS is High) and (CONS is High) and (CO is High)
Then Acquisition is AcquisitionCluster8 = −5.2 * MNG − 6 * BS + 0.8 * INF − 3.4 * HS + 4.3 * CONS + 2.6 * CO − 6.6 * JUR + 12.8 * STP + 8.2 * TMNG − 2

For an easy-to-interpret representation, the fuzzy rule set of above is depicted in graphical form in Fig. 8.21 which shows customer acquisition changes versus each individual satisfaction category. The following conclusions can be drawn from Fig. 8.20

- Increases in MNG, BS, INF, CONS and JUR satisfaction are paralleled by overall decrease in customer acquisition.
- The reverse relationships can be seen for STP and TMNG as increases in these satisfaction categories parallel the average increase in customer acquisition. In particular, as rule 8 reveals, when satisfaction for BS, INF, HS, CONS, and CO is high, then STP and TMNG affect acquisition significantly and have the greatest impact of all.
- The relationship between acquisition and HS cannot be represented in a linear form.

Each of the following diagrams represents the relationship between acquisition and one of the issues for which the abbreviations are shown below:

Category indicator	Abbreviation
Management	MNG
Booking system	BS
Infrastructure	INF
Health and safety	HS
Container sorting	CONS
Cost	CO
Jurisdiction	JUR
Stevedores performance	STP
Time management	TMNG

Based on Fig. 8.21, the horizontal axis represents the issue and the vertical axis represents customer acquisition. The graphs above indicate that management and booking system have similar trends and the lack of improvement in either of these will decrease the level of customer acquisition. In management, after 5 it starts increasing and in booking system it increases after 6. Health and safety issues start their activity just over -1 and this increases sharply to 3 as does acquisition; however, it decreases until 6 and with a simultaneous decrease in customer acquisition. Then from 6, it increases to enhance acquisition. This implies that authorities have taken better care of individuals in that area of operations. Infrastructural issues have a negative and constant trend starting from 5 and ending below -1.

Based on these results and facts gleaned from the questionnaire, the Fremantle port authority seems unable to produce customer and employee satisfaction using the existing infrastructures and it is unlikely to create customer acquisition using their current system. Cost, on the other hand, has a positive impact on customer acquisition up to 3 and after that it starts going down and negatively affects customer acquisition. The graph starts at above 1 and ends at around -3, indicating a slump of 4. This again clearly demonstrates that authorities must carefully scrutinize port activity costs that are incurred by customers, as this current trend in costs will not lead the port to customer acquisition. Container sorting has a negative impact on acquisition, with fluctuation evident in certain parts. The graph shows a general drop, starting at above 7 and ending at below -2. Stevedore performance has a negative inclination up to 3, and then incontrovertibly increases customer acquisition and this occurs due to the integral role that stevedores play in Fremantle port activities. Jurisdiction with respect to acquisition has a negative trend until 4 and then it starts increasing acquisition. In the case of time management with respect to acquisition, when time management has a positive trend, customer acquisition increases. Time management is very fortuitous and sometimes even experts cannot predict the time that needs to be assigned for each process. This implies that improved time management will improve customer acquisition.

Figure 8.22 is produced after running the codes and program, and illustrates level 2.2 and the relationship between customer satisfaction (and its nine issues) and customer acquisition. As can be noted in Fig. 8.22, the customer acquisition

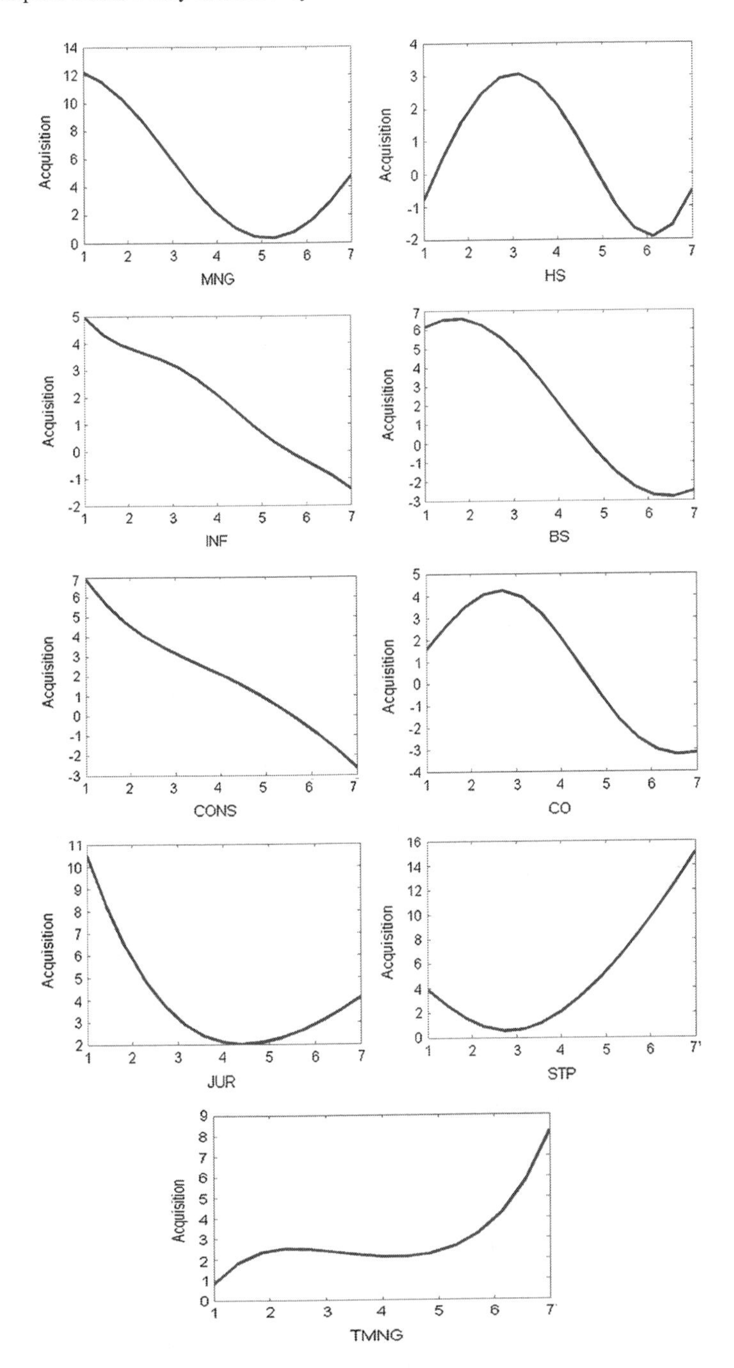

Fig. 8.21 Customer acquisition changes versus each individual satisfaction category

and customer satisfaction completely overlap each other. This alone indicates the appropriateness of the function. Likewise, it shows that the error is almost zero and we are quite sure that the established anfis is correct.

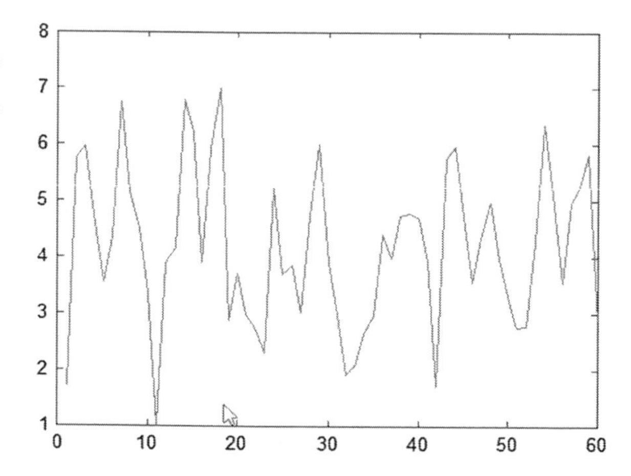

Fig. 8.22 Relationship analysis for level 2.2 which is the relationship between customer issues and customer acquisition

8.4.5 Summary of Level 2.1 (Customer Issues and Loyalty) and Level 2.2 (Customer Issues and Customer Acquisition)

In level 2.1 and 2.2, we established the relationship between customer issues with customer loyalty (level 2.1) and customer acquisition (level 2.2) and the impact of loyalty and customer acquisition on the customer issues. We provide graphs and various figures based on each level and created different rules in accordance with fuzzy inference systems. We also interpreted each of the figures according to the relationship between the issues and both loyalty and customer acquisition. Likewise, we provided recommendations for each level.

8.5 Significance of Utilising Linear and Non-linear Methodology

Previously, researchers have used fuzzy inference systems to address outstanding issues which we touch upon in this section. In short, we have utilised Neuro fuzzy inference systems with respect to customer satisfaction and CRM. Reference [13] propose a methodology to create customer satisfaction using the neuro fuzzy approach. They also generate fuzzy rules based on a survey. As mentioned previously, the extant literature fails to address and ascertain the relationship between customer satisfaction with its antecedents and consequences based on our conceptual framework. To the best of our knowledge, none of the academic works has undertaken research on the relationship between perceived value and interactivity with main issues in a particular industry. Moreover, no research has been done on customer complaints with respect to customer loyalty and customer acquisition. As discussed previously, in the existing literature, no work has been done to examine and evaluate the relationship between various components of CRM. As perceived value, interactivity, customer satisfaction, loyalty and customer acquisition are integral parts of CRM, we attempted to ascertain their relationship and influence in order to evaluate customer satisfaction and decrease the level of complaints using various approaches.

Using both methodologies of linear and non-linear analysis, we overcame the challenges presented by our main variables and were able to establish the relationships between them. To the best of our knowledge, no researchers have utilised a linear methodology and our formulation to test and ascertain such relationships. Those researchers who have utilised a non-linear methodology have established critical ground work which has been extensively cited in research. However, our thesis presents a state-of-the-art feature which comprises a non-linear approach. We introduce various levels by which we can evaluate the relationships using a linear approach and create rules based on a non-linear approach. In the first level, we examine the impacts of perceived value and interactivity on customer satisfaction by proposing research questions which we address using linear methodology. We conclude the first part of this chapter by calculating an overall degree of agreement (ODA) for each customer in the four sections of perceived value, interactivity, customer loyalty, and customer acquisition. In the next stage, using the adaptive Takagi-Sugeno fuzzy inference system (TS-FIS), we presume from a set of inputs that the relationships between output and inputs imply fuzzy IF–THEN rules. Simultaneously, we utilised subtractive clustering to achieve the number of fuzzy sets for each input of ANFIS. Using the aforementioned method, we determined rules that must be taken into consideration by the company. Also, we produced many figures based on the results and facts obtained from software. In the following section, we summarize some of the important rules and recap the main interpretations:

1. When both perceived value and interactivity are very low or low, customer satisfaction is low.
2. When both perceived value and interactivity are high or very high, s significant increase in customer satisfaction is observed.
3. Increases either in perceived value or in interactivity will increase the satisfaction but this increase is not as big as the increase that the concurrent synergy can generate.
4. Management, infrastructure, stevedore performance, container sorting, and time management share a similar relationship with loyalty as increases in these satisfaction categories parallel the average increase in loyalty.
5. The reverse relationships can be seen for booking system, health and safety, and jurisdiction as increases in these satisfaction categories parallel the average decrease in loyalty.

6. The relation between loyalty and cost satisfaction (CO) cannot be presented in a linear form. However, one could say that high satisfaction with cost has a positive impact on loyalty.
7. Increases in management, booking system, infrastructure, container sorting and jurisdiction satisfaction is paralleled by overall decrease in customer acquisition.
8. The adverse relationships can be noted for stevedore performance and time management as increases in these satisfaction categories parallel the average increase in customer acquisition. In particular, as rule 8 reveals, when satisfaction with the booking system, infrastructure, container sorting and cost is high, then stevedore performance and time management affect acquisition significantly and have the greatest impact of all.
9. The relationship between acquisition and HS cannot be presented in a linear form.

8.6 Productivity, Quality and Reputation in I-CRM

Predicated upon our conceptual framework that we introduced and developed, we believe that I-CRM can bring about productivity, quality and reputation for a company. According to [14, 15, 16] and [17], complaints should be considered as an opportunity for improvement. However, most CRM systems fail to address customer issues methodically and do not provide any recommendations. Traditional CRMs fail to provide new business opportunities. Our novel and innovative model increases customer perception and encourages them to lodge complaints which can be responded to and addressed appropriately and promptly. In the following framework which we adopt from Chap. 4, we clearly illustrate that I-CRM variables will turn customer dissatisfaction to customer satisfaction. This will create productivity and increase the quality of the services offered by the company, thereby providing positive word of mouth and enhancing reputation. Together, these three factors are conducive to business improvement. I-CRM variables develop flow in organizational procedures and improve the entire business activities of the organization. Based on [18], components of CRM can improve company productivity and trust. However, Intelligent CRM is more responsive and interactive with customers in terms of developing relationship with customers and generating satisfaction and loyalty. Likewise, I-CRM ensures the long-lasting success of the system as well as increasing the reputation of the company.

8.6.1 Productivity

Due to the increasing level of customer complaints, companies are employing various CRM systems and strategies to decrease the number of complaints, and increase customer satisfaction and company productivity. However, not all CRM systems can successfully deal with complaints and appropriately manage and address them. While the customers and employees continue to complain about issues, productivity cannot improve. Intelligent CRM which is the focus of this thesis has a strategic view in terms of the complaints and the organizational performance intended to capture the customer complaints. Intelligent CRM assists a company to identify any obstacles and increase productivity and efficiency [19]. According to [20], CRM adoption signals a company's commitment to reduce costs and optimise the time and cost needed for the implementation. Also, based on our conceptual framework, intelligent CRM can optimise the issues as well as increasing productivity of the system as it transforms customer complaints into business opportunities on one hand, and on the other hand, it converts customer dissatisfaction to customer satisfaction which enhances productivity.

8.6.2 Quality

Quality is another outcome of I-CRM. Our system is intended to improve the quality of products and services which is also the intention of companies which adopt a CRM software or strategy. However, sometimes, due to the high expectations of the customers, a company may fall short of meeting customer expectations or requirements, and this must be avoided. In Chaps. 7 and 8, we introduce a complete methodology to increase customer satisfaction. Moreover, we also evaluate levels of customer loyalty. In doing so, we established the quality of the services currently offered at the Fremantle port and addressed various issues in order to improve the quality of these services.

Customers have different perceptions about quality of the services. Quality provides convenience for the customers and makes a company trustworthy [21]. The company needs to produce certain perceptions in the minds of customers. Quality can leverage customers and make them more loyal [22]. In our case, the truck drivers become more satisfied and loyal as the quality of the services are being checked using our conceptual framework and methodologies postulated in Chap. 7 and early in Chap. 8, and this alone increases customer satisfaction. Finally, high quality services enhance a company's name and establish it as a competitive brand. A certain level of quality is offered by all companies, and customers respond to this level with either positive or negative behaviours.

8.6.3 Reputation

Once a company reaches its desired productivity level and improves the quality of its services, the customer is likely to remain with the company and begin spreading positive word-of-mouth. Reputation is an intangible quality which may take a long time to establish but can be ruined in an instant. Using the Intelligent CRM framework, a company can improve its reputation and take complete control over performance. I-CRM has a profound impact on a company's reputation as it constantly evaluates each of the variables including perceived value, interactivity, customer satisfaction, loyalty and customer acquisition. Moreover, the system continuously identifies and addresses the issues. The main goal of a company is to retain its customers and keep them satisfied which in turn will positively influence the company's reputation.

According to [23], reputation might be regarded as an invaluable asset that provides feasible competitive advantage for the company. Reputation can be considered as an intangible resource which competitors cannot copy to their own advantage. Reputable companies can greatly influence other companies and their

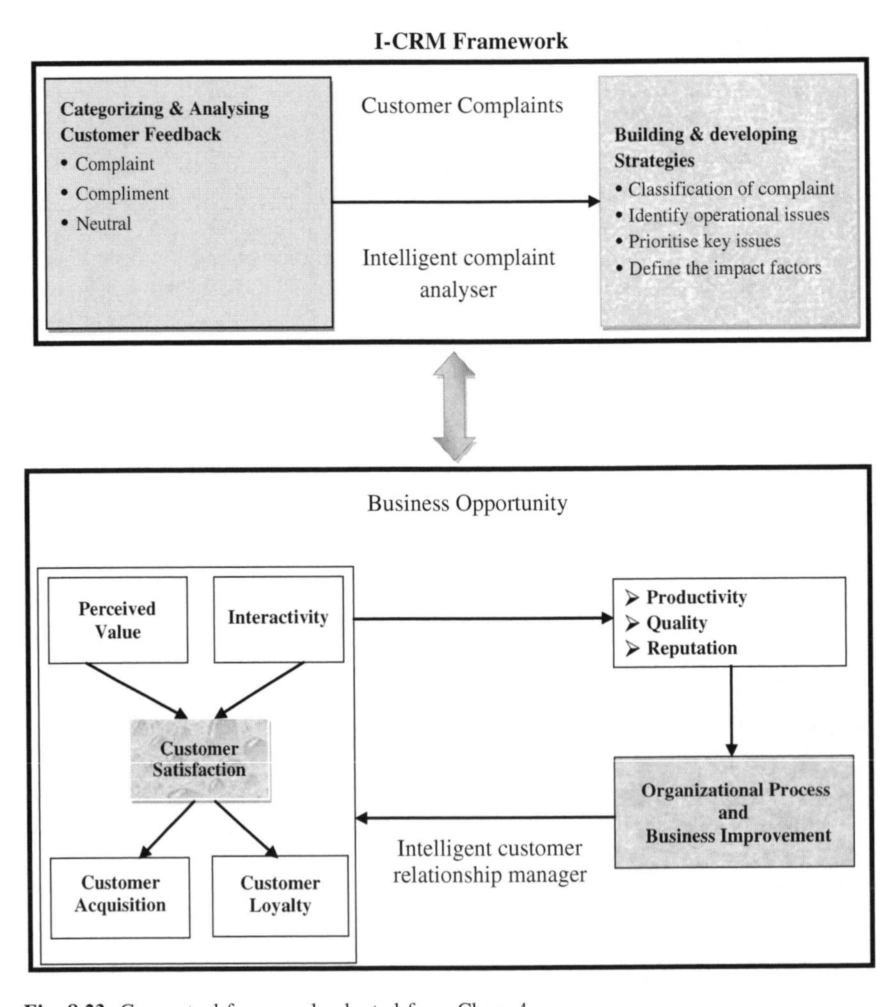

Fig. 8.23 Conceptual framework adopted from Chap. 4

pricing strategies. The reputable companies sometimes become idle for the customers as they start admiring the companies for their services and products provided by them and they hope to support the company in various ways. One of the main goals of I-CRM is to generate reputation for companies. Based on our conceptual framework and according to Walsh et al. [24], customer satisfaction and loyalty are the antecedent and consequent of reputation, respectively. According to [24], "reputation is defined as the ability to identify and truly interpret 'what a company stands for'".

Based on our discussion in Chap. 4, the above conceptual framework clearly depicts two procedures in which an intelligent complaint analyser has the capability of identifying the issues, categorizing them and prioritizing the customer complaints using a qualitative approach. The second procedure which is related to an intelligent customer relationship manager has two different sections which have reciprocal relationships with each other. On the left, the main I-CRM variables (perceived variable, interactivity, customer satisfaction, loyalty and customer acquisition) can be noted. In addition, earlier in Chaps. 6, 7 and 8, we consistently emphasized the positive relationship between these significant factors. As, customer satisfaction was central to other variables, we achieved a result and this led to productivity, quality and reputation of the system as well as the company. These factors lead the company towards business improvement and create new business opportunities. As our process is an iterative one, we expect to increase the satisfaction and decrease the level of dissatisfaction by the frequent repetition of the process (Fig. 8.23).

8.7 Creating Opportunities from Customer Complaints and Generating Business Improvement

Based on our conceptual framework and the aim of this research, we provide the following module. We can simply turn each of the customer complaints into an opportunity for the company. Once we provide new opportunities, we are able to improve productivity and effectiveness, generating greater profits. Productive and efficient companies have knowledgeable staff and managers. As customers become aware of companies' performances and reliability, once a single company can legitimately meet the needs of the customers, they try to be loyal to the company and it is unlikely that they will take their business elsewhere. This creates positive word of mouth and the company will become well-known and reputable. When these three aspects of growth are established, the company may experience improvement. In Fig. 8.24, we will show how customer complaints are converted to business opportunities using our methodology and our framework. To the best of our knowledge, it is first kind in this area of research.

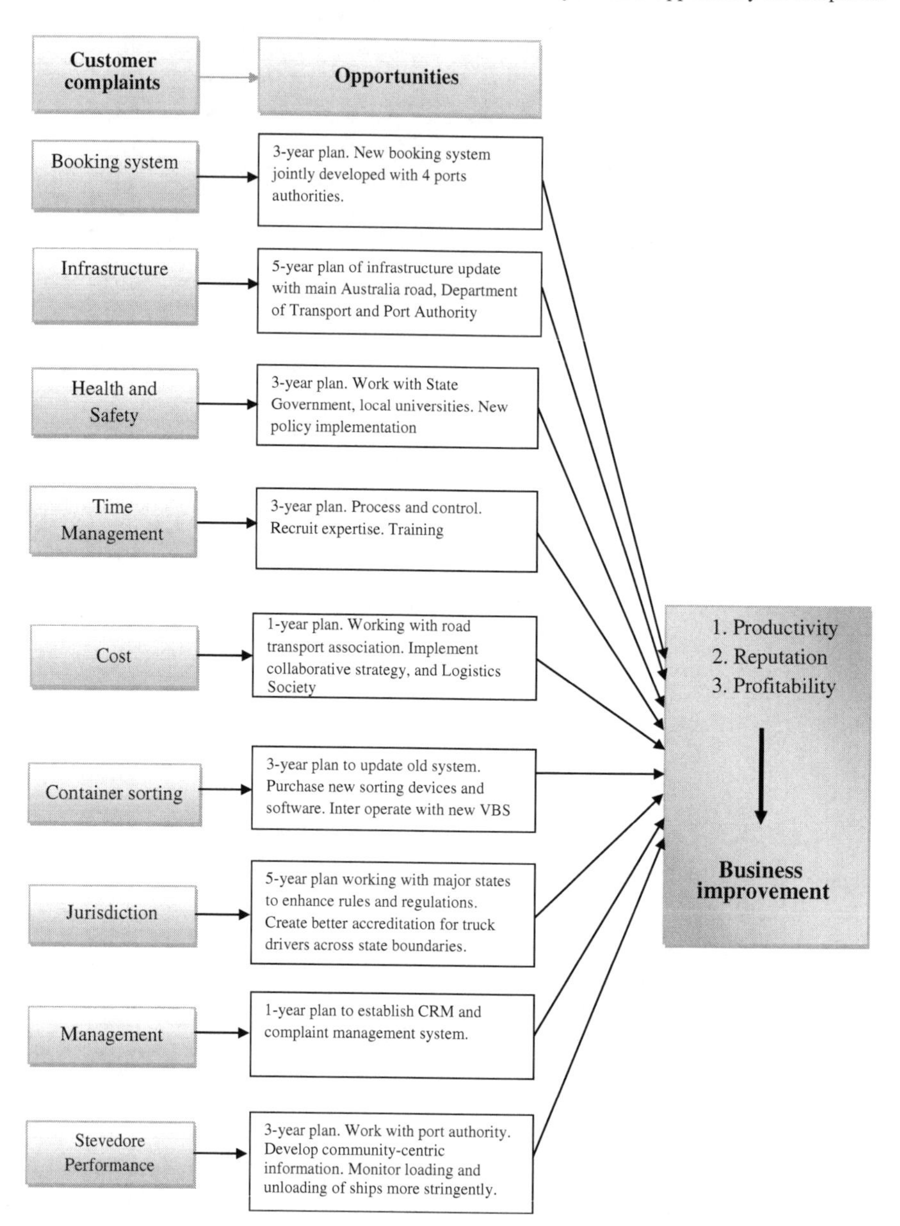

Fig. 8.24 Creating opportunities from customer complaints

8.7.1 Booking System

A new opportunity arises to develop a collaborative system with four other ports namely Brisbane port, Sydney ports, Melbourne port and Adelaide port to jointly

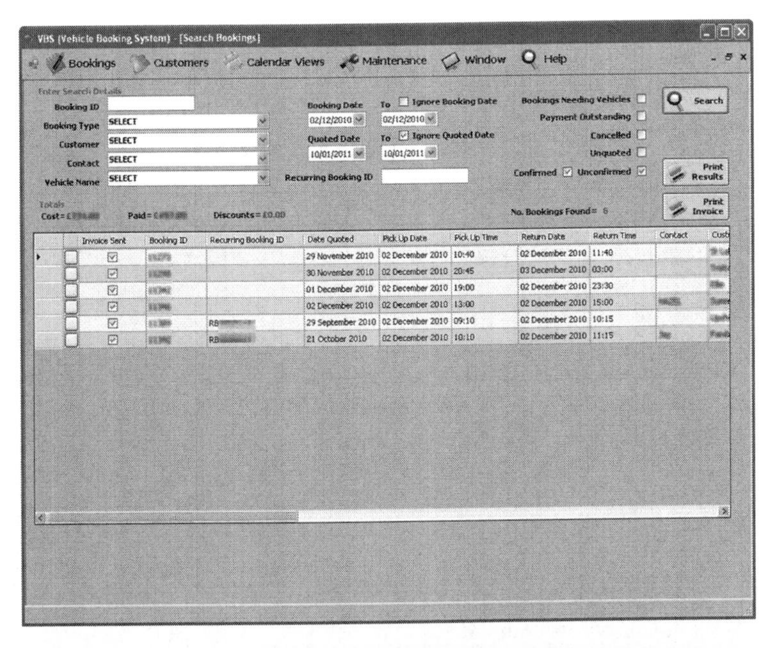

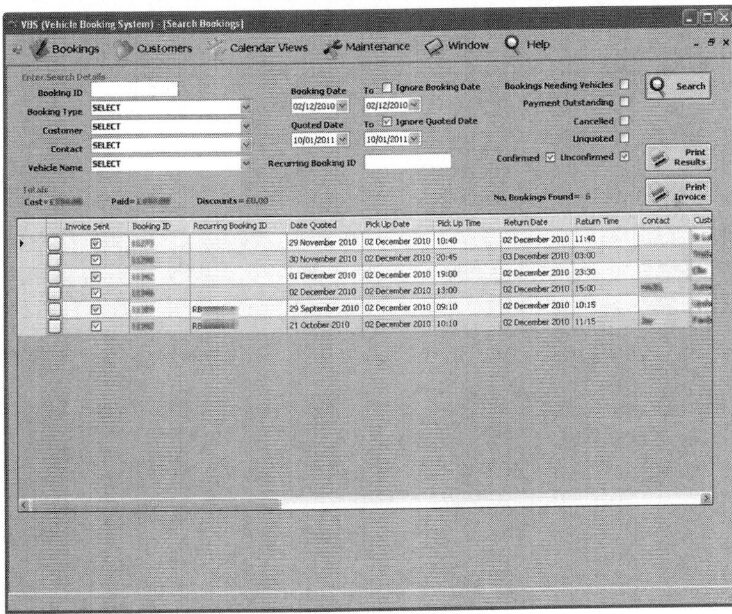

Fig. 8.25 An update of VBS

sponsor a project and develop a new booking system, sharing the cost and generate benefits over three years. Likewise, leverage the joint financial resource support with federal government matching funds namely, Aus-industry. In order to achieve more organized planning and continuity, the Fremantle port must incorporate the latest technology, and adopt a new booking system that integrates road carriers and creates harmony at the port. Simultaneously, they need to provide various time slots (Fig. 8.25).

A vehicle booking system (VBS) must be easy to access and easy to use regardless of the knowledge of the drivers. Drivers must have simple access to the system and make their bookings. Also, the system needs to provide an alternative for drivers to edit or view their profiles. It could also simply generate information regarding their salary. An effective and intelligent VBS should provide information regarding the shifts available to customers in the following days and months. Likewise, drivers may wish to have a link to other drivers and access information which is not confidential. A precise and comprehensive VBS will enable drivers to instantly know where and when to go for their loading and unloading. The VBS software also needs to be user friendly, easy to use and downloadable to mobile phones. Also, system security must be ensured as users will be entering their confidential information into the system. A new booking system will enable more efficient administration of data and will prevent the loss of sensitive data. A new VBS should be directly related to customer expectations; if these expectations are met by saving time and expense, then customer satisfaction will increase. Based on our analysis in Chap. 5, a new VBS is considered to be one of the determinants of overall customer satisfaction at the Fremantle port and therefore must definitely be taken into account. Not only will a new VBS produce customer satisfaction; it will also be able to identify various customers and collect part of their complaints.

8.7.2 Infrastructure

The new opportunity includes engaging with the state government, road authority, port authority and department of transport to put forward a proposal to improve and update port infrastructure within the next five years (Fig. 8.26).

Fig. 8.26 Repair and extend the road infrastructure

The control of congestion has been one of the biggest issues and challenges for many years, and still confronts the Fremantle port today. This issue can be resolved by constructing a separate road for truck drivers. Also, more cameras are needed to control the traffic and congestion. The dearth of signs and arrows has given rise to various issues at the port such as accidents and increased traffic. The Fremantle port authority needs to provide more signage and arrows for customers to better direct the truck drivers, especially new drivers who may be unfamiliar with the port. The construction of more footpaths in addition to a new road must be considered as priorities which will benefit employees and customers at the Fremantle port. Moreover, an extra parking lot for trucks should be constructed. Drivers who are far from home and need to stay overnight require accommodation. Vending machines dispensing cold foods and drinks should be made available. Furthermore, security services at the facilities must be better organised and improved in order to create better customer satisfaction. It is also a good idea to construct a sporting gym for port employees as well as truck drivers. This will help to motivate employees, create better job satisfaction and provide a venue where truck drivers can choose to go when they are required to remain at the port for any length of time.

The port authorities must construct a road and organise working restrictions while this is being done, so that construction companies can build or repair the road. A new road will help to decrease the traffic at peak times.

8.7.3 Health and Safety

As health and safety are of paramount importance for the truck drivers, port authorities need to review and reconsider their current strategies. It is recommended that a three-year plan be adopted whereby port authorities work with the appropriate state government departments and local universities to revise existing health and safety policies and regulations to address the issues of concern that have been identified in our research. Once recommendations have been made, port authorities need to evaluate them and devise strategies for their implementation (Fig. 8.27).

Fig. 8.27 Health and safety signs

A better physical environment must be created at the Fremantle port since truck drivers spend a lot of time in this area and deserve to work in a clean and green environment. Also, to prevent disasters, there should be regulations forbidding drivers to sit inside their cabins while their containers are being loaded or unloaded. Also, to prevent drivers from becoming dehydrated on hot days, free cold water should be supplied to them. During hot days, the air conditioning systems in trucks should be checked and this should be a free service. Likewise, operators can install a refrigerator inside those cabins that are not already equipped with one. Also, in some trucks, the companies must install the air conditioner to create a more comfortable environment for truck drivers.

Crane operators must be careful and vigilant when loading or unloading the containers and ships especially when there are crews working on the ships or drivers standing beside their trucks. To ensure the safety of all concerned, the port authority must install advanced equipment such as cameras enabling the area to be checked before cargo is loaded or unloaded. Another issue that faces the Fremantle port is that currently it is a challenging and complicated task to monitor the traffic. Fumigation of containers needs to be accomplished every time to safeguard the health of the drivers and the employees at the Fremantle port. Adherence to the guidelines in this regard may lessen the peril of entering the containers and exporting the containers. Also, port authorities must act according to an established code of practice and not randomly use man power for cargo sorting. They must use employees only if they follow safety standards and obtain permission from the employees. Likewise, occupational safety must be regularly overseen by the port authority to ensure that safety standards are consistently met. In fact, a blaze at the Fremantle port was started when containers caught fire. This is a real example of safety issues that need to be addressed by having enough safety facilities and equipment. Furthermore, there should have been an expert team on hand that could make vital decisions and take appropriate action to deal with this critical situation, and ensure that there was no repetition of a similar incident [25].

Last, but not the least, port authorities must be genuinely committed to safeguarding the employees, customers and containers at the port.

8.7.4 Time Management

Port authorities need to develop a three-year policy, process and control for time management at the Fremantle port (Fig. 8.28).

Time slots should be guaranteed and split shifts must be assigned to truck drivers. Timely breaks must be assigned as well, as the drivers need to rest to be fresh for another round of work. The time that the drivers sit in their cabins waiting aimlessly is another issue that must be monitored as this affects the driver's health, particularly in hot weather, as well as the efficiency of the port. The port authorities must provide restrictions for working hours and these must be regularly revised.

Fig. 8.28 Controlling the time of loading and unloading cargos

8.7.5 Cost

To be cost effective and decrease various costs at the Fremantle port, responsible people need to be innovative and work together with road transport associations and logistics society to conduct training programs related to collaboration in logistics services and acknowledgement of training programs instead of competing to slash the cost and improve all small and medium enterprises (Fig. 8.29).

Costs must be drastically reduced by the port authority as most of the products and foods available to drivers are prohibitively expensive, as are the costs of accommodation, petrol and technical services. Port authorities need to provide some strategies to decrease the cost of these items but in a way that will avoid confrontation with current providers while benefitting the customers and port employees. An expert team may need to be called into deal with the issue with the assistance of new technology.

To reduce costs, authorities could provide incentive programs such as a 'green reward' to recognise the efforts of employees and drivers who are promoting a better environment at the Fremantle port.

Fig. 8.29 The new opportunity to decrease the Cost

In order to prevent or decrease costs incurred as the result of fire, the port authorities must create fire protection around the Fremantle port. To this end, they need to establish water lines connected to each of the port areas to minimize fire danger.

8.7.6 Container Sorting

The authorities must develop a three-year plan of adopting advanced container sorting systems and inter-operate with new vehicle booking system to enhance productivity and efficiency of the port operations (Fig. 8.30).

To decrease waiting time, loading and unloading needs to be precise and quick using new technologies. The capacity of containers must be increased for some of the trucks according to their carrying capacity. Stickers should be made available for trucks to indicate whether they are old, new, ordinary or powerful. In order to have more effective container sorting, the Fremantle port needs to be equipped with an updated cargo system and software as discussed. Also, they need to use cranes in a productive way and check their systems regularly. The managers of this section must provide better labeling and positioning for the containers. Also, the containers must be laid our based on the date they are received.

Currently, customers are finding it difficult and time-consuming to locate their containers. A new system and effective strategies must be implemented to make it much easier to track and find the containers. Another important point is that the companies operating at the port need to have big depots to separate empty containers from the loaded containers. Container sorting needs more deliberation and flexible arrangement. Also, stevedore companies are in charge of this part of port operations and along with the port authority, they need to be able to accurately monitor the process of container sorting.

Fig. 8.30 The new automated container sorting system

Fig. 8.31 The new Australian fumigation accreditation scheme

8.7.7 *Jurisdiction*

To have a legitimate jurisdiction, the Fremantle port needs to have a five-year plan working together with other major states like Victoria, New South Wales, and Queensland to develop new rules, regulations and legislations. The port needs to establish a close partnership with state governments to implement new rules and regulations (Fig. 8.31).

WA accreditation must be approved inter-state. In one side of the truck, there will have a sticker of vehicle accreditation. The eastern state vehicles have a green sticker (NVA: national) which are difficult to obtain for WA trucks. Trucks from all other states are accepted in WA, but this is not reciprocated. So, trucks from other states are recognized nationally, but WA vehicles are recognized only in WA. Furthermore, stevedores must meet the criteria and required international standards and work under the supervision of the Fremantle port. Also, authorities at the Fremantle port must provide an appropriate link between logistics companies and stevedores.

Furthermore, the process of employing new truck drivers and organising their wages must be facilitated throughout an intelligent system. The Fremantle port also must generate new vacancies and job opportunities for members of the public including researchers, students and experienced individuals. They should be encouraged to seek a career in transport and make a contribution to the port. Each department in the port should be required to produce monthly or quarterly publications showing their contribution to the port and any improvements that they have made to port operations. This makes staff and departments more transparent and accountable, and will increase efficiency and productivity.

8.7.8 Management

A one-year plan must be adopted to establish and integrate CRM and complaint management in order to have intelligent CRM at the Fremantle port. This system will enable management to collect information about customer dissatisfaction and to address customer issues using a complete methodology. Moreover, it evaluates their satisfaction and decreases their dissatisfaction. The management must act in a transparent way and act as an effective hub within the port to create trust and ease of mind. Port authorities must provide flow in the procedures establishing new departments such as an I-CRM department and an operational management department. The decisions provided by each of the departments may have an effective impact on the Fremantle port performance. CRM managers are responsible for acquiring and retaining customers by developing appropriate customer service strategies. Furthermore, as discussed in this thesis, an I-CRM department must be able to manage customer complaints. They must accurately receive the complaints on time, and categorise and analyse them to enhance productivity and quality of the services. Likewise, this department can increase customer satisfaction using methodologies that have been employed throughout this thesis. The staff must have the skills to build relationships with international partners. Changing managerial positions or what is known as re-positioning in management increases the efficiency of the managers. A chain of responsibilities must be defined and provided to ensure all employees and customers are aware of the positions. The management team needs to conduct training courses for the employees as well as truck drivers to increase their knowledge and skills (Fig. 8.32).

The port authority must have an effective strategy in order to initiate the business opportunities and prevent risks and issues involved within procedures. They also need to map all procedures from the beginning by thoroughly examining all issues and outlining solutions and strategies for them. The Fremantle port authority, to perform more effectively, should create an environmental management department to monitor the work environment in order to identify and address risks and challenges, accurately control missions, control regulations and ensure that they are

Fig. 8.32 The new collaborative management method and approaches

being adhered to. Management also needs to adopt advertising strategies to promote the Fremantle port via national channels. Apart from partnering with universities, the port authorities can engage schools by inviting students to the port, introducing them to various departments and port activities, and encouraging interest in the port as a pathway to future studies and a possible career.

In the next phase, the Fremantle port could offer educational programs to their staff and drivers; academics or student groups can be invited to conduct these courses. A new management team must be created to separate identified areas of responsibility for outstanding circumstances and functions related to various customers and employees. Rotation of responsibilities and duties must be taken into account to develop productivity and improve customer satisfaction. Rotation of duties generates perceived value as the customers experience new faces as well as new services. This alone increases customer loyalty as the company is perceived as one that is making improvements. Additionally, appropriate management will ensure the upgrade of various kinds of programs within the working environment such as software, security and control programs. It also protects ongoing procedures.

An updated management system of training activities can be investigated. The effectiveness of training programs for both employees and truck drivers can be monitored to determine whether there has been any improvement in performance, especially with respect to the customers. Effective management can increase the awareness of the individuals who are dealing directly with the Fremantle port. In addition to implementing appropriate training programs, they may have cross training programs in order to ensure that each of the individuals and employees can perform part of the duty within the procedure to generate improvement and effectiveness throughout the company.

The establishment of hotlines at the Fremantle port will allow port authorities to collect and respond to customer feedback in a timely fashion, particularly in the case of emergencies. This needs to be made a task of the customer service department.

The port management team must be mindful of nearby residential areas, ensuring that resident safety is maximized and that pollution of any kind, including noise pollution, is kept to a minimal level in consideration of nearby residents.

The management team should revise and update the port's mission, incorporating new visions and strategies to improve productivity and enhance the port's reputation.

The Fremantle port could raise its public profile by founding or sponsoring a sporting event. The port could be made the starting point of a competition such as a cycling event.

Last but not the least, the port authorities must apply the most effective and relevant ethical standards to the WA environment based on the culture, and fairness and equity should be the basis of all actions and decisions. They must respect each individual and all clients. As one of the most significant facets of each work setting is its teamwork, the Fremantle port must attempt to foster an atmosphere of team work, and thereby improve its operations.

Fig. 8.33 New stevedore performance

8.7.9 Stevedore Performance

The port must embark on a three-year plan working with authorities, stevedores and port logistics to develop a community-based information-sharing portal showing container movement, track queuing situation and port operation, that will enable the community to collaborate together to help and monitor each other and provide recommendations and solutions to the port community (Fig. 8.33).

The port authority must take up the challenge of developing honesty and trust among operators. This is one way of improving the effectiveness of operations at the Fremantle port.

The duopoly should be broken and new stevedoring companies should be encouraged to use this venue and create healthy competition which in turn should lead to improvement. As stevedores are the first to sort out the containers, they need to create the best impression for the ships which need to be unloaded. To this end, their job must be done precisely, honestly and without any damage or error. Likewise, stevedore performance must be regularly monitored and assessed in order to make them more efficient. The stevedores must show transparency and reliability throughout their performance in order to compete with employees in other ports across Australia. The haphazard working arrangements of stevedores at the Freemantle port also cause problems; whereas flexible planning and programming of their work schedules would alleviate many stevedore-related issues, allow performance to be better monitored, and in turn increase productivity.

Stevedoring companies must adopt regular programs to clean up the waterfront and provide safety using initiatives from the management authorities at the Fremantle port.

Those individuals who are recruited by the Fremantle port must be screened based on their abilities and qualifications. They need to be employed based on international criteria that are used by all ports worldwide when recruiting employees. In this case, different ports around the world can dispatch their intelligent employees by selection in order to develop their capabilities and intelligence and this may help to improve the quality of staff expertise at Fremantle port. On top of that, it increases the reputation of the company.

8.8 Conclusion

In the first section of this chapter, we analysed the relationship between the main variables. To do so, we tested a linear modelling using PCA and the result we obtained from data envelopment analysis and our key customers; we ascertained and analysed the relationship between perceived value and interactivity with customer satisfaction. Prior to that, we used Alpha Cronbach to verify the validity of the data and to ensure that there was no need to collect additional data. Then, using a neuro-fuzzy approach, we tested the relationship between complaints and customer loyalty and acquisition to verify the relationship between the hypotheses, especially the one that had been rejected previously in Chap. 6 using structural equation modelling. Likewise, using a diagram, we outlined how I-CRM can change customer complaints into new business opportunities. Ultimately, we provided remedial solution to the issues by converting customer complaints into business opportunities to generate productivity, reputation, profitability and business improvement. The solution and recommendations that we provided in Chap. 8 have a major influence on the system as well as improving the satisfaction of employees. With our proposed solutions, we can effectively identify the preferences of the customers and decrease their complaints over time, thereby increasing customer loyalty.

References

1. Kwong, C., Wong, T., & Chan, K. (2009). A methodology of generating customer satisfaction models for new product development using a neuro-fuzzy approach. *Expert Systems with Applications, 36*, 11262–11270.
2. Faed, A., Hussain, O. K., Faed, M., & Saberi, Z. (2012). *Linear modelling and optimization to evaluate customer satisfaction and loyalty*. Presented at the The 9th IEEE International Conference on e-Business Engineering.
3. A. Faed, Hussain, O. K., & Chang, E. (2012). *Linear and fuzzy approaches for customer satisfaction analysis*. Service Oriented Computing and Applications, Springer, Under review.
4. Mamdani, E. H., & Assilian, S. (1975). An experiment in linguistic synthesis with a fuzzy logic controller. *International Journal of Man-Machine Studies, 7*, 1–13.
5. Tsukamoto, Y. (1979). An approach to fuzzy reasoning method. In M. M. Gupta, R. K. Ragade, & R. R. Yager (Eds.), *Advances in fuzzy set theory and application*. Amsterdam: Holland.
6. Takagi, T., & Sugeno, M. (1985). Fuzzy identification of system and its applications to modelling and control. *Transactions on Systems, Man, and Cybernetics, 1*, 5.

7. Li, J. A., Liu, K., Leung, S. C. H., & Lai, K. K. (2004). Empty container management in a port with long-run average criterion*. *Mathematical and Computer Modelling, 40*, 85–100.
8. Jang, J. S. R. (1993). ANFIS: Adaptive-network-based fuzzy inference system. *Systems, Man and Cybernetics, IEEE Transactions on, 23*, 665–685.
9. Jang, S. R., Sun, T., & Mizutani, E. (1997). Neuro-fuzzy and soft computing: A computational approach to learning and machine intelligence. Upper Saddle River, NJ: Prentice Hall.
10. Bao, Y., Bao, Y., & Sheng, S. (2011). Motivating purchase of private brands: Effects of store image, product signatureness, and quality variation. *Journal of Business Research, 64*, 220–226.
11. Yager, R., & Filev, D. (1994). Generation of fuzzy rules by mountain clustering. *Journal of Intelligent & Fuzzy Systems, 2*(3), 209–219.
12. Faed, A., & Chang, E. (2012). *Adaptive neuro-fuzzy inference system based approach to examine customer complaint issues*. Presented at the Second World Conference on Soft Computing, Baku, Azerbaijan.
13. Kwong, C. K., Wong, T. C., & Chan, K. Y. (2009). A methodology of generating customer satisfaction models for new product development using a neuro-fuzzy approach. *Expert Systems with Applications, 36*, 11262–11270.
14. Coussement, K., & Van den Poel, D. (2008). Improving customer complaint management by automatic email classification using linguistic style features as predictors. *Decision Support Systems, 44*, 870–882.
15. Namkung, Y., Jang, S., & Choi, S. K. (2011). Customer complaints in restaurants: Do they differ by service stages and loyalty levels? *International Journal of Hospitality Management, 30*, 495–502.
16. Ro, H., & Wong, J. (2012). Customer opportunistic complaints management: A critical incident approach. *International Journal of Hospitality Management, 31*, 419–427.
17. Galitsky, B. A., González, M. P., & Chesñevar, C. I. (2009). A novel approach for classifying customer complaints through graphs similarities in argumentative dialogues. *Decision Support Systems, 46*, 717–729.
18. Faed, A. R., Ashouri, A., & Wu, C. (2011). Maximizing productivity using CRM within the context of M-commerce.
19. Elmuti, D., Jia, H., & Gray, D. (2009). Customer relationship management strategic application and organizational effectiveness: An empirical investigation. *Journal of Strategic Marketing, 17*, 75–96.
20. Krasnikov, A., Jayachandran, S., & Kumar, V. (2009). The impact of customer relationship management implementation on cost and profit efficiencies: Evidence from the US commercial banking industry. *Journal of Marketing, 73*, 61–76.
21. Richards, K. A., & Jones, E. (2008). Customer relationship management: Finding value drivers. *Industrial Marketing Management, 37*, 120–130.
22. Chang, H. H., Wang, Y. H., & Yang, W. Y. (2009). The impact of e-service quality, customer satisfaction and loyalty on e-marketing: Moderating effect of perceived value. *Total Quality Management, 20*, 423–443.
23. Keh, H. T., & Xie, Y. (2009). Corporate reputation and customer behavioral intentions: The roles of trust, identification and commitment. *Industrial Marketing Management, 38*, 732–742.
24. Walsh, G., Mitchell, V. W., Jackson, P. R., & Beatty, S. E. (2009). Examining the antecedents and consequences of corporate reputation: A customer perspective. *British Journal of Management, 20*, 187–203.
25. News, A. (2012, May 2012). Containers catch fire at Port of Fremantle. Available: http://www.abc.net.au/news/2012-04-08/containers-ablaze-at-port-of-fremantle/3938518

Chapter 9
Conclusion of the Thesis and Future Works

9.1 Introduction

Customer relationship management has already become one of the most cutting edge solutions extensively implemented by organizations. CRM has lately attracted much attention from researchers around the globe. It focuses on building the most effective and strategic relationships with customers. Currently, numerous researchers are working on CRM and its components such as customer satisfaction, perceived value, interactivity, customer acquisition and loyalty. However, their research outcomes cannot fully address all customer issues. In this thesis, we discuss the issue of customer complaints and have developed ways of addressing these. The literature review provided in Chap. 2 indicates that much improvement has been made in this area by various studies. Simultaneously, the drawbacks of these studies have been outlined and we illustrated that none of the studies presented in the current literature provides a comprehensive methodology for intelligent customer relationship management. The reason for this is that none of the studies provides a consolidated approach to finding a solution that a new researcher can use to provide recommendations on how to deal with customer complaints. Moreover, extant studies fail to tackle one or more of the following: the relationship between perceived value and interactivity with customer satisfaction, and a methodical approach to evaluating the level of customer satisfaction and determining the effect of various levels of customer satisfaction on loyalty and customer acquisition. In order to propose a comprehensive and complete recommendation and solution to the issues, in this thesis, we identified eight research directions which were addressed in separate chapters.

9.2 Problems Addressed in this Thesis

In this thesis, we introduced and discussed eight important stages that we completed on our way to addressing the Fremantle port issues using I-CRM. These stages consisted of the following:

A. R. Faed, *An Intelligent Customer Complaint Management System with Application to the Transport and Logistics Industry*, Springer Theses, DOI: 10.1007/978-3-319-00324-5_9,
© Springer International Publishing Switzerland 2013

1. Define the concept of customer relationship management, customer complaint, perceived value, interactivity, customer satisfaction, loyalty and customer acquisition. These were considered in relation to a real-world case and each was analysed separately.
2. Propose a means to categorize and analyse the texts of collected interviews using software. This part was the first step of the analysis where we qualitatively analysed the interview texts and prioritized the complaints in their order of importance.
3. Propose a means for comparison of satisfaction between I-CRM variables, testing two sets of hypotheses to find the relationship between variables, using our data collected from Fremantle port survey (questionnaires). We proposed nine hypotheses regarding the nine acquired variables based on the Fremantle ports issues. We then tested and proved the relationship between dissatisfaction and the issues raised.
4. Propose a means to test the second set of hypotheses using software, based on the conceptual framework. Here, we proposed another set of hypotheses which are related to I-CRM variables. We then tested and ascertained the relationship between variables.
5. Propose a means to analyse customer satisfaction by introducing a five-step algorithm. As the number of customer was 60, and they had various requirements and cause of dissatisfaction, we grouped them into three categories to identify the key customers.
6. Propose a means to provide recommendations to reduce failings and strengthen best practices. There is not enough potential for improvement of customer satisfaction since the company has harmed truck drivers, sometimes irreparably.
7. Propose a means to find the relationship between customer satisfaction, its antecedents and consequences to increase the likelihood of potential improvement.
8. Propose a means to obtain an optimal solution for relationship analysis. The relationship is between perceived value and interactivity with nine issues of customer satisfaction; and in the final stage, finding the legitimate relationship between those nine issues and issues of loyalty and customer acquisition. We intended to solve these issues to find the sensitivities and finalize the optimization process.

9.3 Recapitulations

Customer relationship management is not a new field of knowledge. It is a significant tool for national and international organizations including public and private sectors and small to medium enterprises or large corporations. However, the majority of organizations do not have a CRM system and if they own one, it is either outdated or it does not have adequate functionality to handle various issues. With the growing economy and new business requirements, CRM needs to be tailored to customers' needs, especially since existing CRM does not have automation and intelligence. Also, it does not include all the mathematical tools needed to analyse customer complaints. Therefore, we want to include automated features to enable a CRM system to more accurately and efficiently analyse the issues.

After data has been collected, it needs to be tested for clarity, consistency and stability using validity and reliability approaches. These factors are the cornerstone of the research and unless the reliability and validity of each data set is established, the results will not be precise and logical. As our data was to be collected at Fremantle port, we targeted the drivers for our specific sample and distributed questionnaires accordingly. It was understood that each driver would respond differently to our questions, according to their differences in knowledge, experience, age, mood at the time, and interest in helping the researcher.

In this thesis, Chaps. 5, 6, 7 and 8 explain the experimentation used to validate the variables, proof of concepts and solution to the issues.

9.3.1 I-CRM Analytic Framework

In this thesis, we acquired and analysed the data from a unique case in an area which has not attracted many researchers due to its difficulties and complexity. However, we were extremely interested in the logistics and transport industry, and keen to establish a reciprocal relationship with the Fremantle port authorities. Prior to collecting the data whether by interviews or questionnaire survey, we deployed our main variables and set up our conceptual framework. We obtained a 360-degree view of the model and of our customers based on the CRM concept in order to deal specifically with customer complaints. To the best of our knowledge, our conceptual framework is unique and applicable to various industries regardless of the services they provide. The I-CRM framework which is created through the work in this thesis is customer-friendly and decreases the number of lost opportunities and client defections. In addition, I-CRM enhances customer service and decreases the need for intellectual re-working. It engages the customer and allows them to talk and offer their proposition or their complaint. The I-CRM generates an effective experience and perceived value for the customers. Then, using interactivity methods, experiences will be converted to customer satisfaction. If the company can keep the customers contented, customers in response will be committed and feel an obligation to stay with the company. By that commitment, customers might remain loyal and provide positive word of mouth. The customer retention developed by I-CRM will enhance the company's reputation and this is followed by the company attracting new customers. Simultaneously, it continuously handles the customer complaints and addresses their issues using the conceptual framework and the complete methodologies introduced in this thesis.

Furthermore, in the first part of our conceptual framework, we have shown some unique features of our research, in which we categorised and analysed customer complaints and created optimization for prioritizing customer complaints and for prioritizing our customers. Additionally, I-CRM has the ability to provide solutions to the issues and subsequent recommendations.

9.3.2 Qualitative and Quantitative Research

Qualitative data are the data represented in document format. Qualitative data can be obtained from different sources such as primary or secondary, publications, and company records. Primary data are data that are gathered from respondents by means of interviews and questionnaires, and secondary data are the kind that already exist. Also, the essence of this research is causal and we look for variables that produce particular results. This is why we propose several hypotheses in this thesis, each of which must be validated. To perform qualitative analysis, we need to collect samples on a regular basis to ensure their validity [1, 2]. Also, the analysis of qualitative data is not simple as there are insufficient established guidelines for qualitative data analysis. In the first part of our data analysis, we utilise qualitative analysis and we start off with data reduction using text mining analysis. Using this, we create data nodes and data codes to display essential words and to obtain the most important complaints. Nodding is the process that we give to a specific group of related statements in a section. Coding is an analytical procedure whereby we reduce the amount of data. And in the final step, we performed categorization which is the procedure of arranging and grouping our codes into different sections [1].

Quantitative data are gathered using various approaches to prove the stringency of the data. Also, we apply quantitative analysis to derive accurate results from collected data [1]. In this thesis, we also use quantitative analysis to arrive at a stringent result. In the very first section of our quantitative analysis, we use multivariate techniques to ascertain the validity of our hypotheses followed by structural equation modelling for testing the relationships between main variables. Furthermore, we utilise linear modelling and non-linear modelling to find new relationships between hypotheses, revalidate hypotheses, and re-investigate hypotheses using other tools and methods.

9.3.3 Text Analysis to Prioritize Complaints

In Chap. 5, text mining analysis was conducted of the transcribed interview surveys to obtain the prioritised customer issues at the Fremantle port. Using our lab tool, we successfully analysed the interviews in detail. This chapter provides details of the interviews collected, and their subsequent categorisation and analysis using the text-mining tool. Several advanced concepts used in text mining analysis are defined, followed by qualitative data analysis. Issues are identified in detail and a scientific approach is presented to justify the issues identified as being those relevant to the port in terms of logistics and transport.

In this chapter, we categorized and analysed the interview texts which were collected. We utilised the text-mining tool to assist analysis. We defined several advanced concepts used in text mining analysis followed by qualitative data analysis. We then identified the key issues and used the scientific approach to justify the issues identified. This is related to the case study in the port and pertains to logistics and transport. We concluded the chapter by discussing the significance of the analysis in terms of real-world problem definition.

9.3.4 *Validation of Identified Issues and Their Impact Factors Through Hypotheses Formulation and Proof of Concept*

In the Chap. 8, through text mining analysis of interview data, we determined the main issues that resulted in customer dissatisfaction with the Fremantle port operations. Furthermore, we ranked the major issues by determining their importance and weights. In this chapter, we validate the issues identified and their impact factors through hypotheses formulation and proof of concept. We carry out analysis using variance and structural equation modelling. In the previous Chap. 8, we presented nine ranked issues which were considered as variables and we used these nine variables as inputs. We carried out statistical testing to ascertain the impact factors. We used these variables to identify those factors that impact on CRM performance.

9.3.5 *Improve Customer Satisfaction Through Customer Type Mapping and Strategies for I-CRM Development*

In this chapter, we carried out the modelling between dissatisfaction to satisfaction through the five steps of customer screening, data reduction, principal component analysis, data envelopment analysis, achieving potential improvement and providing recommendations in order to address the issues at the Fremantle port. We also re-validated our hypotheses for major variables using a complete methodology. This process assisted us to separate the customers into clusters and we mapped the issues and the impact factors to the customer clusters. This in turn helped us to determine and prioritise the significant issues. These should be addressed first in order to improve customer satisfaction in the shortest possible time with maximum results. This concept also can be proven by an analytical hierarchy process and sensitivity analysis which provide mapping between customer type and satisfaction level. This process helps determine the significant issues that should be addressed first to obtain maximum results of customer satisfaction in a timely fashion.

The following framework clearly demonstrates the five steps that were taken to address the issues at the Fremantle port. Additionally, using formulations, we revalidate the relationship between the main variables or what we call I-CRM variables (Fig. 9.1).

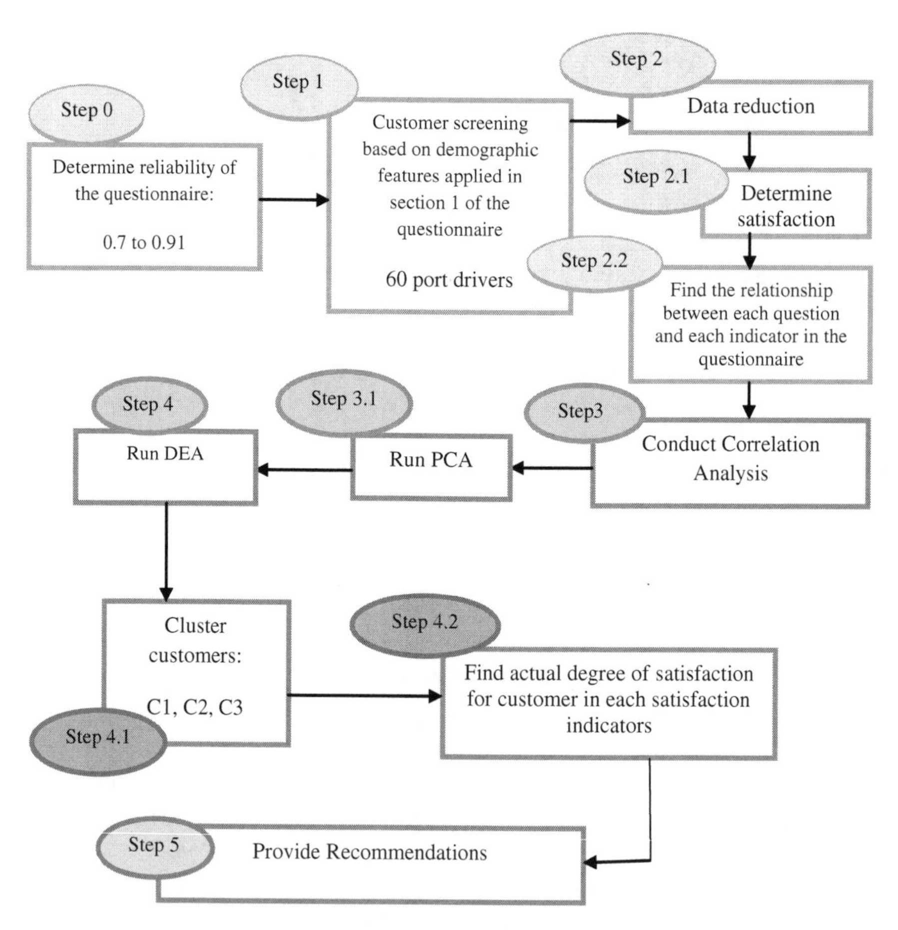

Fig. 9.1 An Intelligent customer relationship management module

9.3.6 *Analysis of Relationship Between Customer Satisfaction and Related Variables*

In Chap. 8, to validate the relationship between our variables, a comprehensive approach is provided. We employed linear modelling to formulate the relationship between customer satisfaction and its antecedents and consequences. Likewise, we adopted non-linear modelling to structure and evaluate the impact of perceived value and interactivity on customer satisfaction issues and then we estimated the ways in which customer satisfaction issues influence customer loyalty and customer acquisition. Using fuzzy inference systems, we ascertained in a detailed and methodical way that our customer relationship management is intelligent. Finally, based on the Takagi-Sugeno type fuzzy inference system, we presented the various rules for three layers of our variables. In the first layer, we provided rules for interactivity, perceived value and customer satisfaction. In layer 2.1, we generated rules that considered the relationship between nine satisfaction issues and loyalty and in layer 2.2; we established rules for the relationship between those nine satisfaction issues with customer acquisition.

One of the main significant aspects of I-CRM is that it is clearly intelligent, because by using an adaptive fuzzy inference system, we made I-CRM learner and adaptive followed by generating rules. And, using those if–then rules, we developed more recommendations for improvement.

We also proposed a detailed model to generate opportunities from customer complaints for each of the satisfaction issues at the Fremantle port.

9.4 Validation and Reliability of the Research

To achieve reliability of the research, we must consider and define the endogenous and exogenous aspects of reliability. Following this, we elaborate on the main factors affecting reliability and consider the way to manipulate them.

9.4.1 *Surveys and Samples*

In this thesis, we used the data from Fremantle port located near the city of Perth in Western Australia. We collected our data during three different periods. First, we began collecting the data by conducting interviews after having obtained approval from the Fremantle port authority. We interviewed truck drivers in December 2011 and January 2011. After finishing the first part of our analysis, in August and September 2011, we conducted another survey using interviews and questionnaires [3].

Prior to that, we obtained approval to interview managers at Fremantle port in order to obtain their viewpoints and their problems with Fremantle port, the variety of activities at the port, and the logistic companies with whom they are dealing.

Furthermore, we held meetings at Curtin University with key managers of Fremantle port to discuss the port issues at Fremantle, and hear their perspective regarding the existing conditions and operations.

After completing our questionnaire, we pre-tested the questions. We conducted pre-testing to ensure that our questions were valid and significant for the purposes of this study.

Following this, we received questionnaire approval and ethical clearance from Curtin University enabling us to proceed with data collection. We conducted interviews at Fremantle port with truck drivers who were considered as customers. Basically, we conducted the second interviews to validate our first previously recorded interviews. We distributed 90 questionnaires and received 60 in return. Of these, some were answered on the spot, others were answered and posted later, and some respondents offered to collect questionnaires from their friends.

All of the sample respondents were selected randomly from truck drivers. The majority of our respondents were male. We were gratified to see various nationalities represented among the truck drivers from different logistics companies.

In the questionnaire, we also included two psychological questions which pertained to familiarity with the culture and characteristics of the truck driver.

9.4.2 Validity Analysis

Validity is established to allow the researcher to collect, analyse and interpret data after eliminating any irrelevant or interfering factors. In the very first stage and to enhance the validity of this research, questionnaires are distributed among a number of selected samples in order to find and eliminate ambiguities. Finally, after providing transparency and addressing any vagueness, the final questionnaires are distributed. In this thesis, to enhance the validity of the questionnaire's contents we use the following tools:

1. Expert opinions.
2. Review of similar questionnaires.
3. Preliminary distribution among selected sample.

9.4.3 Reliability Analysis

Reliability implies addressing the requirement to demonstrate that it pertains to distinctive features of a system. The reliability of data is evaluated from extant or obtained information [4].

A reliability measure must be obtained in order to determine the consistency of the data, in that consistency illustrates the goodness of measuring a concept. To estimate the reliability of a questionnaire, we must have a precise estimation using an Alpha scale (Cronbach alphas) which is a reliability coefficient. It is advisable to use Cronbach alphas if there are multiple answers in a questionnaire. The highest Alpha score is 1 which indicates the internal consistency [1]. We must note that reliability is related to the score, not to the individuals.

It is critical to know the level of reliability prior to obtaining the outcomes. Generally, if the result of Alpha is less that 0.6 it is regarded as having poor reliability, results within a 0.7 range are acceptable with good consistency, and results above 0.8 are considered to have good consistency and stability. Those above 0.9 are the most reliable [1].

In the following section, we discuss the reliability of each variable based on the real-world quantitative data that was gathered using a questionnaire survey at Fremantle port. The reliability test is necessary as it establishes the consistency of the measurement. The reliability of the construct is evaluated on Cronbach alphas using SPSS 13.0. The results show that all criteria were met and data was ready for analysis [5], and [6]. All values are above 0.70, thereby demonstrating consistency. In the next section, each value is considered separately.

9.4.4 *Summary of Questionnaire Reliability*

As demonstrated in Chap. 6, we achieved the Cronbach's alpha which is a coefficient of reliability. In this thesis, we estimated that for each section of our questionnaire data to measure reliability.

For customer satisfaction, we obtained 0.809, which highlighted strong reliability. Regarding perceived value, we obtained 0.74 which shows good amount of reliability. For the interactivity section, we obtained 0.884 which is very strong. Regarding customer acquisition, we obtained 0.7 which is reliable, and regarding loyalty we obtained 0.91 which shows a great amount of reliability and consistency.

9.5 Contributions of this Thesis to the Existing Literature

The significant contribution of this thesis to the extant literature is that it proposes a complete methodology for customer relationship and its consequences based upon our conceptual framework. In this thesis, we made various major contributions to the literature as we have utilised many tools and approaches to address the issues in the exclusive customer relationship management area which is the logistics and transport industry at the Fremantle port. In this thesis, we proposed six methods of analysis.

1. Analytics methodology for categorizing a range of CRM literature to provide a new concept and definition using current literature.
2. Methodology for categorizing and analyzing customer complaints in the first part of our qualitative analysis using interview surveys.
3. Methodology for validation of the Fremantle issues and their impact factors on the hypotheses formulation and proof of concept (quantitative analysis).
4. Methodology for developing customer satisfaction through customer type mapping and providing recommendations and strategies for I-CRM enhancement.
5. Methodology for analyzing the relationship between main variables using linear modelling and mathematically formulating them.
6. Methodology for analyzing the relationship between customer satisfaction and related variables and re-justification of hypotheses using non-linear modelling.

9.5.1 Contribution 1: Comprehensive Survey of Current Literature

In Chap. 2, we carried out a comprehensive survey of existing literature in CRM and its main variables, examining many well-known and extensively cited researches [7, 8]. The conceptual framework that we present in this thesis is state of the art which can be adopted and applied in B2B, and B2C CRM environments. In Chap. 2, the literature survey yielded various definitions of customer relationship management and its major components. In this thesis, we categorised various approaches based upon this literature research. We also critically evaluated the current literature in terms of categorising customer complaints in I-CRM and assessed customer satisfaction in relation to its antecedents and its outcomes.

9.5.2 Contribution 2: Definitions of Advanced I-CRM and Creating an Intelligent CRM Framework

Previously, various researchers defined CRM in different formats and for various work settings. However, due to the dynamic nature of our subject, we introduce the new concept of Intelligent CRM, and define that in a new way based on previous definitions. Also, although there are various definitions for perceived value which are all correct, in this thesis, we have defined perceived value as an experience that a customer can obtain from a system.

Although numerous researches have been undertaken on CRM and customer satisfaction and other variables pertaining to CRM, we could not find a conceptual model which handles the customer complaints with respect to our main variables such as perceived value, interactivity, customer satisfaction, loyalty and customer acquisition. Also, none of the current method provides a complete methodology to create satisfaction followed by loyalty and customer satisfaction. Hence, we proposed our state-of-the-art conceptual framework and presented it in Chap. 4.

9.5.3 Contribution 3: An Analytic Approach Through Text Mining

The third outstanding contribution of thesis is that it proposes a methodology for text mining analysis based on qualitative research. The methodology was presented in Chap. 5. To the best of our knowledge, the extant literature has not as yet provided such a methodology in the customer relationship and complaint management area. The prominent features of this methodology are as follows:

1. In the first stage, we collected the interviews from truck drivers and transcribed them all. Also, we categorized our interviews that we conducted with managers of the port authority, truck drivers. Also, we transcribed the meetings we had with key stakeholders and we derived the major issues.
2. After transcribing the interviews into a text format, we employed software to mine and analyse the text to discover the main issues before prioritizing them.
3. Next, we used excel software and imported the data we obtained using the text-mining tool in order to draw charts and graphs to be able to interpret each of the issues separately.

9.5.4 Contribution 4: New Approaches for Testing the Hypotheses

1. Following the prioritisation of customer complaints, we established new hypotheses regarding the acquired issues. We had thirteen hypotheses regarding customer dissatisfaction and introduced a methodology to ascertain them.
2. Also, based on our conceptual framework, we established four more hypotheses that we tested and proved utilizing software.

9.5.5 A Five-Step Algorithm for the Analysis of Customer Satisfaction and Analytic Approach

This is a fundamental contribution of this thesis in that it proposes a methodology for finding key customers and providing recommendations and solutions to the issues. The prominent features of this methodology are as follows:

1. Introducing algorithm to categorize the customers into different groups based on their importance. Conducting correlation analysis to evaluate and recognise the relationship between the questions in each category.
2. Reducing the number of data, using principal component analysis to reduce the number of variables under analysis of decision-making units.
3. Employing data envelopment analysis to find the key customers which in this study we name them as category one, two and three customers.
4. Providing recommendations, strategies and solutions based on the importance of the customers.
5. Addressing sensitivity analysis using analytical hierarchy process to generate the necessary objective to facilitate the priorities and to create the best decision-making process.

We conduct an analytic approach followed by optimization. The prominent features of this methodology are as follows:

1. It proposes linear modelling to evaluate the relationship between variables and to achieve the most significant relationship.
2. We re-validate previous outcomes as we could not ascertain the relationship between customer satisfaction and loyalty.
3. We use data envelopment analysis to find the most efficient way to produce customer satisfaction and optimize cost and time to improve customer satisfaction.
4. Key customer recognition optimization.
5. Key customer complaint recognition optimization.

9.5.6 A New Rule-Based Methodology for Customer Complaints, Antecedents and Consequences: Three-Level Analytics Approach

The final contribution of this thesis to the existing literature is that it proposes non-linear modelling using an adaptive fuzzy inference system to analyse the relationship between dissatisfaction issues which are considered in the sections on customer satisfaction and perceived value, interactivity, loyalty and customer acquisition. Furthermore, it presents various rules for customer satisfaction and its relevant variables. Using software, we established rules on three levels that we identified based on our conceptual framework.

1. The rules for the relationship between perceived value, interactivity and customer satisfaction.
2. The rules for the relationship between customer dissatisfaction variables and loyalty.
3. The rules for the relationship between customer dissatisfaction variables and customer acquisition.

The work that we undertook in this thesis has been published as a part of proceedings in peer reviewed international conferences and journals. We have attached some selected publications in the appendix 8.

9.6 New Features of I-CRM

As discussed earlier, the proposed I-CRM methodology and system has many capabilities, such as obtaining complaints and performing analysis using the qualitative method. It gives a positive experience to the customers using various interactivity methods. I-CRM finds key customers and key complaints. I-CRM evaluates customer satisfaction and decreases the level of complaints. I-CRM optimizes costs and time in order to achieve customer satisfaction. Likewise, using fuzzy approaches, I-CRM generates rules to address major complaints of the key customers and as, the system is learner-responsive and adaptive, and we can strongly claim that it is intelligent. The proposed I-CRM has various capabilities, and can be employed by many industries to generate their effectiveness, reputation, productivity and improve their performance such as Airlines, service industries, banks, hospitality, consumer goods, retail, manufacturing, entertainment, health care, and tourism industry. The new I-CRM is user friendly and easier to be employed. Also, the proposed I-CRM can be implemented in the above mentioned industries to elicit and analyse customers' negative feedback, categorise them followed by addressing and solving their issues.

Our I-CRM framework can be employed into various phases of CRM such as technology, procedures, consolidation and various customer centric approaches in marketing, management, customer satisfaction and loyalty. Our study was focused on analytical CRM, however, it can be implemented into operational and collaborative CRM. In this study we have chosen a logistics and transport industry and based upon that we have selected our customers at the Fremantle port, however, the process of customer complaint handling can be also conducted over other groups of customers. The new set of customers can be different managers of the company, stevedores and security people who play a big role in safety of the employees. We can extract their concerns and viewpoints and evaluate their feedback followed by their satisfaction. In all three aforementioned cases, the proposed I-CRM can categorise, analyse and address all sorts of customer complaints and viewpoints.

9.7 Future Work

The proposed strategies, recommendations and decision-making methods of intelligent customer relationship management which are the focuses of this thesis are the first of their type and represent an initiative in the area of CRM and complaint management systems. This thesis provides an avenue for companies to recognize the importance of acknowledging customer complaints and attempts to classify them and categorize their customers utilizing the methods mentioned in this thesis.

We believe that there is a broad scope for a plethora of future work on the same research topic which can be extended to the following areas:

1. Automated evaluation for customer complaints through I-CRM on the cloud with meaningful evaluation of customer service based on customer feedback.

It allows I-CRM to analyse complaints using open access and transparent evaluation of customer services on the cloud. Hence, one needs to employ a cloud service because cloud computing provides a low cost solution and is an open platform. Using a cloud as a shared knowledge, we can help all of the companies deal with their customer issues.

I-CRM on the cloud is focused on shared knowledge. Due to the duration of time for completion of this thesis, the researcher cannot completely concentrate on cloud computing to provide full I-CRM on the cloud platform, the main points of the cloud, cloud services and open source. Semi-automatic methods must be developed to categorize feedback based on the proposed classification scheme. In this section, an attempt is made to define the cloud and include it in our thesis but only as a concept due to time constraints. In future, researchers could extend our idea by integrating our conceptual framework into cloud and cloud computing.

The development of an open-source I-CRM system will allow large and medium-sized enterprises to access I-CRM services throughout the cloud. Cloud is defined as a key differentiating element of successful information technology with its ability to become a legitimate and economical contributor to cyber-infrastructure. It acts as a data center which includes expensive data and can be customized or personalized. Cloud computing is cost-free and there are no impediments to its implementation. Furthermore, the environment in which it stores data is very secure. Cloud computing is measurable and multi-tenanted. "Cloud computing" is "the next natural stage in the evolution of on-demand IT goods and services. To a large extent, Cloud computing will be based on virtualized resources" [9]. Likewise, [10] defined cloud computing as "a computing platform that exists in a huge data-centre and capable of generating an aptitude to address the requirements". Using a cloud-based Customer Relationship Management (CRM) system, the client data can be collected, stored and analysed. CRM software allows the storage of records of interaction between customer and the company.

2. Ontological representation of I-CRM system.

Sometimes, companies intend to address complaints but the recommendations and evaluations are inaccurate due to their inefficiencies and lack of knowledge in that specific domain. Lack of knowledge creates bias in the work setting which may decrease the number of customers and increase negative word of mouth. Thus, apart from the methodologies that have been developed to determine the level of customer satisfaction and other variables, a new platform can be constructed and developed utilizing ontology. As everyone expresses complaints differently, manually responding to the complaints may offend some customers. So, ontology clarifies the true nature and implication of customer complaints. It is one of the options that makes the system interactive and auto-populates the complaints.

The ontological representation of the I-CRM system may lead the company to realize the major and significant assessment of customer satisfaction issues, since ontology enables issues to be effectively and thoroughly investigated. Also, in-depth analysis reveals new customer perspectives and new customer issues. Utilising the customer complaints ontology, knowledge can be obtained from customers in addition to the semantics of the customer complaints. With ontology, new models and categorization of customers can be developed using an ontological method, followed by refining the complaints. Once the customer complaints have been categorized and prioritised, the system can use the vocabularies for each group that were obtained earlier in Chap. 5. To implement ontology in our research, we need to have a complete view and complete profile of the customers including details regarding each complainant, complaint, customer's address, contact details of each customer, and the recommendation provided for each customer.

It is critically important to involve ontology-based complaint handling because it can reveal knowledge of which people are unaware. Also, with ontology, customers can better express themselves. Burrel and Morgan [11] believe that ontology creates new queries about the reality of the issues.

Ontology options could be developed to mediate the process so that customers are appropriately engaged and resolutions of priorities are optimised.

3. To further improve, a fuzzy-based approach through text analysis and computation of the words will allow timely feedback in response to customer complaints.

As customer complaints are always combined with uncertainty and inconsistencies in terms of being legitimate and valid knowledge, fuzzy systems might be further utilised to better address the complaints. With fuzzy systems, the results can be enhanced effectively. Likewise, using a fuzzy system, after obtaining the outcomes, the results could be simulated and a new complaint management system could be implemented. This thesis used an adaptive fuzzy inference system to acquire an understanding about various relationships between interactivity and perceived value with the customer issues, and the relationship between customer issues with loyalty and customer acquisition. There are also various fuzzy methods that can be used for decision modelling and for complaint categorization. Fuzzy systems can also be used to rank and assess the complaints, followed by providing various recommendations to the issues and developing automated customer complaint handling.

In Chap. 5, text analysis and prioritization of the customer complaints was conducted. After qualitative data were obtained, qualitative analysis produced the outcomes. A fuzzy system can be utilized as well to monitor customer requirements. Also, the prioritized complaints can be transformed to create new qualitative values and optimize the process of customer satisfaction. In addition, a fuzzy system can help to predict the occurrence of customer complaints and the system would be able to prevent it.

Furthermore, using a fuzzy neural network, the company can effectively target those customers that have complaints, and using a fuzzy algorithm, the issues can be adequately addressed. The collected data from the Fremantle port can also be processed and converted into quantitative data. In so doing, all the relationships between the inputs and outputs will be shifted to fuzzy relationships. Then, utilizing a fuzzy judgment method, new rules and recommendations to improve the system can be postulated. Utilizing fuzzy systems, the company can evaluate service quality and increase its productivity via survey questionnaire data. This method promotes a productive decision-making procedure and the researchers can recognize the significant variables which are influential in the procedure. Furthermore, the fuzzy systems and approaches can be employed to detect those customers who are likely to leave the company and predict the likelihood of their complaints and preempt their decision to leave. The fuzzy approach creates a platform and a model to accurately identify the important drivers for customers to churn and also find the customers who may become dissatisfied in future. It also creates strategies, rules and recommendations to prevent customer angst and retain their loyalty to the company.

References

1. Sekaran, U. (2006). *Research methods for business: A skill building approach.* Wiley-India.
2. Elliott, R. & Waller. (2011). *Marketing*. Wiley, Ltd.
3. Arheart, K. L., Sly, D. F., Trapido, E. J., Rodriguez, R. D., & Ellestad, A. J. (2004). Assessing the reliability and validity of anti-tobacco attitudes/beliefs in the context of a campaign strategy. *Preventive Medicine, 39*, 909–918.
4. Papazoglou, I. A. (1999). Bayesian decision analysis and reliability certification. *Reliability Engineering and System Safety, 66*, 177–198.
5. Wallenburg, C. (2009). Innovation in logistics outsourcing relationships: proactive improvement by logistics service providers as a driver of customer loyalty. *Journal of Supply Chain Management, 45*, 75–93.
6. Ryu, K., Han, H., & Kim, T. H. (2008). The relationships among overall quick-casual restaurant image, perceived value, customer satisfaction, and behavioral intentions. *International Journal of Hospitality Management, 27*, 459–469.
7. Payne, A., & Frow, P. (2005). A strategic framework for customer relationship management. *Journal of Marketing, 69*, 167–176.
8. Reynolds, K. E., & Beatty, S. E. (1999). Customer benefits and company consequences of customer-salesperson relationships in retailing. *Journal of Retailing, 75*, 11–32.
9. Vaquero, L. M. (2008). A break in the clouds: towards a cloud definition. *Computer communication review, 39*, 50.
10. Jaeger, P. T., Lin, J., & Grimes, J. M. (2008). Cloud computing and information policy: Computing in a policy cloud? *Journal of Information Technology and Politics, 5*, 269–283.
11. Morgan, G., & Smircich, L. (1980). The case for qualitative research. *Academy of Management Review, 5*, 491–500.

Chapter 10
Appendix

10.1 Appendix 1: Covariance Matrix for Chapter 6 and 7

	Manageme	Runningt	Organize	Manage_A	Transpor	Traffic
Manageme	1.83					
Runningt	1.19	2.07				
Organize	0.91	1.26	1.91			
Manage_A	0.90	1.04	0.87	1.54		
Transpor	0.44	0.27	0.03	0.08	1.88	
Traffic	0.74	0.78	0.33	0.52	0.15	1.77
Waitingt	0.44	0.33	0.23	0.19	−0.11	0.40
Insideca	0.63	0.92	1.16	0.70	0.46	0.40
Queuetim	−0.11	−0.17	0.04	−0.29	−0.15	0.05
Cost	0.43	0.41	0.19	0.18	0.25	0.47
Health	0.26	0.30	0.49	0.16	0.53	−0.40
Road	0.67	0.21	0.54	0.49	0.47	0.08
Competet	1.05	0.77	0.68	0.44	0.53	0.59
Vbs	0.64	0.44	0.48	0.44	0.18	0.30
Is	−0.03	−0.01	0.21	−0.11	0.30	0.01
Stevedor	0.58	0.63	0.74	0.34	0.10	0.41
Containe	0.26	0.15	0.35	0.13	0.46	0.17
Access	0.25	0.53	0.37	0.66	0.43	0.30
Sign	−0.13	−0.02	0.15	−0.02	−0.30	0.02
Forklift	0.12	−0.31	−0.37	−0.20	−0.08	−0.03
Separate	0.36	0.29	0.34	0.48	−0.08	0.28
Danger	−0.26	−0.04	0.03	−0.21	0.30	−0.13
Newstore	−0.14	−0.18	−0.42	−0.13	0.13	−0.29
Faciliti	0.20	−0.11	−0.04	−0.36	0.22	0.28
Parking	0.07	0.07	−0.02	0.09	0.28	0.08
Pavement	−0.03	0.29	0.56	0.10	0.18	0.16
Cameras	0.22	0.53	0.20	0.22	0.41	0.20
Tobemoti	0.53	0.57	0.45	0.16	−0.26	0.30
Workingp	0.07	0.34	0.19	0.26	0.04	0.36

(continued)

A. R. Faed, *An Intelligent Customer Complaint Management System with Application to the Transport and Logistics Industry*, Springer Theses, DOI: 10.1007/978-3-319-00324-5_10,
© Springer International Publishing Switzerland 2013

(continued)

	Manageme	Runningt	Organize	Manage_A	Transpor	Traffic
Petrolst	−0.34	0.13	0.07	−0.19	0.00	−0.27
Accomoda	−0.24	0.33	0.55	0.04	−0.12	0.08
Responsi	0.27	0.55	0.60	0.40	0.09	−0.05
Technica	−0.01	0.15	0.22	0.04	0.20	−0.12
Respectt	0.51	0.17	0.28	−0.10	0.19	−0.13
Timeslot	0.45	0.85	0.48	0.53	0.24	0.54
Shift	0.71	0.27	0.30	0.15	0.42	0.17
Recommen	0.27	0.11	0.16	−0.18	0.51	0.27
Switch	0.41	0.26	0.08	−0.42	1.01	0.44
Havingth	0.06	−0.13	−0.13	0.11	−0.08	−0.04
Tendency	0.29	0.09	0.06	−0.40	0.70	0.31
Wordofmo	0.00	−0.04	−0.06	−0.43	0.01	0.30
Attracti	0.56	0.44	0.48	0.26	0.41	0.59
Incentiv	0.46	0.04	0.55	0.19	0.47	0.23
Respon_A	0.63	−0.04	0.23	0.06	0.42	0.18
Numberof	0.12	0.06	0.12	0.05	0.08	0.22
Knowledg	0.36	0.34	0.44	0.17	0.01	−0.16
Necessar	0.45	0.29	0.24	0.29	0.09	0.01
Cleannes	0.52	0.05	0.42	0.25	0.38	−0.03
Salary	0.41	0.34	0.48	0.24	0.74	0.09
Services	−0.03	0.24	0.16	0.28	0.13	0.21
Quality	0.37	0.72	0.56	0.28	0.44	0.30
Improvem	0.49	0.62	0.64	0.44	0.26	0.03
Relation	0.37	0.33	0.40	0.48	0.08	0.17
Workatmo	0.61	0.32	0.39	0.49	0.31	−0.10
Workat_A	0.01	0.29	0.11	0.16	0.16	0.38
Noenough	0.60	0.54	0.26	0.28	0.41	0.03
Quickfee	0.35	0.10	0.38	0.18	0.08	0.20
Interact	0.55	0.46	0.27	0.34	0.68	−0.06

Covariance Matrix

	Waitingt	Insideca	Queuetim	Cost	Health	Road
Waitingt	1.34					
Insideca	0.25	2.41				
Queuetim	0.26	−0.03	0.97			
Cost	0.25	0.44	−0.16	1.44		
Health	−0.06	0.27	0.09	−0.20	1.55	
Road	0.14	0.54	−0.01	0.13	0.45	1.95
Competet	0.31	0.47	−0.03	0.51	0.34	0.73
Vbs	0.37	0.56	−0.11	0.40	0.45	0.64
Is	0.10	0.29	0.16	0.32	0.08	0.16
Stevedor	0.36	0.55	−0.05	0.21	0.37	0.92
Containe	0.12	0.18	0.12	0.07	0.49	0.61

(continued)

(continued)

	Waitingt	Insideca	Queuetim	Cost	Health	Road
Access	−0.03	0.79	−0.24	0.34	0.38	0.49
Sign	0.16	0.08	0.26	−0.02	0.04	0.57
Forklift	0.34	0.02	0.32	−0.25	0.10	0.46
Separate	−0.06	0.29	−0.35	0.33	0.04	0.19
Danger	−0.01	−0.10	0.15	−0.37	−0.02	0.40
Newstore	0.24	−0.21	−0.03	0.06	0.27	−0.14
Faciliti	0.47	−0.20	0.43	0.20	0.01	0.07
Parking	0.19	0.37	0.17	0.34	0.28	0.13
Pavement	0.12	0.38	0.25	0.21	0.35	0.30
Cameras	−0.11	−0.25	0.04	−0.16	0.22	−0.04
Tobemoti	0.59	0.51	0.46	0.25	0.35	−0.08
Workingp	−0.05	0.31	0.06	0.25	−0.06	0.01
Petrolst	0.04	0.08	0.11	−0.07	0.29	−0.09
Accomoda	0.24	0.19	0.06	0.00	0.02	−0.03
Responsi	0.07	0.14	−0.11	−0.24	0.26	0.43
Technica	−0.18	−0.01	−0.31	0.17	0.08	−0.22
Respectt	0.32	−0.19	0.03	−0.11	0.52	0.31
Timeslot	0.38	0.43	0.23	0.02	0.16	0.60
Shift	−0.05	0.32	0.06	0.17	0.10	0.70
Recommen	−0.19	0.48	0.07	0.32	0.27	0.78
Switch	−0.10	0.60	0.23	0.21	0.51	0.83
Havingth	−0.17	0.02	0.10	−0.14	0.01	−0.11
Tendency	−0.06	0.33	0.30	0.13	0.18	0.37
Wordofmo	0.01	0.03	0.43	−0.12	−0.07	0.16
Attracti	−0.02	0.53	0.06	−0.02	0.19	0.74
Incentiv	−0.07	0.39	−0.25	−0.12	0.41	0.27
Respon_A	0.09	0.19	0.10	0.14	0.15	0.56
Numberof	0.06	0.11	0.11	0.04	−0.09	0.05
Knowledg	0.04	0.25	0.24	−0.24	0.11	0.30
Necessar	−0.08	0.11	0.07	−0.45	0.06	0.62
Cleannes	−0.08	0.55	−0.12	−0.02	0.38	0.82
Salary	0.19	0.67	0.12	0.73	0.55	0.65
Services	0.56	0.45	0.00	0.21	−0.08	0.28
Quality	0.34	0.45	0.23	−0.04	0.20	0.64
Improvem	−0.01	0.77	0.19	0.04	0.18	0.82
Relation	0.07	0.56	−0.04	−0.03	0.26	0.75
Workatmo	0.18	0.50	−0.10	0.20	0.36	0.94
Workat_A	0.51	0.25	0.05	−0.13	0.63	0.35
Noenough	0.67	0.22	0.12	−0.04	0.65	0.89
Quickfee	0.41	0.42	0.29	0.10	0.37	0.69
Interact	0.11	0.49	−0.16	−0.34	0.83	0.80

Covariance Matrix

	Competet	Vbs	Is	Stevedor	Containe	Access
Competet	1.74					
Vbs	0.66	1.42				
Is	−0.05	0.12	1.41			
Stevedor	0.26	0.53	0.51	1.63		
Containe	0.73	0.53	0.43	0.68	2.04	
Access	0.08	0.25	0.21	0.47	0.30	1.90
Sign	−0.34	0.06	0.39	0.49	0.13	0.30
Forklift	0.17	0.36	−0.05	0.37	0.12	0.02
Separate	0.37	0.54	−0.06	0.13	0.16	0.67
Danger	−0.25	−0.43	0.03	0.25	0.08	−0.27
Newstore	−0.16	0.16	0.19	−0.05	0.12	−0.07
Faciliti	0.22	0.02	0.39	0.17	0.13	−0.34
Parking	0.01	−0.02	−0.15	0.13	0.35	0.26
Pavement	0.14	0.16	0.70	0.73	0.52	0.13
Cameras	0.01	−0.02	0.06	0.16	0.33	0.09
Tobemoti	0.71	0.48	0.12	0.26	0.33	−0.17
Workingp	0.32	0.20	0.15	0.34	0.13	0.03
Petrolst	−0.17	−0.19	0.33	0.22	0.54	0.36
Accomoda	0.02	−0.26	0.03	0.42	0.11	0.06
Responsi	0.45	0.60	0.25	0.51	0.71	0.40
Technica	0.02	0.18	0.10	0.01	0.09	0.50
Respectt	0.51	0.39	0.19	0.60	−0.04	−0.02
Timeslot	0.45	0.41	−0.06	0.64	0.40	0.25
Shift	0.66	0.30	0.17	0.29	0.15	0.22
Recommen	0.67	0.33	0.19	0.49	0.34	0.53
Switch	0.34	0.33	0.36	0.85	0.41	0.42
Havingth	0.06	−0.10	0.07	−0.29	−0.03	−0.02
Tendency	0.19	0.37	0.37	0.36	0.13	0.46
Wordofmo	−0.08	0.18	0.27	0.27	0.02	0.07
Attracti	0.71	0.36	−0.19	0.50	0.50	0.20
Incentiv	0.43	0.12	0.09	0.15	0.08	0.27
Respon_A	0.85	0.33	0.31	0.06	0.45	0.31
Numberof	0.09	0.02	−0.06	0.05	−0.02	0.11
Knowledg	0.39	0.07	−0.04	0.21	−0.04	−0.03
Necessar	0.48	0.23	0.04	0.49	0.53	−0.18
Cleannes	0.49	0.16	−0.06	0.31	0.26	0.21
Salary	0.87	0.47	0.23	0.11	0.41	0.74
Services	0.04	0.36	0.29	0.14	0.04	0.44
Quality	0.68	0.27	0.49	0.41	0.75	0.41
Improvem	0.32	0.48	0.23	0.84	0.65	0.11
Relation	0.54	0.38	0.09	0.45	0.05	0.91
Workatmo	0.53	0.50	0.01	0.55	0.19	0.71
Workat_A	0.30	0.41	0.46	0.62	0.20	0.78
Noenough	0.70	0.53	0.01	0.56	0.34	0.50

(continued)

(continued)

	Competet	Vbs	Is	Stevedor	Containe	Access
Quickfee	0.69	0.34	0.21	0.53	0.65	0.62
Interact	0.83	0.53	−0.14	0.28	0.52	0.86

Covariance Matrix

	Sign	Forklift	Separate	Danger	Newstore	Faciliti
Sign	2.08					
Forklift	0.36	1.97				
Separate	0.14	0.17	2.22			
Danger	0.35	0.32	−0.25	2.22		
Newstore	−0.03	0.12	0.01	−0.50	1.67	
Faciliti	0.69	0.51	−0.06	0.26	−0.13	2.10
Parking	0.01	0.24	0.18	0.25	0.10	0.18
Pavement	0.53	−0.10	0.40	0.76	−0.31	0.70
Cameras	0.28	0.25	0.05	0.39	−0.21	0.44
TOBEMOTI	−0.03	0.19	0.18	−0.49	0.29	0.01
Workingp	−0.26	0.08	0.60	0.34	−0.01	−0.15
Petrolst	0.45	−0.03	−0.27	0.17	0.87	0.12
Accomoda	0.12	−0.36	−0.28	0.44	−0.49	−0.14
Responsi	0.37	0.44	0.77	−0.03	0.11	0.14
Technica	−0.25	−0.53	0.37	−0.56	0.09	−0.09
Respectt	−0.23	0.81	0.06	0.07	0.01	0.47
Timeslot	0.32	0.17	−0.04	0.11	−0.12	0.31
Shift	0.09	0.20	0.35	0.17	−0.10	0.68
Recommen	0.45	0.81	0.22	0.05	0.05	0.30
Switch	0.79	0.49	0.11	0.06	0.42	0.49
Havingth	−0.01	0.07	−0.12	−0.17	0.05	−0.05
Tendency	0.50	0.56	0.32	−0.17	0.31	0.20
Wordofmo	0.65	0.47	0.25	−0.19	0.46	0.01
Attracti	0.22	0.24	−0.06	−0.17	−0.43	0.28
Incentiv	−0.12	−0.20	0.54	−0.19	−0.18	0.38
Respon_A	−0.24	0.17	−0.12	−0.55	0.11	0.12
Numberof	−0.22	0.03	0.17	−0.04	−0.17	0.04
Knowledg	0.05	0.32	0.30	0.30	−0.16	0.14
Necessar	0.02	0.31	0.00	0.23	−0.20	0.43
Cleannes	0.27	0.03	0.15	−0.01	−0.10	0.01
Salary	−0.69	−0.27	0.75	−0.04	0.08	0.14
Services	0.72	0.20	0.27	0.30	0.05	0.47
Quality	0.33	0.05	−0.01	0.19	−0.39	0.28
Improvem	0.60	0.41	0.16	0.67	−0.52	0.26
Relation	0.74	0.53	0.27	−0.02	−0.26	0.16
Workatmo	0.57	0.54	0.05	0.09	−0.03	0.27
Workat_A	0.59	0.44	0.32	−0.22	0.09	0.27
Noenough	0.08	0.71	0.17	0.40	0.24	0.06

(continued)

(continued)

	Sign	Forklift	Separate	Danger	Newstore	Faciliti
Quickfee	0.55	0.49	0.16	0.24	−0.19	0.44
Interact	0.10	0.56	−0.09	−0.19	0.09	−0.32

Covariance Matrix

	Parking	Pavement	Cameras	Tobemoti	Workingp	Petrolst
Parking	1.87					
Pavement	0.22	2.21				
Cameras	0.22	0.17	1.63			
Tobemoti	0.26	0.32	0.39	2.19		
Workingp	−0.10	0.68	0.12	0.49	1.58	
Petrolst	0.34	0.30	0.05	0.13	−0.11	2.07
Accomoda	0.07	0.52	−0.25	0.00	−0.13	0.07
Responsi	0.10	0.18	0.61	−0.03	0.14	0.36
Technica	−0.14	−0.10	−0.06	−0.49	−0.05	0.04
Respectt	−0.15	−0.16	0.11	−0.21	0.06	−0.55
Timeslot	0.47	0.42	0.30	0.48	0.33	−0.06
Shift	−0.10	0.23	−0.22	−0.44	0.31	−0.17
Recommen	0.18	0.45	0.38	0.59	0.41	0.17
Switch	−0.07	0.53	0.13	0.33	0.19	0.44
Havingth	−0.19	−0.08	−0.19	0.10	−0.10	0.04
Tendency	−0.20	0.34	0.16	0.56	0.30	0.18
Wordofmo	−0.47	0.00	0.21	0.59	0.35	0.20
Attracti	−0.19	0.04	0.04	−0.24	0.08	0.01
Incentiv	0.06	0.72	−0.01	−0.27	0.33	−0.31
Respon_A	−0.29	−0.19	0.13	0.16	0.18	0.09
Numberof	0.16	−0.04	0.07	−0.01	0.03	−0.24
Knowledg	−0.03	0.02	−0.03	−0.06	0.36	−0.11
Necessar	0.03	0.34	0.09	−0.23	0.52	0.01
Cleannes	0.06	0.14	−0.17	−0.01	−0.01	0.00
Salary	0.48	0.25	−0.17	0.09	0.67	−0.05
Services	0.12	0.09	0.64	0.36	0.12	0.04
Quality	0.09	0.23	0.52	0.40	−0.01	0.21
Improvem	0.02	0.60	0.32	0.02	0.27	−0.15
Relation	0.11	0.09	0.17	−0.05	0.20	0.26
Workatmo	0.22	0.03	0.03	−0.38	−0.04	0.29
Workat_A	−0.30	0.51	0.21	0.55	0.32	0.45
Noenough	0.16	0.37	0.26	0.26	0.15	0.36
Quickfee	0.21	0.62	−0.04	0.28	0.19	0.29
Interact	−0.26	−0.07	−0.13	−0.15	0.06	0.40

10.2 Appendix 2: Correlation Analysis (minitab® release 14.1)

Results for: Management

Correlation analysis: Q2-1, Q2-2, Q2-3, Q2-4, Q2-5, Q2-6, Q2-10, Q2-12, Q2-15, Q2-16, Q2-19, Q2-20, Q2-21, Q2-28, Q2-29, Q2-31, Q2-32, Q2-33, Q2-34, Q2-35, Q2-36

	Q2-1	Q2-2	Q2-3	Q2-4	Q2-5	Q2-6	Q2-10	Q2-12	Q2-15
Q2-2	0.613								
	0.000								
Q2-3	0.489	0.634							
	0.000	0.000							
Q2-4	0.535	0.584	0.511						
	0.000	0.000	0.000						
Q2-5	0.237	0.135	0.015	0.047					
	0.069	0.303	0.910	0.724					
Q2-6	0.411	0.408	0.179	0.313	0.083				
	0.001	0.001	0.170	0.015	0.528				
Q2-10	0.264	0.239	0.117	0.119	0.150	0.295			
	0.042	0.066	0.374	0.366	0.254	0.022			
Q2-12	0.353	0.107	0.279	0.282	0.245	0.044	0.075		
	0.006	0.417	0.031	0.029	0.060	0.737	0.567		
Q2-15	−0.018	−0.006	0.128	−0.076	0.183	0.008	0.226	0.097	
	0.891	0.963	0.328	0.564	0.162	0.949	0.082	0.462	
Q2-16	0.335	0.343	0.421	0.212	0.055	0.241	0.134	0.518	0.337
	0.009	0.007	0.001	0.103	0.677	0.063	0.306	0.000	0.008
Q2-19	−0.064	−0.007	0.076	−0.010	−0.151	0.010	−0.009	0.284	0.229
	0.625	0.955	0.564	0.938	0.248	0.942	0.946	0.028	0.079
Q2-20	0.062	−0.151	−0.193	−0.117	−0.044	−0.018	−0.151	0.234	−0.031
	0.635	0.249	0.141	0.373	0.738	0.890	0.249	0.072	0.817
Q2-21	−0.178	−0.135	−0.165	−0.258	0.038	−0.143	−0.185	−0.090	0.033
	0.173	0.303	0.207	0.047	0.775	0.275	0.157	0.494	0.800
Q2-28	0.265	0.266	0.219	0.087	−0.129	0.153	0.143	−0.037	0.067
	0.041	0.040	0.092	0.508	0.325	0.242	0.275	0.777	0.608
Q2-29	0.043	0.188	0.111	0.167	0.022	0.216	0.167	0.006	0.101
	0.747	0.150	0.399	0.202	0.866	0.097	0.201	0.965	0.443
Q2-31	−0.138	0.180	0.318	0.028	−0.070	0.047	−0.002	−0.020	0.021
	0.292	0.168	0.013	0.831	0.594	0.722	0.988	0.880	0.875
Q2-32	0.130	0.254	0.289	0.213	0.045	−0.023	−0.131	0.205	0.142
	0.322	0.050	0.025	0.102	0.735	0.859	0.320	0.116	0.280
Q2-33	−0.006	0.082	0.125	0.022	0.113	−0.071	0.109	−0.121	0.066
	0.965	0.533	0.339	0.866	0.391	0.592	0.405	0.357	0.616
Q2-34	0.264	0.117	0.202	0.011	0.174	−0.028	−0.091	0.216	0.090
	0.042	0.374	0.122	0.936	0.183	0.834	0.488	0.098	0.493

(continued)

(continued)

	Q2-1	Q2-2	Q2-3	Q2-4	Q2-5	Q2-6	Q2-10	Q2-12	Q2-15
Q2-35	0.241	0.426	0.250	0.308	0.126	0.292	0.015	0.310	−0.037
	0.063	0.001	0.054	0.017	0.336	0.024	0.912	0.016	0.780
Q2-36	0.373	0.134	0.155	0.085	0.215	0.092	0.101	0.356	0.103
	0.003	0.306	0.239	0.517	0.099	0.487	0.441	0.005	0.435

	Q2-16	Q2-19	Q2-20	Q2-21	Q2-28	Q2-29	Q2-31	Q2-32	Q2-33
Q2-19	0.268								
	0.038								
Q2-20	0.208	0.176							
	0.111	0.178							
Q2-21	−0.070	−0.066	−0.081						
	0.596	0.614	0.538						
Q2-28	0.136	−0.015	0.090	−0.081					
	0.300	0.908	0.494	0.539					
Q2-29	0.215	−0.144	0.048	−0.318	0.264				
	0.099	0.273	0.715	0.013	0.042				
Q2-31	0.262	0.065	−0.202	0.148	−0.002	−0.080			
	0.043	0.623	0.121	0.260	0.990	0.542			
Q2-32	0.262	0.170	0.208	−0.342	−0.015	0.073	−0.016		
	0.043	0.193	0.111	0.007	0.909	0.581	0.902		
Q2-33	0.009	−0.134	−0.293	−0.192	−0.261	−0.028	0.054	0.263	
	0.948	0.307	0.023	0.142	0.044	0.831	0.684	0.042	
Q2-34	0.357	−0.057	0.337	0.059	−0.147	0.023	0.067	0.268	0.195
	0.005	0.666	0.008	0.657	0.261	0.861	0.611	0.039	0.136
Q2-35	0.361	0.160	0.087	0.020	0.236	0.190	0.183	0.102	−0.094
	0.005	0.221	0.506	0.879	0.070	0.146	0.161	0.439	0.474
Q2−36	0.162	0.043	0.102	−0.167	−0.209	0.175	−0.240	0.078	0.195
	0.215	0.745	0.436	0.201	0.109	0.182	0.065	0.555	0.136

	Q2-34	Q2-35
Q2-35	0.103	
	0.433	
Q2-36	0.364	0.283
	0.004	0.029

CellContents:Pearsoncorrelation
P-Value

Results for: Booking system

Correlations: Q2-5, Q2-6, Q2-7, Q2-8, Q2-9, Q2-14, Q2-17, Q2-18, Q2-35, Q2-36

	Q2-5	Q2-6	Q2-7	Q2-8	Q2-9	Q2-14	Q2-17	Q2-18	Q2-35
Q2-6	0.083								
	0.528								
Q2-7	−0.068	0.260							
	0.608	0.045							
Q2-8	0.215	0.193	0.138						
	0.100	0.140	0.292						
Q2-9	−0.115	0.040	0.228	−0.018					
	0.382	0.759	0.080	0.890					
Q2-14	0.112	0.188	0.270	0.306	−0.096				
	0.394	0.150	0.037	0.018	0.464				
Q2-17	0.234	0.088	0.075	0.081	0.083	0.314			
	0.071	0.505	0.568	0.539	0.527	0.015			
Q2-18	0.226	0.162	−0.018	0.371	−0.178	0.150	0.151		
	0.082	0.216	0.893	0.004	0.174	0.252	0.251		
Q2-35	0.126	0.292	0.240	0.199	0.172	0.246	0.201	0.133	
	0.336	0.024	0.065	0.128	0.189	0.058	0.124	0.311	
Q2-36	0.215	0.092	−0.028	0.144	0.047	0.176	0.074	0.111	0.283
	0.099	0.487	0.835	0.272	0.723	0.179	0.574	0.397	0.029

Cell Contents: Pearson correlation

P-Value

Results for: Infrastructure

Correlations: Q2-8, Q2-10, Q2-15, Q2-19, Q2-23, Q2-24, Q2-25, Q2-26, Q2-27, Q2-29, Q2-30, Q2-31, Q2-33, Q2-34

	Q2-8	Q2-10	Q2-15	Q2-19	Q2-23	Q2-24	Q2-25	Q2-26	Q2-27
Q2-10	0.237								
	0.068								
Q2-15	0.157	0.226							
	0.231	0.082							
Q2-19	0.037	−0.009	0.229						
	0.781	0.946	0.079						
Q2-23	0.107	−0.036	−0.125	0.014					
	0.418	0.785	0.342	0.918					
Q2-24	−0.089	0.116	0.225	0.329	0.071				
	0.501	0.379	0.083	0.010	0.588				
Q2-25	0.174	0.208	−0.093	0.004	−0.057	0.092			
	0.183	0.111	0.480	0.978	0.667	0.482			
Q2-26	0.167	0.116	0.396	0.247	0.159	0.324	0.107		
	0.203	0.379	0.002	0.058	0.226	0.012	0.415		
Q2-27	−0.125	−0.107	0.037	0.153	0.124	0.238	0.126	0.091	

(continued)

(continued)

	Q2-8	Q2-10	Q2-15	Q2-19	Q2-23	Q2-24	Q2-25	Q2-26	Q2-27
	0.343	0.414	0.780	0.244	0.344	0.067	0.336	0.490	
Q2-29	0.159	0.167	0.101	−0.144	0.006	−0.083	−0.060	0.366	0.077
	0.225	0.201	0.443	0.273	0.962	0.528	0.649	0.004	0.559
Q2-30	−0.035	0.042	−0.196	−0.216	0.466	−0.057	−0.172	−0.141	−0.026
	0.793	0.750	0.134	0.098	0.000	0.666	0.188	0.281	0.844
Q2-31	0.099	−0.002	0.021	0.065	0.300	−0.077	0.042	0.280	−0.157
	0.452	0.988	0.875	0.623	0.020	0.558	0.748	0.030	0.230
Q2-33	−0.007	0.109	0.066	−0.134	−0.056	−0.051	−0.082	−0.055	−0.037
	0.957	0.405	0.616	0.307	0.669	0.699	0.535	0.677	0.778
Q2-34	−0.005	−0.091	0.090	−0.057	0.058	0.251	−0.094	0.009	0.064
	0.972	0.488	0.493	0.666	0.661	0.053	0.474	0.947	0.629

	Q2-29	Q2-30	Q2-31	Q2-33
Q2-30	0.059			
	0.655			
Q2-31	−0.080	−0.036		
	0.542	0.782		
Q2-33	−0.028	−0.024	0.054	
	0.831	0.857	0.684	
Q2-34	0.023	0.159	0.067	0.195
	0.861	0.225	0.611	0.136

Cell Contents: Pearson correlation
P-Value

Results for: Health and safety

Correlations: Q2-11, Q2-21, Q2-22, Q2-28

	Q2-11	Q2-21	Q2-22
Q2-21	−0.019		
	0.886		
Q2-22	0.009	−0.111	
	0.949	0.399	
Q2-28	0.192	−0.081	0.223
	0.141	0.539	0.087

Cell Contents: Pearson correlation
P-Value

Results for: Container Sorting

Correlations: Q2-17, Q2-18

Pearson correlation of Q2-17 and Q2-18 = 0.151
P-Value = 0.251

Results for: Cost

Correlations: Q2-10, Q2-23, Q2-24, Q2-25, Q2-30

	Q2-10	Q2-23	Q2-24	Q2-25
Q2-23	−0.036			
	0.785			
Q2-24	0.116	0.071		
	0.379	0.588		
Q2-25	0.208	−0.057	0.092	
	0.111	0.667	0.482	
Q2-30	0.042	0.466	−0.057	−0.172
	0.750	0.000	0.666	0.188

Cell Contents: Pearson correlation
P-Value

Results for: Jurisdiction

Correlations: Q2-4, Q2-12, Q2-13

	Q2-4	Q2-12
Q2-12	0.282	
	0.029	
Q2-13	0.264	0.381
	0.042	0.003

Cell Contents: Pearson correlation
P-Value

Results for: Stevedores performance

Correlations: Q2-13, Q2-16, Q2-20

	Q2-13	Q2-16
Q2-16	0.145	
	0.270	
Q2-20	0.092	0.208
	0.484	0.111

Cell Contents: Pearson correlation
P-Value

Results for: Time management

Correlations: Q2-6, Q2-7, Q2-20, Q2-21, Q2-23, Q2-35, Q2-36

	Q2-6	Q2-7	Q2-20	Q2-21	Q2-23	Q2-35
Q2-7	0.260					
	0.045					
Q2-20	−0.018	0.208				

(continued)

(continued)

	Q2-6	Q2-7	Q2-20	Q2-21	Q2-23	Q2-35
	0.890	0.110				
Q2-21	−0.143	0.036	−0.081			
	0.275	0.785	0.538			
Q2-23	0.169	−0.158	−0.065	0.006		
	0.198	0.227	0.619	0.963		
Q2-35	0.292	0.240	0.087	0.020	0.064	
	0.024	0.065	0.506	0.879	0.624	
Q2-36	0.092	−0.028	0.102	−0.167	0.056	0.283
	0.487	0.835	0.436	0.201	0.673	0.029

Cell Contents: Pearson correlation
P-Value

10.3 Appendix 3: Principal Component Analysis (minitab® release 14.1 output): Results for: Management

Principal Component Analysis: Q2-1, Q2-2, Q2-3, Q2-4, Q2-5, Q2-6, Q2-10, Q2-12, Q2-15, Q2-16, Q2-19, Q2-20, Q2-21, Q2-28, Q2-29, Q2-31, Q2-32, Q2-33, Q2-34, Q2-35, Q2-36

Eigenanalysis of the Correlation Matrix

Eigenvalue	4.2701	2.1655	1.8262	1.6824	1.4752	1.4339	1.1653	0.9972
Proportion	0.203	0.103	0.087	0.080	0.070	0.068	0.055	0.047
Cumulative	0.203	0.306	0.393	0.474	0.544	0.612	0.668	0.715
Eigenvalue	0.8681	0.8057	0.7574	0.6682	0.5067	0.4820	0.3960	0.3783
Proportion	0.041	0.038	0.036	0.032	0.024	0.023	0.019	0.018
Cumulative	0.756	0.795	0.831	0.863	0.887	0.910	0.929	0.947
Eigenvalue	0.3095	0.2810	0.2403	0.1639	0.1270			
Proportion	0.015	0.013	0.011	0.008	0.006			
Cumulative	0.961	0.975	0.986	0.994	1.000			

Variable	PC1	PC2	PC3	PC4	PC5	PC6	PC7	PC8
Q2-1	−0.363	0.068	0.096	−0.176	−0.178	−0.169	−0.104	0.339
Q2-2	−0.369	0.252	0.034	0.105	−0.018	−0.134	0.035	0.080
Q2-3	−0.345	0.121	0.011	0.301	0.094	−0.081	−0.007	0.162
Q2-4	−0.315	0.203	0.051	0.002	0.075	−0.252	−0.248	0.021
Q2-5	−0.115	−0.124	0.284	−0.067	−0.391	0.142	0.029	0.117

(continued)

(continued)

Variable	PC1	PC2	PC3	PC4	PC5	PC6	PC7	PC8
Q2-6	−0.234	0.235	−0.049	−0.174	−0.134	0.082	−0.018	−0.184
Q2-10	−0.149	0.201	0.139	−0.137	−0.110	0.492	−0.123	0.079
Q2-12	−0.256	−0.323	−0.128	−0.034	−0.150	−0.005	−0.309	−0.036
Q2-15	−0.089	−0.161	−0.018	0.131	0.011	0.631	0.061	0.251
Q2-16	−0.319	−0.202	−0.221	0.168	−0.003	0.195	0.146	−0.028
Q2-19	−0.068	−0.227	−0.346	0.135	0.161	0.205	−0.481	−0.131
Q2-20	−0.034	−0.387	−0.292	−0.318	0.098	−0.145	0.183	0.136
Q2-21	−0.153	0.029	0.172	−0.249	0.544	0.061	−0.156	−0.132
Q2-28	−0.128	0.271	−0.347	−0.187	0.107	0.085	0.273	0.315
Q2-29	−0.154	0.110	0.002	−0.304	0.227	0.216	0.434	−0.365
Q2-31	−0.059	0.124	−0.169	0.567	−0.060	0.031	0.202	−0.231
Q2-32	−0.188	−0.235	0.082	0.128	0.488	−0.108	0.026	0.158
Q2-33	−0.043	−0.046	0.530	0.260	0.204	0.079	0.067	−0.116
Q2-34	−0.164	−0.389	0.141	0.094	−0.097	−0.160	0.426	0.131
Q2-35	−0.267	0.001	−0.223	0.004	−0.182	−0.113	0.081	−0.486
Q2-36	−0.200	−0.287	0.278	−0.209	−0.163	0.004	−0.054	−0.318

Variable	PC9	PC10	PC11	PC12	PC13	PC14	PC15	PC16
Q2-1	0.161	−0.102	0.050	−0.029	−0.053	−0.169	−0.032	−0.059
Q2-2	−0.010	0.065	0.224	0.008	−0.096	0.271	−0.227	0.122
Q2-3	−0.056	−0.301	−0.078	0.054	−0.276	−0.071	−0.055	0.146
Q2-4	−0.139	0.023	−0.163	0.327	0.198	0.333	0.237	−0.440
Q2-5	−0.471	0.490	−0.085	−0.154	−0.283	−0.055	−0.208	−0.160
Q2-6	0.419	0.456	0.169	0.317	0.012	−0.392	0.032	0.121
Q2-10	0.330	0.009	−0.183	−0.366	0.134	0.493	0.026	0.210
Q2-12	−0.190	−0.097	−0.436	−0.117	0.267	−0.195	−0.013	0.173
Q2-15	−0.172	−0.061	0.224	0.353	−0.051	0.010	0.437	−0.122
Q2-16	0.126	−0.010	−0.211	0.093	0.341	−0.234	−0.158	−0.048
Q2-19	0.100	0.011	0.303	−0.042	−0.232	0.113	−0.463	−0.256
Q2-20	0.230	0.225	−0.048	−0.108	−0.043	0.235	0.167	−0.131
Q2-21	0.048	0.095	−0.249	−0.207	−0.357	−0.217	0.229	−0.213
Q2-28	−0.171	−0.184	0.193	−0.414	0.044	−0.265	−0.001	−0.135
Q2-29	−0.227	−0.120	−0.198	0.265	−0.015	0.146	−0.377	0.011
Q2-31	0.135	0.141	−0.302	−0.185	−0.332	−0.007	0.216	0.001
Q2-32	−0.186	0.340	0.180	−0.014	0.114	0.090	0.049	0.538
Q2-33	0.118	0.007	0.199	−0.268	0.408	−0.173	−0.108	−0.320
Q2-34	0.299	−0.058	−0.004	0.035	−0.118	0.157	−0.097	−0.220
Q2-35	−0.241	0.073	0.339	−0.282	0.169	0.130	0.300	−0.067
Q2-36	0.059	−0.424	0.225	−0.013	−0.259	−0.055	0.174	0.211

Variable	PC17	PC18	PC19	PC20	PC21
Q2-1	−0.132	0.013	0.398	0.255	−0.567
Q2-2	−0.427	0.008	0.026	−0.603	0.107
Q2-3	0.116	−0.259	−0.622	0.230	−0.066
Q2-4	0.198	−0.071	0.151	0.130	0.299
Q2-5	−0.017	−0.061	−0.058	0.149	0.152
Q2-6	0.284	−0.091	−0.123	−0.082	0.079
Q2-10	0.117	0.064	−0.072	0.133	0.038
Q2-12	0.249	−0.084	0.006	−0.469	−0.122
Q2-15	−0.038	−0.023	0.017	−0.190	−0.170
Q2-16	−0.483	0.204	−0.009	0.297	0.300
Q2-19	0.153	−0.012	0.099	0.069	−0.026
Q2-20	−0.190	−0.550	−0.144	−0.014	−0.026
Q2-21	−0.173	0.309	−0.112	−0.123	−0.026
Q2-28	0.272	−0.038	0.145	−0.061	0.316
Q2-29	0.108	−0.149	0.136	0.026	−0.245
Q2-31	0.058	−0.191	0.412	−0.013	−0.036
Q2-32	0.173	0.067	0.209	0.188	0.034
Q2−33	0.023	−0.359	−0.018	−0.088	−0.088
Q2-34	0.382	0.440	−0.076	−0.153	0.016
Q2-35	−0.020	0.217	−0.228	0.118	−0.281
Q2-36	−0.038	−0.173	0.229	0.076	0.397

Results for: Booking system

Principal Component Analysis: Q2-5, Q2-6, Q2-7, Q2-8, Q2-9, Q2-14, Q2-17, Q2-18, Q2-35, Q2-36

Eigenanalysis of the Correlation Matrix

Eigenvalue	2.3918	1.5007	1.1034	1.0045	0.8618	0.8140	0.7131	0.6036
Proportion	0.239	0.150	0.110	0.100	0.086	0.081	0.071	0.060
Cumulative	0.239	0.389	0.500	0.600	0.686	0.768	0.839	0.899

Eigenvalue	0.5782	0.4289
Proportion	0.058	0.043
Cumulative	0.957	1.000

Variable	PC1	PC2	PC3	PC4	PC5	PC6	PC7	PC8
Q2-5	−0.281	0.381	−0.322	−0.060	0.265	−0.221	0.689	−0.169
Q2-6	−0.331	−0.199	0.267	0.267	0.054	−0.688	−0.008	0.476
Q2-7	−0.247	−0.514	0.314	−0.087	−0.030	0.079	0.373	−0.311

(continued)

(continued)

Variable	PC1	PC2	PC3	PC4	PC5	PC6	PC7	PC8
Q2-8	−0.391	0.161	0.291	0.212	0.195	0.516	0.156	0.196
Q2-9	−0.030	−0.551	−0.346	0.088	0.503	0.309	−0.011	0.269
Q2-14	−0.408	−0.031	0.171	−0.392	−0.509	0.213	0.024	0.158
Q2-17	−0.305	0.011	−0.282	−0.701	0.217	−0.119	−0.275	0.164
Q2-18	−0.319	0.388	0.251	0.149	0.364	0.084	−0.442	−0.155
Q2-35	−0.400	−0.249	−0.232	0.204	−0.075	−0.147	−0.299	−0.631
Q2-36	−0.276	0.087	−0.544	0.395	−0.436	0.148	−0.032	0.251

Variable	PC9	PC10
Q2-5	−0.009	0.216
Q2-6	−0.065	0.019
Q2-7	0.506	−0.264
Q2-8	−0.430	−0.374
Q2-9	0.011	0.384
Q2-14	−0.097	0.557
Q2-17	0.036	−0.414
Q2-18	0.484	0.263
Q2-35	−0.409	0.017
Q2-36	0.377	−0.212

Results for: Infrastructure

Principal Component Analysis: Q2-8, Q2-10, Q2-15, Q2-19, Q2-23, Q2-24, Q2-25, Q2-26, Q2-27, Q2-29, Q2-30, Q2-31, Q2-33, Q2-34

Eigenanalysis of the Correlation Matrix

Eigenvalue	2.1782	1.6818	1.5979	1.3523	1.2619	1.1445	0.9839	0.8292
Proportion	0.156	0.120	0.114	0.097	0.090	0.082	0.070	0.059
Cumulative	0.156	0.276	0.390	0.486	0.577	0.658	0.729	0.788
Eigenvalue	0.7904	0.6029	0.5316	0.4574	0.3261	0.2619		
Proportion	0.056	0.043	0.038	0.033	0.023	0.019		
Cumulative	0.844	0.887	0.925	0.958	0.981	1.000		

Variable	PC1	PC2	PC3	PC4	PC5	PC6	PC7	PC8
Q2-8	−0.214	−0.178	−0.422	0.114	0.001	0.134	−0.096	0.490
Q2-10	−0.236	−0.060	−0.395	−0.110	0.175	0.376	−0.390	−0.330
Q2-15	−0.440	0.074	−0.071	−0.312	−0.127	−0.169	−0.225	−0.030
Q2-19	−0.357	0.176	0.284	0.219	−0.192	−0.094	−0.291	−0.039
Q2-23	0.024	0.609	−0.195	−0.275	−0.003	−0.044	0.071	0.170

(continued)

(continued)

Variable	PC1	PC2	PC3	PC4	PC5	PC6	PC7	PC8
Q2-24	−0.384	0.028	0.413	−0.103	0.075	0.259	−0.187	0.061
Q2-25	−0.160	0.134	−0.163	0.371	0.195	0.584	0.316	0.097
Q2-26	−0.526	−0.183	−0.038	0.050	−0.005	−0.249	0.220	−0.085
Q2-27	−0.157	0.038	0.408	0.035	0.424	0.095	0.358	−0.227
Q2-29	−0.177	−0.211	−0.274	−0.214	0.464	−0.395	0.285	0.015
Q2-30	−0.223	0.541	−0.121	0.053	−0.232	−0.080	0.347	0.043
Q2-31	−0.132	−0.327	−0.082	0.267	−0.572	−0.010	0.296	−0.115
Q2-33	0.038	−0.056	−0.105	−0.506	−0.282	0.314	0.255	−0.464
Q2-34	−0.055	−0.236	0.265	−0.475	−0.132	0.248	0.181	0.563

Variable	PC9	PC10	PC11	PC12	PC13	PC14
Q2-8	−0.571	0.088	−0.164	−0.223	−0.232	0.023
Q2-10	0.182	0.010	0.419	−0.242	0.077	−0.255
Q2-15	−0.123	−0.655	−0.065	0.250	0.142	0.269
Q2-19	−0.292	0.478	0.352	0.372	0.066	0.035
Q2-23	0.175	0.042	0.194	0.117	−0.629	0.002
Q2-24	0.309	0.198	−0.333	−0.376	−0.165	0.381
Q2-25	0.162	−0.083	−0.083	0.467	0.133	0.169
Q2-26	0.220	0.069	−0.317	0.153	−0.193	−0.595
Q2-27	−0.433	−0.277	0.273	−0.194	−0.232	−0.088

Results for: Health and safety
Principal Component Analysis: Q2-11, Q2-21, Q2-22, Q2-28
Eigenanalysis of the Correlation Matrix

Eigenvalue	1.3455	1.0303	0.9150	0.7092
Proportion	0.336	0.258	0.229	0.177
Cumulative	0.336	0.594	0.823	1.000

Variable	PC1	PC2	PC3	PC4
Q2-11	0.396	0.720	0.345	0.454
Q2-21	0.349	−0.535	0.769	−0.040
Q2-22	−0.543	0.403	0.498	−0.543
Q2-28	0.653	0.184	−0.205	−0.706

Results for: Container Sorting
Principal Component Analysis: Q2-17, Q2-18

Eigenanalysis of the Correlation Matrix

Eigenvalue	1.1506	0.8494
Proportion	0.575	0.425
Cumulative	0.575	1.000

Variable	PC1	PC2
Q2-17	0.707	0.707
Q2-18	0.707	−0.707

Results for: Cost
Principal Component Analysis: Q2-10, Q2-23, Q2-24, Q2-25, Q2-30
Eigenanalysis of the Correlation Matrix

Eigenvalue	1.5274	1.2547	0.9308	0.8026	0.4846
Proportion	0.305	0.251	0.186	0.161	0.097
Cumulative	0.305	0.556	0.743	0.903	1.000

Variable	PC1	PC2	PC3	PC4	PC5
Q2-10	0.149	0.620	−0.432	0.596	−0.228
Q2-23	0.628	−0.292	−0.073	0.336	0.634
Q2-24	0.083	0.500	0.834	0.130	0.175
Q2-25	0.359	0.493	−0.266	−0.713	0.219
Q2-30	0.669	−0.191	0.205	−0.082	−0.684

Results for: Jurisdiction
Principal Component Analysis: Q2-4, Q2-12, Q2-13
Eigenanalysis of the Correlation Matrix

Eigenvalue	1.6210	0.7610	0.6180
Proportion	0.540	0.254	0.206
Cumulative	0.540	0.794	1.000

Variable	PC1	PC2	PC3
Q2-4	−0.528	0.847	−0.068
Q2-12	−0.605	−0.319	0.729
Q2-13	−0.596	−0.426	−0.681

Results for: Stevedores performance
Principal Component Analysis: Q2-13, Q2-16, Q2-20
Eigenanalysis of the Correlation Matrix

Eigenvalue	1.3019	0.9154	0.7827
Proportion	0.434	0.305	0.261
Cumulative	0.434	0.739	1.000

Variable	PC1	PC2	PC3
Q2-13	−0.488	−0.840	0.238
Q2-16	−0.642	0.161	−0.750
Q2-20	−0.591	0.519	0.617

Results for: Time management
Principal Component Analysis: Q2−6, Q2-7, Q2-20, Q2-21, Q2-23, Q2-35, Q2-36
Eigenanalysis of the Correlation Matrix

Eigenvalue	1.6807	1.2576	1.1252	0.9608	0.8736	0.5691	0.5330
Proportion	0.240	0.180	0.161	0.137	0.125	0.081	0.076
Cumulative	0.240	0.420	0.580	0.718	0.843	0.924	1.000

Variable	PC1	PC2	PC3	PC4	PC5	PC6	PC7
Q2-6	0.506	−0.184	−0.328	−0.442	−0.027	0.180	0.612
Q2-7	0.425	0.532	−0.243	−0.185	0.016	0.359	−0.560
Q2-20	0.259	0.393	0.416	0.153	0.696	−0.098	0.296
Q2-21	0.190	-0.230	0.599	−0.631	−0.073	−0.253	−0.290
Q2-23	−0.083	0.619	0.273	−0.052	−0.632	−0.075	0.357
Q2-35	0.558	−0.048	−0.173	0.370	−0.208	−0.681	−0.111
Q2-36	0.375	−0.301	0.444	0.458	−0.257	0.544	−0.004

10.4 Appendix 4: Questionnaire Used in Chapter 6 and 7

A Questionnaire to Determine Customer Satisfaction

Purpose of questionnaire:
This set of questionnaires will be distributed to the drivers at Port of Fremantle in Perth, WA to determine level of their satisfaction for the services that are being provided to them. Using the feedback, we will be able to carry out an analysis

followed by solution and recommendation to enhance the productivity at the Fremantle port and generate drivers' satisfaction and loyalty.

How are we going to protect anonymity?
This is an anonymous questionnaire. Please **do not include your name** anywhere on the questionnaire.
Although we do not know the precise number of respondents, we estimate to have at least 40 respondents. Should you choose to respond, your questionnaire will be one of the many received. To help maintain your anonymity, **please return the questionnaire in an unmarked envelope**. All received envelop will be opened by Alireza Faed, PhD student at Curtin University. Once all questionnaires are received, they will be categorized and analysed.

Please return this questionnaire with the responses to the following address:

PhD Student
Alireza Faed
Curtin University
Digital Ecosystems and Business Intelligence Institute
Curtin University of Technology
Enterprise Unit 4, De Laeter Way, Technology Park
Bentley WA 6102

Should you require any further information about this project, kindly email me at Alireza.Faed@postgrad.curtin.edu.au, or A.R.Faed@gmail.com
Mobile phone: 0433117587

Please return questionnaire by November 15th 2011. Thank you.
This survey is 9 pages long and should take only 20 min to be completed.
Please answer every question.

10.4.1 Demographics, Please Circle the Appropriate Response

1. Age Group:
   (a) 21–30
   (b) 31–40
   (c) 41–50
   (d) 51–60
2. Gender:
   (a) Male
   (b) Female

3. What is the level of your education?

    (a) Some high school
    (b) High school graduate
    (c) Some college/technical
    (d) College graduate

4. What is your position?

    (a) A novice
    (b) Driver and operator
    (c) Experienced driver
    (d) Driver and owner

5. Please indicate on what basis you are **employed** in your current job(s) by selecting the appropriate box:

    (a) Full time
    (b) Part time
    (c) Casual

6. How close is your home to the port?

    (a) 1–10 K
    (b) 10–20 k
    (c) More than 20 K

7. How familiar are you with the working culture in WA?

    (a) Poor
    (b) Fair
    (c) Good
    (d) Very good
    (e) Excellent

8. Which characteristic best describes you as a heavy vehicle driver? (Circle the word please)

    (a) Relax and calm
    (b) Easy going and friendly
    (c) Easily distracted
    (d) Easily frustrated

10.4.2 Customer Satisfaction

How satisfied are you with/by________[the Fremantle Port] in each of the following cases?

Numbers	Level of satisfaction based on 7 point likert scale
1	Strongly dissatisfied
2	Mostly dissatisfied
3	Somewhat dissatisfied
4	Neutral
5	Moderately satisfied
6	Very satisfied
7	Extremely satisfied

1. The current management at the port.

1	2	3	4	5	6	7

2. Are you happy with the way that the port is run?

1	2	3	4	5	6	7

3. Management satisfaction in terms of being organized and professional?

1	2	3	4	5	6	7

4. Management strategies at the port.

1	2	3	4	5	6	7

5. Transportation status for your shifts.

1	2	3	4	5	6	7

6. Current traffic and congestion at the port.

1	2	3	4	5	6	7

7. Current waiting time while you get to the queue which picking up the container.

1	2	3	4	5	6	7

8. Staying in your cabin while you are in a queue at Patrick and at DP world.

1	2	3	4	5	6	7

9. How long are you willing to stay in a queue while you are waiting?
   (a) 15–20 min
   (b) 30 min
   (c) More than 30 min
   (d) More than One hour

10. The current cost of food available at Fremantle port.

1	2	3	4	5	6	7

11. The existing health and safety support at Fremantle port.

1	2	3	4	5	6	7

12. The road regulation and the sufficiency and flexibility of operational rules and regulations at Fremantle port.

1	2	3	4	5	6	7

13. Competition between the companies at the Fremantle port.

1	2	3	4	5	6	7

14. The current booking system of the Fremantle port.

1	2	3	4	5	6	7

15. Intelligent system and devices employed in the Fremantle port.

1	2	3	4	5	6	7

16. Port operators and Stevedore performance within Fremantle port.

1	2	3	4	5	6	7

Comment:

____

____

____

17. Current container sorting and labelling procedures.

1	2	3	4	5	6	7

18. The speed of accessing and finding your container.

1	2	3	4	5	6	7

19. Signs and arrows to find your direction.

1	2	3	4	5	6	7

20. The performance of forklift drivers.

1	2	3	4	5	6	7

21. Providing separate road for the trucks.

7	6	5	4	3	2	1

22. The likely danger created by the presence of heavy vehicle on the roads.

1	2	3	4	5	6	7

23. If new store (Supermarket) built at the Fremantle port.

7	6	5	4	3	2	1

24. The current Rest room facilities, café, restaurant, and the services provided to you now.

1	2	3	4	5	6	7

Comment:

____

____

____

25. Current parking at the Fremantle port.

1	2	3	4	5	6	7

26. Pavement and walking spaces at Fremantle port.

1	2	3	4	5	6	7

27. The number of cameras at the Fremantle port.

1	2	3	4	5	6	7

28. To be monitored by security cameras while you are in the Fremantle port.

1	2	3	4	5	6	7

29. The working procedures at the Fremantle port.

1	2	3	4	5	6	7

30. If you have a petrol station at the Fremantle port.

1	2	3	4	5	6	7

31. Accommodation in the Fremantle port.

1	2	3	4	5	6	7

32. Staff responsiveness in dealing with you.

1	2	3	4	5	6	7

33. Technical competence of engineers and their response time.

1	2	3	4	5	6	7

34. Consideration and respecting cultural values of each individual by the people at the Fremantle port.

1	2	3	4	5	6	7

35. Availability of requested time slots.

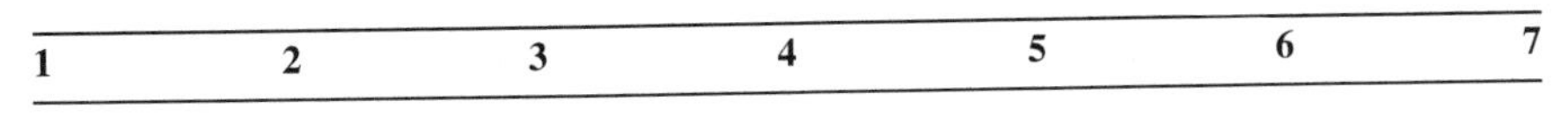

36. Pick up shifts assigned to you.

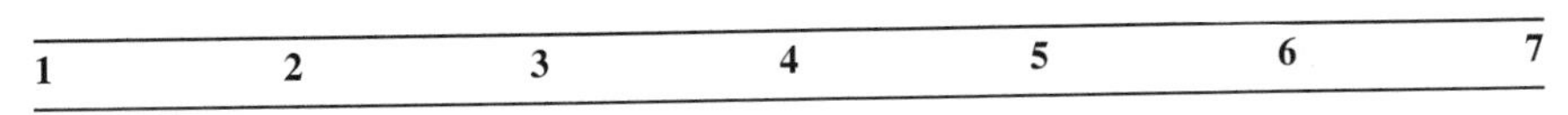

37. Have you ever lodged a complaint? If the answer is yes, how satisfied were you by the feedback of your complaint?

1	2	3	4	5	6	7

10.4.3 Customer Loyalty Section

Number	Customer Loyalty
1	Extremely unlikely
2	Very unlikely
3	Somewhat unlikely
4	Neutral
5	Somewhat likely
6	Very Likely
7	Extremely likely

1. Would you recommend the stevedores to your friends who are seeking a career?

1	2	3	4	5	6	7

2. How likely are you to switch from your present job?

7	6	5	4	3	2	1

3. Are you employed? How long have you been with your current employer/3rd transportation provider?

   (a) less than 1 year
   (b) 1–4 years
   (c) 5–10 years
   (d) More than 10 years.

4. How likely are you to continue working in the same company?

1	2	3	4	5	6	7

5. Will you provide word of mouth regarding the company you are working with?

1	2	3	4	5	6	7

10.4.4 Perceived Value Section

Number	Perceived value
1	Totally unacceptable
2	Unacceptable
3	Slightly unacceptable
4	Neutral
5	Slightly acceptable
6	Acceptable
7	Perfectly Acceptable

1. The personnel at the port have enough knowledge about their jobs.

1	2	3	4	5	6	7

2. The personnel provide necessary information to the customers and make them content.

1	2	3	4	5	6	7

3. How clean and spacious is your work space.

1	2	3	4	5	6	7

4. You have a very good salary and you are happy with it.

1	2	3	4	5	6	7

5. The services provided at the port are satisfactory and all of the drivers are happy about them.

1	2	3	4	5	6	7

6. The level of quality is acceptable in comparison with other entities in the same field.

1	2	3	4	5	6	7

7. There is an improvement by the port, in terms of the services that they provide since the last year or two.

1	2	3	4	5	6	7

10.4.5 Interactivity Section

Number	Interactivity
1	Strongly disagree
2	Disagree
3	Somewhat disagree
4	Neither agree or disagree
5	Somewhat agree
6	Agree
7	Strongly agree

1. The relationship between you and stevedore personnel/other employees in your organization is good.

1	2	3	4	5	6	7

2. Collegiality/work atmosphere at Fremantle port.

1	2	3	4	5	6	7

3. Collegiality/work atmosphere at your employer.

1	2	3	4	5	6	7

4. The current interaction and communication on behalf of the booking system

7	6	5	4	3	2	1

with the drivers is not enough.

5. I receive a quick and appropriate feedback, if I have a question or a complaint.

1	2	3	4	5	6	7

6. The company interacts with employees beyond the standards.

1	2	3	4	5	6	7

10.4.6 Customer Acquisition Section

Number	Customer acquisition
1	Strongly disagree
2	Disagree
3	Somewhat disagree
4	Neither agree or disagree
5	Somewhat agree
6	Agree
7	Strongly agree

1. Your organization endeavours to attract more drivers.

1	2	3	4	5	6	7

2. Your organization provides incentive while it hires a new driver.

1	2	3	4	5	6	7

3. My organization works or endeavours to take my concerns into accounts.

1	2	3	4	5	6	7

4. The current numbers of drivers in comparison with last year have decreased.

   (a) Yes
   (b) No
   (c) I do not know

   General Comment: ______________________________

____

____

____

THANK YOU for taking the time to complete this questionnaire.

10.5 Appendix 5: Prototype Implementation Sources

```
clear all
clc
%read data from
excel**********************************************
management=xlsread('Alirezadata.xlsx','Management','
c3:w62');
BookingSystem=xlsread('Alirezadata.xlsx','Booking
System','c3:l62');
Infrastructure=xlsread('Alireza
data.xlsx','Infrastructure','c3:p62');
HealthandSafety=xlsread('Alirezadata.xlsx','Health  and
Safety','c3:f62');
ContainerSorting=xlsread('Alirezadata.xlsx','Container
Sorting','c3:d62');
Cost=xlsread('Alirezadata.xlsx','Cost','c3:g62');
Jurisdiction=xlsread('Alirezadata.xlsx','
Jurisdiction','c3:e62');
Stevedoresperformance=xlsread('Alireza data.xlsx','Ste-
vedores performance','c3:e62');
Timemanagement=xlsread('Alirezadata.xlsx','Time
management','c3:e62');
```

```
%calculate PCA
**************************************************
****
[management_pc,                      management_zscores,
management_pcvars]=princomp(management);
[BookingSystem_pc,                BookingSystem_zscores,
BookingSystem_pcvars]=princomp(BookingSystem);
[Infrastructure_pc,              Infrastructure_zscores,
Infrastructure_pcvars]=princomp(Infrastructure);
[HealthandSafety_pc,            HealthandSafety_zscores,
HealthandSafety_pcvars]=princomp(HealthandSafety);
[ContainerSorting_pc,          ContainerSorting_zscores,
ContainerSorting_pcvars]=princomp(ContainerSorting);
[Cost_pc, Cost_zscores, Cost_pcvars]=princomp(Cost);
[Jurisdiction_pc,                  Jurisdiction_zscores,
Jurisdiction_pcvars]=princomp(Jurisdiction);
[Stevedoresperformance_pc,             Stevedoresperfor-
mance_zscores, Stevedoresperformance_pcvars]=
princomp(Stevedoresperformance);
[Timemanagement_pc,              Timemanagement_zscores,
Timemanagement_pcvars]=princomp(Timemanagement);

%calculate new normalized variables (customer satisfaction
indicators
% management
00000000000000000000000000000000000000000000000000000
00000000
management_new=(management_zscores*management_pcvars)/
(sum(management_pcvars));
management_new=((management_new-min(management_new))/
(max(management_new)-min(management_new)))*6+1;

% BookingSystem
00000000000000000000000000000000000000000000000000000
000000000
BookingSystem_new = (BookingSystem_zscores*BookingSys-
tem_pcvars)/(sum(BookingSystem_pcvars));
BookingSystem_new = ((BookingSystem_new-min(BookingSys-
tem_new))/(max(BookingSystem_new)-
min(BookingSystem_new)))*6 + 1;

% Infrastructure
00000000000000000000000000000000000000000000000000000
0000000000
```

```
Infrastructure_new=(Infrastructure_zscores*Infrastruc-
ture_pcvars)/(sum(Infrastructure_pcvars));
Infrastructure_new=((Infrastructure_new-min(Infra-
structure_new))/(max(Infrastructure_new)-
min(Infrastructure_new)))*6 + 1;

% HealthandSafety
0000000000000000000000000000000000000000000000000000000
000000000
HealthandSafety_new = (HealthandSafety_zscores*Healt-
handSafety_pcvars)/(sum(HealthandSafety_pcvars));
HealthandSafety_new = ((HealthandSafety_new-min(Healt-
handSafety_new))/(max(HealthandSafety_new)-
min(HealthandSafety_new)))*6+1;

% ContainerSorting
0000000000000000000000000000000000000000000000000000000
000000000
ContainerSorting_new=(ContainerSorting_zscores*Con-
tainerSorting_pcvars)/(sum(ContainerSorting_pcvars));
ContainerSorting_new=((ContainerSorting_new-min(Con-
tainerSorting_new))/(max(ContainerSorting_new)-
min(ContainerSorting_new)))*6+1;

% Cost
0000000000000000000000000000000000000000000000000000000
000000000
Cost_new=(Cost_zscores*Cost_pcvars)/
(sum(Cost_pcvars));
Cost_new = ((Cost_new-min(Cost_new))/(max(Cost_new)-
min(Cost_new)))*6+1;

% Jurisdiction
0000000000000000000000000000000000000000000000000000000
000000000
Jurisdiction_new=(Jurisdiction_zscores*Jurisdic-
tion_pcvars)/(sum(Jurisdiction_pcvars));
Jurisdiction_new=((Jurisdiction_new-min(Jurisdic-
tion_new))/(max(Jurisdiction_new)-
min(Jurisdiction_new)))*6+1;

% Stevedoresperformance
0000000000000000000000000000000000000000000000000000000
000000000
```

```
Stevedoresperformance_new = (Stevedoresperfor-
mance_zscores*Stevedoresperformance_pcvars)/
(sum(Stevedoresperformance_pcvars));
Stevedoresperformance_new=((Stevedoresperformance_new-
min(Stevedoresperformance_new))/(max(Stevedoresperfor-
mance_new)-min(Stevedoresperformance_new)))*6+1;

% Timemanagement
00000000000000000000000000000000000000000000000000
000000000
Timemanagement_new=(Timemanagement_zscores*Timemanage-
ment_pcvars)/(sum(Timemanagement_pcvars));
Timemanagement_new=((Timemanagement_new-min(Timeman-
agement_new))/(max(Timemanagement_new)-
min(Timemanagement_new)))*6+1;

newdata = [management_new BookingSystem_new Infrastruc-
ture_new HealthandSafety_new ContainerSorting_new...
Cost_new Jurisdiction_new Stevedoresperformance_new
Timemanagement_new];
xlswrite('Alirezadata.xlsx',newdata,'newdata',
'b2:j61');

%codes for part
2 ====================================================
======
loyalty=xlsread('Alirezadata.xlsx','Sections 3,4,5,6
(2)','b3:e62');
perceived=xlsread('Alirezadata.xlsx','Sections 3,4,5,6
(2)','f3:l62');
interactivity=xlsread('Alirezadata.xlsx','Sec-
tions 3,4,5,6 (2)','m3:r62');
aquisition=xlsread('Alirezadata.xlsx','Sections 3,4,5,6
(2)','s3:u62');
satisfaction=newdata;

%calculate PCA
**************************************************
***
[loyalty_pc,                    loyalty_zscores,
loyalty_pcvars]=princomp(loyalty);
[perceived_pc,                  perceived_zscores,
perceived_pcvars]=princomp(perceived);
[interactivity_pc,              interactivity_zscores,
interactivity_pcvars]=princomp(interactivity);
```

```
[aquisition_pc,                     aquisition_zscores,
aquisition_pcvars]=princomp(aquisition);
[satisfaction_pc,                 satisfaction_zscores,
satisfaction_pcvars]=princomp(satisfaction);
%calculate new normalized variables

% loyalty
0000000000000000000000000000000000000000000000000000000
000000000
loyalty_new=(loyalty_zscores*loyalty_pcvars)/
(sum(loyalty_pcvars));
loyalty_new=((loyalty_new-min(loyalty_new))/(max(loy-
alty_new)-min(loyalty_new)))*6+1;

% perceived
0000000000000000000000000000000000000000000000000000000
000000000
perceived_new=(perceived_zscores*perceived_pcvars)/
(sum(perceived_pcvars));
perceived_new=((perceived_new-min(perceived_new))/
(max(perceived_new)-min(perceived_new)))*6+1;

% interactivity
0000000000000000000000000000000000000000000000000000000
0000000000
interactivity_new=(interactivity_zscores*interactiv-
ity_pcvars)/(sum(interactivity_pcvars));
interactivity_new=((interactivity_new-min(interactiv-
ity_new))/(max(interactivity_new)-
min(interactivity_new)))*6+1;

% aquisition
000000000000000000000000000000000000000000000000000000
0000000000
aquisition_new=(aquisition_zscores*aquisition_pcvars)/
(sum(aquisition_pcvars));
aquisition_new=((aquisition_new-min(aquisition_new))/
(max(aquisition_new)-min(aquisition_new)))*6+1;

% satisfaction
0000000000000000000000000000000000000000000000000000000
000000000
satisfaction_new=(satisfaction_zscores*satisfac-
tion_pcvars)/(sum(satisfaction_pcvars));
```

```
satisfaction_new=((satisfaction_new-min(satisfac-
tion_new))/(max(satisfaction_new)-
min(satisfaction_new)))*6+1;

newdatasections=[loyalty_new perceived_new interactiv-
ity_new aquisition_new satisfaction_new];
xlswrite('Alireza data.xlsx',newdatasections,'newdata-
sections','
c2:g61');
```

10.6 Appendix 6: Representation of Variables Derived from Structural Equation Modelling (Chapter 6)

10.6.1 Customer Satisfaction and its Observed Variables

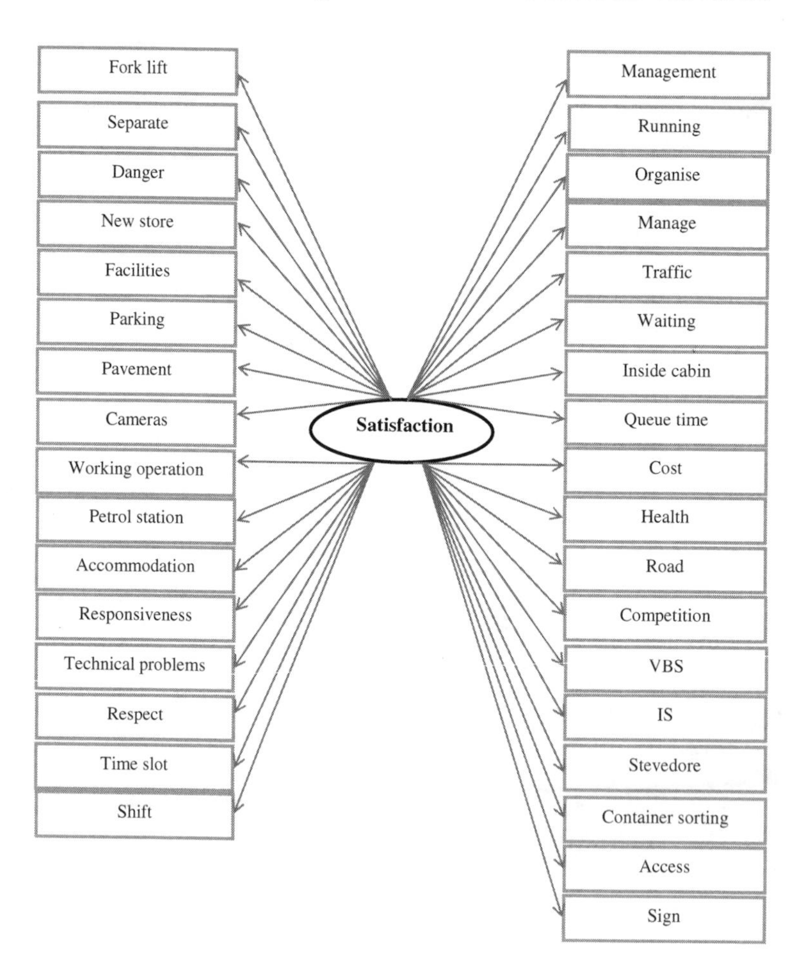

10.6.2 The Relationship Between Antecedents of Customer Satisfaction and Their Observed Variables

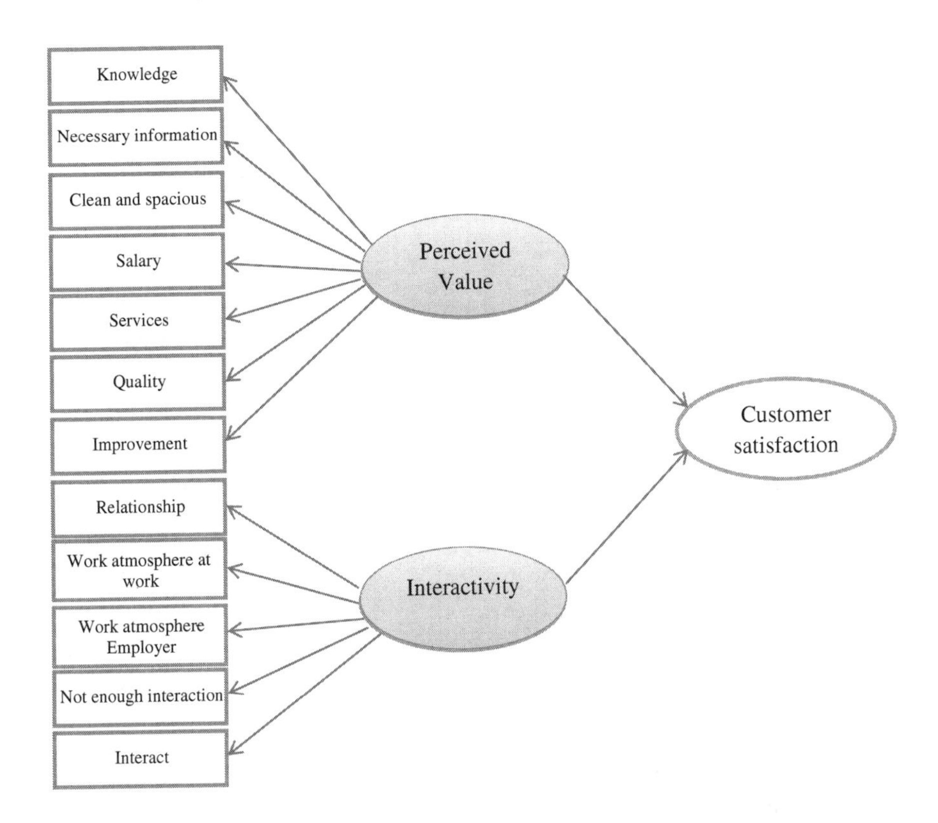

10.6.3 The Relationship Between Consequences of Customer Satisfaction and Their Observed Variables

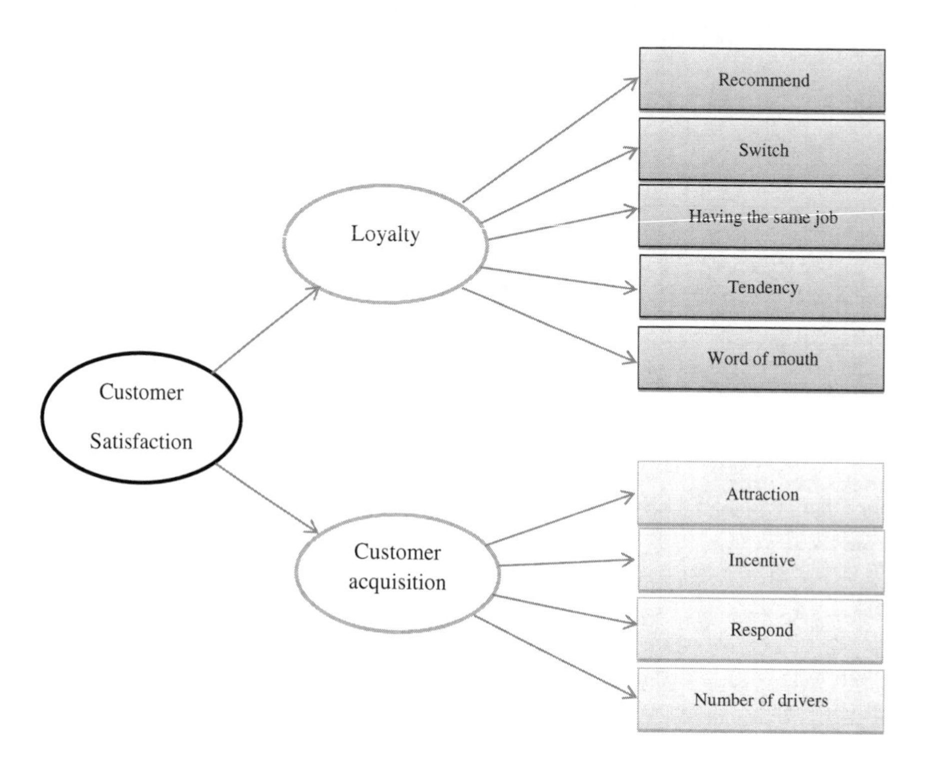

Short Biography of the Author

Alireza Faed obtained his Ph.D. in the area of e-CRM from Curtin University, WA, Australia. He achieved his M.Sc. in E-Commerce and Marketing from Sweden and Iran, respectively. He holds a B.Sc. degree in Industrial Management from Iran. Currently, he is working as an associate Lecturer at Curtin University and teaches Marketing and Communication in Business.

Alireza is a faculty member and has over 7 years' experience teaching range of academic environments as a lecturer. Prior to that, he used to work as an instructor for academic and vocational institutes. His research and teaching interests include embedded academic support, CRM, Complaint Management Systems, e-Marketing, Sport sponsorship, e-commerce and Communication in business. He is a member of IEEE and IEEE Intelligent Transportation Systems Society. He has published his research in reputable International Journals and Conferences.

A. R. Faed, *An Intelligent Customer Complaint Management System with Application to the Transport and Logistics Industry*, Springer Theses, DOI: 10.1007/978-3-319-00324-5,
© Springer International Publishing Switzerland 2013

Printed by Publishers' Graphics LLC
LMO130701.15.15.25